# Fields Institute Communications

Volume 78



The Communications series features conference proceedings, surveys, and lecture notes generated from the activities at the Fields Institute for Research in the Mathematical Sciences. The publications evolve from each year's main program and conferences. Many volumes are interdisciplinary in nature, covering applications of mathematics in science, engineering, medicine, industry, and finance.

More information about this series at http://www.springer.com/series/10503

Wai Keung Li · David A. Stanford
Hao Yu
Editors

# Advances in Time Series Methods and Applications

## The A. Ian McLeod Festschrift

FIELDS The Fields Institute for Research in the Mathematical Sciences

Springer

*Editors*
Wai Keung Li
Department of Statistics and Actuarial Science
University of Hong Kong
Hong Kong

David A. Stanford
Department of Statistics and Actuarial Science
University of Western Ontario
London, ON
Canada

Hao Yu
Department of Statistics and Actuarial Science
University of Western Ontario
London, ON
Canada

ISSN 1069-5265 ISSN 2194-1564 (electronic)
Fields Institute Communications
ISBN 978-1-4939-6567-0 ISBN 978-1-4939-6568-7 (eBook)
DOI 10.1007/978-1-4939-6568-7

Library of Congress Control Number: 2016949559

Mathematics Subject Classification (2010): 62M10, 62P20, 62P12



Cover illustration: Drawing of J.C. Fields by Keith Yeomans

Printed on acid-free paper

This Springer imprint is published by Springer Nature
The registered company is Springer Science+Business Media LLC
The registered company address is: 233 Spring Street, New York, NY 10013, U.S.A.

# Preface

Time series analysis took off with a burst of activity in the 1970s due to a number of publications employing time domain techniques. No doubt, the pivotal event in this advance can be ascribed to the book by Box and Jenkins which appeared in 1970. The so-called Box and Jenkins's approach for autoregressive integrated moving-average (ARIMA) models was made popular by many researchers who refined and expanded its initial framework.

Among these researchers, A. Ian McLeod stands out as one who has contributed to almost every aspect of the Box–Jenkins framework both in theory and in practice. His method in deriving diagnostic statistics via the asymptotic distribution of ARMA model residuals is versatile and applicable to nearly all kinds of new time series models. His work in long memory time series with Keith Hipel was truly ahead of time. Professor McLeod was one of the early advocates of the uses of inverse and inverse partial autocorrelations. The McLeod–Hipel time series package was one of the few available comprehensive computer softwares for time series analysis in the 1970s and 1980s.

Ian McLeod's research interests cover also random number generation and environmental statistics, especially on water resources issues. His many influential contributions are summarized by a review article in this monograph.

Since the 1980s time series analysis has grown in many different directions. The new areas that have appeared include, among other topics, nonstationary time series, nonlinear models and conditional heteroscedasticity models. Despite the range of these new developments, the papers in this volume testify to the impact that Ian McLeod's influence is still being felt widely.

This volume arises as a consequence of a Festschrift in Ian McLeod's honour held at the University of Western Ontario June 2–3, 2014 that was partially supported by the Fields Institute. Participants of the Festschrift were invited to submit works to form this volume. The resulting peer-reviewed monograph consists of 13 technical papers and one review on Ian McLeod's work. The papers reflect the diversity of time domain time series analysis since its infancy in the 1970s. The topics covered include diagnostic checks for duration time series models, partially nonstationary vector time series, methodology for ordered categorical data, a new

C(α) test for estimating equations, model testing using wavelets, an adaptive Lasso approach to vector autoregressions, identification of threshold nonlinear models, graphical methods, as well as business and environmental applications. We believe that the papers in this volume shed light on a variety of areas in time series analysis, and are hopeful that it will be useful to both theorists and practitioners.

The editors would like to take this opportunity to thank the Fields Institute, the University of Western Ontario and the University of Hong Kong, and the participants of the Festschrift in 2014 for their support. Thanks are also due to the authors and referees for papers of this volume for their effort and hard work.

Hong Kong — Wai Keung Li
London, ON, Canada — David A. Stanford
London, ON, Canada — Hao Yu

# Contents

# Ian McLeod's Contribution to Time Series Analysis—A Tribute

**W.K. Li**

**Abstract** Ian McLeod's contributions to time series are both broad and influential. His work has put Canada and the University of Western Ontario on the map in the time series community. This article strives to give a partial picture of McLeod's diverse contributions and their impact by reviewing the development of portmanteau statistics, long memory (persistence) models, the concept of duality in McLeod's work, and his contributions to intervention analysis.

**Keywords** Asymptotic distributions · Box–Jenkins approach · Duality · Intervention analysis · Long memory models · Residual autocorrelations

## 1 Introduction

The "Big Bang" in time domain time series was triggered by the monograph, "Time Series Analysis: forecasting and control" by George Box and Gwilym Jenkins in 1970. With the advance of computer technology came the dawn of time domain time series. Soon thereafter a series of papers by Ian and several co-authors appeared on expanding, developing and explaining the methodology. Four papers by Ian stand out in expounding the Box–Jenkins approach. These include the "Derivation of the theoretical autocovariance function of ARMA time series" (*Applied Statistics* [50]); "Intervention Analysis in Water Resources" (Hipel et al., *Water Resources Research* [21]); "Advances in Box–Jenkins Modelling (I): Model Construction" (Hipel et al.—*Water Resources Research* [24]) and "Advances in Box–Jenkins Modelling (II): Applications" (McLeod et al.—*Water Resources Research* [59]).

The first of these papers is important as it gives an algorithm to calculate the theoretical autocovariance of the autoregressive moving average (ARMA) model in terms of the model parameters. This result is needed if one wants to calculate the exact likelihood of an ARMA model. Back in the Seventies, evaluation of the exact

W.K. Li (✉)
Department of Statistics and Actuarial Science,
University of Hong Kong, Pokfulam Road, Hong Kong
e-mail: hrntlwk@hku.hk



W.K. Li et al. (eds.), *Advances in Time Series Methods and Applications*,
Fields Institute Communications 78, DOI 10.1007/978-1-4939-6568-7_1

likelihood was a major research activity in time series analysis. I will comment upon the second paper later. The third and the fourth papers give a detailed exposition of the Box–Jenkins methodology. In addition to the skeleton of the original Box–Jenkins proposal, these papers present more discussion of a variety of practical issues that are useful to a practitioner. These include choice of the Box–Cox transformation, testing for skewness and kurtosis for time series, and the exploitation of the duality property of ARMA models in model identification, by introducing the inverse autocorrelation functions and the inverse partial autocorrelation function. All these and some later developments were powerfully implemented in the McLeod–Hipel time series package, which in those days was the most comprehensive Box–Jenkins time series package. It was also one of the earliest. It appeared around 1977, well before the official date mentioned on Ian's website. The package was followed by an all encompassing treatise: Time Series Modelling of Water Resources and Environmental systems by Hipel and McLeod [23] which ably summarized the progress by 1994 of the Box–Jenkins approach to time series analysis.

Needless to say, Ian's contributions are much broader and deeper than the papers mentioned above. In the following I would try to give a snapshot on Ian McLeod's contribution to time series analysis in the following four areas: (1) the portmanteau test; (2) the persistence (long memory) phenomenon; (3) the role of duality and (4) the intervention analysis.

## 2 The Asymptotic Distribution of Residual Autocorrelation of the ARMA Models and the Portmanteau Test

An important stage in the Box–Jenkins approach to time series modelling is model diagnostic checking. The residuals of a good time series fit should appear approximately as if they were white noise. The joint distribution of residual autocorrelation from a fitted autoregressive moving average model [4] is therefore important. This leads to the portmanteau goodness of fit test.

In 1978 Ian published the paper "On the distribution of residual autocorrelations in Box–Jenkins Models" in *Journal of the Royal Statistical Society, Series B*. It gives a somewhat different derivation in the ARMA case based on martingale differences. This approach avoids the least squares properties used in Box–Pierce [4], and can be applied to other situations such as the multiplicative Seasonal ARMA models. As we shall see, in fact, its impact goes far beyond the domain of ARMA models. Because of its importance I present its ideas according to the following development.

Suppose that the time series $z_t$, $t = 1, \ldots, n$, is generated by the ARMA time series model

$$\phi(B)z_t = \theta(B)a_t, \qquad (1)$$

where

$$\phi(B) = 1 - \phi_1 B - \cdots - \phi_p B^p, \quad \theta(B) = 1 - \theta_1 B - \cdots - \theta_q B^q,$$

and $B$ is the backshift operator on $t$. The white noise series, $a_t$, is assumed to be independent and identically distributed with mean 0 and variance $\sigma^2$, and the ARMA model is assumed to be stationary, invertible and not redundant. Let $\boldsymbol{\beta} = (\phi_1, \ldots, \phi_p, \theta_1, \ldots, \theta_q)$ be the $p + q$ dimensional vector of the true model parameters.

Let $\hat{a}_t$ be the residuals from the fitted autoregressive moving average model. Let the lag $l$ residual autocorrelations be

$$\hat{r}(l) = \frac{\sum_{t=l+1}^{n} \hat{a}_t \hat{a}_{t-l}}{\sum_{t=1}^{n} \hat{a}_t^2} \qquad l = 1, 2, \ldots$$

and let $\hat{\boldsymbol{r}} = \left(\hat{r}(1), \ldots, \hat{r}(m)\right)$. Define $r(l)$ as the white noise counterpart of $\hat{r}(l)$ with $a_t$, the white noise process driving the ARMA model, replacing $\hat{a}_t$. Let $\boldsymbol{\beta}$ be the vector of ARMA parameters and $\hat{\boldsymbol{\beta}}$ an estimator of $\boldsymbol{\beta}$ based on minimizing say, the sum of squares

$$S = \sum_{t=p+q+1}^{n} a_t^2,$$

where $p, q$ are the AR and MA orders respectively.

There are three steps in McLeod's approach:

$$\text{Step 1 : Show } (\hat{\boldsymbol{\beta}} - \boldsymbol{\beta}) \cong \mathrm{I}^{-1} \frac{\partial \mathrm{I}}{\partial \boldsymbol{\beta}}, \qquad (2)$$

$$\text{Step 2 : Show } \sqrt{n}(\hat{\boldsymbol{\beta}} - \boldsymbol{\beta}, \boldsymbol{r}) \overset{\text{asym}}{\sim} MVN(0, \mathrm{V}), \qquad (3)$$

where

$$\mathrm{V} = \left( \frac{\mathrm{I}^{-1}}{-\mathrm{X}\mathrm{I}^{-1}} \middle| \frac{-\mathrm{I}^{-1}\mathrm{X}^{\mathrm{T}}}{1_{m \times m}} \right),$$

$$\mathrm{I} \cong \frac{1}{2n} \mathrm{E}\left( \frac{\partial^2 S}{\partial \boldsymbol{\beta} \partial \boldsymbol{\beta}^{\mathrm{T}}} \right),$$

$$1_{m \times m} = m \times m \text{ identity matrix},$$

$$n\,\mathrm{E}\left[(\hat{\boldsymbol{\beta}}-\boldsymbol{\beta})r^{\mathrm{T}}\right]=-\mathrm{I}^{-1}\mathrm{X}^{\mathrm{T}}.$$

$$\text{Step 3 : Show } \hat{r}\cong r+\mathrm{X}(\hat{\boldsymbol{\beta}}-\boldsymbol{\beta})\ .$$

Combining the above gives

$$n\,\mathrm{var}(\hat{r})=(1_{m\times m}-\mathrm{X}\mathrm{I}^{-1}\mathrm{X}^{\mathrm{T}}).\tag{4}$$

For ARMA models when $n\gg m\gg 0$,

$$\mathrm{X}^{\mathrm{T}}\mathrm{X}\cong\mathrm{I}.$$

Therefore the right hand side of (4) is idempotent. This leads to the portmanteau test

$$Q_m=n\sum_{l=1}^{m}\hat{r}^2(l)$$

and the modification [45]

$$Q_m^*=n(n+2)\sum_{l=1}^{m}\hat{r}^2(l)/(n-l),$$

is known as the Ljung–Box–Pierce portmanteau test.

If the ARMA model is adequate in describing the data, then both statistics will be asymptomatically $\chi^2$ with $m-p-q$ degrees of freedom if $n\gg m\gg 0$. The result also implies that $\mathrm{var}(\hat{r}(l))\ll\frac{1}{n}$ for $l$ small (see [12] for a review). There are many extensions of the above approach. A similar approach has been used in [53] on the distribution of residual cross correlations in univariate ARMA models. This leads to tests of the so-called Granger causality between two time series under contemporaneous correlation, extending Haugh [20]. A robust versions were derived by Li and Hui [37] and Duchesne and Roy [10]. The approach is very general and can be applied to many other situations (models). Extension to $k$ dimensional multivariate ARMA models was done in [39]. The modified portmanteau statistic by Li and McLeod takes the form

$$Q^*(m)=n\sum_{l=1}^{m}\mathrm{tr}\left(\hat{C}_l^{\mathrm{T}}\hat{C}_0^{-1}\hat{C}_l\hat{C}_0^{-1}\right)+\frac{k^2m(m+1)}{2n},\tag{5}$$

where $\hat{C}_l$ is the lag $l$ residual autocovariance matrix. Hosking [25] independently obtained the portmanteau statistic

$$\tilde{Q}(m) = n^2 \sum_{k=1}^{m} \frac{1}{n-l} \mathrm{tr}\left(\hat{C}_l^{\mathrm{T}} \hat{C}_0^{-1} \hat{C}_l \hat{C}_0^{-1}\right).$$

Kheoh and McLeod [30] showed that the Li–McLeod test (5) is closer to the nominal significance level in moderate to large samples.

Periodic time series analysis is another area to which Ian (and Keith Hipel) have made important theoretical and applied contributions. A univariate approach to periodic ARMA models where model parameters are seasonally dependent (i.e., periodic) is difficult. However, it can be tackled in an easier way by interpreting periodic ARMA models as special multivariate ARMA models. McLeod [55] showed that for periodic autoregressions, a proper portmanteau test can be defined in terms of the autocorrelations of the residuals $\hat{a}_{t(r,m)}$ where $m$ is the period, $m = 1, \ldots, s$, so that $t(r, m) = (r-1)s + m$.

Denote by $\hat{r}_{l,m}$ the residual autocorrelation for lag $l$ and period $m$. Let $\hat{r}_m = (\hat{r}_{1,m}, \ldots, \hat{r}_{L,m})$. A suitable portmanteau test is then

$$\tilde{Q}_{L,m} = \sum_{l=1}^{L} \frac{\hat{r}_{l,m}^2}{\mathrm{var}(r_{l,m})},$$

where

$$\mathrm{var}(r_{l,m}) = \frac{n - l/s}{n(n+2)}.$$

Towards the end of the Seventies, interest in nonlinear time series models began to emerge. Among these are bilinear models [17] and threshold models [69]. It was suggested that nonlinearity could be revealed by higher order moments. McLeod pioneered the use of the autocorrelations of squared residuals from univariate ARMA models (See the McLeod–Hipel Time Series Package). This resulted in the McLeod–Li [60] paper and the portmanteau test

$$Q_m = n(n+2) \sum_{k=1}^{m} \hat{r}_{aa}^2(k) / (n-k),$$

where

$$\hat{r}_{aa}(k) = \frac{\sum_{t=k+1}^{n} \left(\hat{a}_t^2 - \hat{\sigma}^2\right)\left(\hat{a}_{t-k}^2 - \hat{\sigma}^2\right)}{\sum_{t=1}^{n} \left(\hat{a}_t^2 - \hat{\sigma}^2\right)^2},$$

where $\hat{\sigma}^2 = \sum_{t=p+q+1}^{n} \hat{a}_t^2 / n$.

As far as I know, this test has been included in the computer packages SAS; ITSM [5] and Cryer and Chan [7]. A distinct feature is that

$$Q_m \overset{\text{asym}}{\sim} \chi^2_m.$$

That is, it is not necessary to subtract $p$ and $q$ from the degrees of freedom $m$. This is so because

$$\frac{\partial r_{aa}(l)}{\partial \boldsymbol{\beta}} = O_p\left(\frac{1}{\sqrt{n}}\right)$$

and hence

$$\hat{r}_{aa}(l) = r_{aa}(l) + O_p\left(\frac{1}{n}\right).$$

Unfortunately, this has been overlooked by many! The test is also wrongly ascribed as the Ljung–Box test despite the fact that they have never shown any results for the distribution of $\hat{r}_{aa}(k)$.

Engle [13, 14] proposed the so-called autoregressive conditional heteroscedastic (ARCH) model to model changing conditional variance in economic/financial time series. Luukkonen et al. [46] pointed out that the McLeod–Li test is asymptotically equivalent to the Lagrange-multiplier test for ARCH effect proposed by Engle. However, the test has been wrongly applied by many without pre-whitening by first fitting a univariate ARMA model (private communication, Mike McAleer). Hence, $Q_m$ is significant because of the underlying ARMA structure, and not due to the presence of ARCH effect or other types of deviations from linearity. A ranked (robust) version of McLeod–Li was proposed in [70]. The result was motivated by Dufour and Roy [11].

Since Engle [13, 14] ARCH or GARCH [3] type models have been very popular in financial econometrics. A natural question is "What sort of diagnostics can be done on such models?". A natural response is "What about using standardized squared residuals from a GARCH model?". For $\varepsilon_t$ satisfying a pure GARCH model, the $\hat{r}_{aa}(l)$ in McLeod–Li test is modified as

$$\hat{r}_l = \frac{\sum_{l+1}^{n}\left(\frac{\hat{\varepsilon}_t^2}{\hat{h}_t} - \overline{\varepsilon}\right)\left(\frac{\hat{\varepsilon}_{t-l}^2}{\hat{h}_t} - \overline{\varepsilon}\right)}{\sum\left(\frac{\hat{\varepsilon}_t^2}{\hat{h}_t} - \overline{\varepsilon}\right)^2}, \quad l = 1, \ldots, m,$$

where $\hat{h}_t$ is the conditional variance estimated by a GARCH model and $\overline{\varepsilon} = n^{-1}\sum \hat{\varepsilon}_t^2/\hat{h}_t$. Define $\hat{\boldsymbol{r}}^{\mathrm{T}} = (\hat{r}_1, \hat{r}_2, \ldots, \hat{r}_m)$ then under Gaussianity and other regular conditions it can be shown [38] that (using the same idea of McLeod [52])

$$\sqrt{n} \cdot \hat{\boldsymbol{r}} \overset{\text{asym}}{\sim} MVN(0, \mathrm{V}),$$

if the GARCH model is adequate and

$$V = 1_{m\times m} - \frac{1}{4}XI^{-1}X^{T},$$

for certain matrices X and I the Fisher information matrix. Note that we can estimate quantities in V by the corresponding sample values, and that V is not idempotent. We can nevertheless define the test statistic

$$Q(m) = n\hat{r}^{T}\hat{V}^{-1}\hat{r}$$

which is asymptotically $\chi^2_m$ distributed.

Nearly all previous results are based on existence of moments up to certain order. There are situations, e.g. in finance, where moments may not exist. Lin and McLeod [42] considered portmanteau tests for ARMA models with infinite variance. Here the ARMA processes are driven by innovations with a stable Paretian distribution. See further works by Lin and McLeod [41] and Mahdi and McLeod [47]. Wong and Li [71] also extend McLeod [51] to the ARCH case.

## 3 Persistence—Long Memory Time Series Models

McLeod and Hipel are among the first handful of authors to pay attention to the presence of the long memory feature in hydrological time series. McLeod and Hipel [57] and Hipel and McLeod [57] are two of the earliest papers that give a comprehensive coverage of long memory time series models during their early days. Note that these *Water Resources Research* papers have the title "Preservation of the Rescaled Adjusted Range I, II and III", respectively. Motivation is based on the so-called Hurst phenomenon in hydrologic time series. Let $z_1, \ldots, z_N$ be a time series. Define the partial sum

$$S_k^* = S_{k-1}^* + (z_k - \bar{z}_N),$$

$S_0 = 0$, $S_N^* = 0$, $\bar{z}_N$ the sample mean of $\{z_i\}$. The adjusted range

$$R_N^* = M_N^* - m_N^*,$$

where

$$M_N^* = \max(0, S_1^*, \ldots, S_N^*), \quad m_N^* = \min(0, S_1^*, \ldots, S_N^*).$$

The rescaled adjusted range is

$$\bar{R}_N^* = \frac{R_N^*}{D_N^*},$$

where $D_N^* =$ sample standard deviated of $\{z_i\}$. Hurst [27, 28] studied long-range storage requirements of the Nile River based on 690 annual time series. Hurst observed

$$\bar{R}_N^* \propto N^H,$$

where $H$ is the Hurst coefficient.

He showed by means of a coin-tossing experiment that

$$E(\bar{R}_N^*) = 1.2533N^{1/2},$$

i.e. $H = 0.5$. The same result can be obtained using probability theory [15]. However, Hurst observed the average value of $H = 0.73$ instead of 0.5. There have been debates on the possible causes of such deviation. A possible explanation is the Fractional Gaussian Noise (FGN), $B_H(t)$, proposed by Mandelbrot [48] and Mandelbrot and Wallis [49]. $B_H(t)$ is required to satisfy the self-similarity property

$$B_H(t+\tau) - B_H(t) \sim \frac{B_H(t+\tau\varepsilon) - B_H(t)}{\varepsilon^H},$$

and this implies

$$E(\bar{R}_N^*) = a\,N^H, \quad 0 < H < 1.$$

The theoretical autocovariance of FGN is not summable if

$$\frac{1}{2} < H < 1.$$

In contrast the autocovariances of the usual ARMA model are summable. The non-summabilty of the autocovariance suggests a long-memory process. A detailed discussion of statistical inference for FGN was provided in [57]. Hipel and McLeod [22] went on to provide evidence that ARMA models fitted to 23 geophysical datasets do have a Hurst coefficient greater that 0.5. Fractionally differenced time series, the discrete time analog of FGN was advocated by Granger and Joyeux [18] and Hosking [26]. In the simplest case, the pure fractionally differenced time series $X_t$ is defined by

$$(1-B)^d X_t = a_t,$$

where $-\frac{1}{2} < d < \frac{1}{2}$, $B$ is the backshift operator, and $a_t$ is white noise. The Hurst coefficient $H = d + \frac{1}{2}$ and $X_t$ is persistent if $d > 0$. Clearly this can be easily extended to incorporate AR and MA components resulting in the so-called Fractional ARMA or Fractionally Integrated ARMA models (FARMA or FARIMA resp.). Exact and approximate MLE for FARMA models were considered in the 1981 UWO Ph.D. Thesis by W.K. Li [36] and the asymptotic distribution of the residual autocorrelations was also obtained. See also [40, 64]. The portmanteau statistic

$$Q_m \overset{\text{asym}}{\sim} \chi^2_{m-p-q-1}$$

as one more parameter, $d$, has to be estimated. Long memory models have found applications to financial time series. Ding et al. [9] and Granger et al. [19] suggested that absolute returns of daily financial series exhibit long memory behavior. Long memory in the conditional variance of returns financial assets has been observed. The so-called Fractionally Integrated GARCH models have been proposed by borrowing the idea from FARMA models. Baillie et al. [1]. It is also natural to combine FARMA with GARCH. In other words long memory in the conditional mean and conditional heteroscedasticity in the variance. See [43, 44]. See also [2]. Two portmanteau tests can be derived: one test for the mean (FARMA) component, and the other for the GARCH component. Portmanteau tests for least absolute deviation (LAD) estimation of FARMA-GARCH have been obtained in [32, 33]. More recently, more general long memory GARCH models have been proposed; for example, [16, 67], Hyperbolic GARCH (HYGARCH) in [8, 31]. Mixture Memory GARCH in [34] and the Rescaled GARCH in [35].

## 4 The Role of Duality in McLeod's Work

The duality property of ARMA models has been cleverly exploited in some of McLeod's work. If the time series $z_t$ satisfies the multiplicative seasonal-moving average model

$$\boldsymbol{\Phi}(B^s)\phi(B)z_t = \boldsymbol{\Theta}(B^s)\theta(B)a_t,$$

$$\phi(B) = 1 - \phi_1 B - \cdots - \phi_p B^p,\ \theta(B) = 1 - \theta_1 B - \cdots - \theta_q B^q,$$

$$\boldsymbol{\Phi}(B^s) = 1 - \boldsymbol{\Phi}_1 B^s - \cdots - \boldsymbol{\Phi}_{p_s} B^{sp_s},\ \boldsymbol{\Theta}(B^s) = 1 - \boldsymbol{\Theta}_1 B^s - \cdots - \boldsymbol{\Theta}_{q_s} B^{sq_s},$$

where $B$ is the backshift operator, $s$ the seasonal period and $a_t$ a sequence of independent normal variables with mean zero and variance $\sigma^2$. Consider the models

$$\boldsymbol{\Phi}(B^s)\phi(B)z_t = \boldsymbol{\Theta}(B^s)\theta(B)a_t,\ \boldsymbol{\Theta}(B^s)\theta(B)y_t = \boldsymbol{\Phi}(B^s)\phi(B)a_t,$$

$$\boldsymbol{\Xi}(B^s)\xi(B)x_t = a_t,\quad w_t = \boldsymbol{\Xi}(B^s)\xi(B)a_t,$$

where

$$\xi(B) = 1 - \Sigma\xi_i B^i = \phi(B)\theta(B),\ \boldsymbol{\Xi}(B^s) = 1 - \Sigma\boldsymbol{\Xi}_i B^{si} = \boldsymbol{\Phi}(B^s)\boldsymbol{\Theta}(B^s).$$

These four models may be referred to as the primal, the dual, the autoregressive adjoint and the moving average adjoint respectively. Duality for pure AR and pure MA models was first noticed by Pierce [66], where it was shown that the nonlinear least squares estimator $\hat{\theta}$ of the moving average parameter $\theta$ of an MA model and the least squares estimator $\tilde{\theta}$ of its AR dual each satisfy the relation

$$\theta - \hat{\theta} = -(\theta - \tilde{\theta}).$$

One implication of this result is that the asymptotic covariance matrix of $\hat{\theta}$ is just that of $\tilde{\theta}$. This insight allows us to obtain the asymptotic variance of estimators of more complicated ARMA models from that of the AR model. McLeod [54] extends Pierce's result to the case of multiplicative models. Let

$$\boldsymbol{\beta} = (\phi_1, \ldots, \phi_p, \theta_1, \ldots, \theta_q, \boldsymbol{\Phi}_1, \ldots, \boldsymbol{\Phi}_{p_s}, \boldsymbol{\Theta}_1, \ldots, \boldsymbol{\Theta}_{q_s}),$$
$$\alpha = (\xi_1, \ldots, \xi_{p+q}, \boldsymbol{\Xi}_1, \ldots, \boldsymbol{\Xi}_{p_s+q_s}).$$

Given a series of $n$ observations from each model, let $\hat{\boldsymbol{\beta}}_z$, $\hat{\boldsymbol{\beta}}_y$, $\hat{\alpha}_x$ and $\hat{\alpha}_w$ denote the corresponding efficient approximate maximum likelihood estimates and denote the corresponding residuals by $\hat{a}_{z,t}$, $\hat{a}_{y,t}$, $\hat{a}_{x,t}$ and $\hat{a}_{w,t}$.

**Theorem 1** [54] *Apart from a quantity which is $O_p(1/n)$,*

$$\hat{a}_{z,t} = \hat{a}_{y,t} = \hat{a}_{x,t} = \hat{a}_{w,t}, \tag{6}$$

$$\hat{\alpha}_w - \alpha = -(\hat{\alpha}_x - \alpha) = -J'(\hat{\boldsymbol{\beta}}_y - \boldsymbol{\beta}) = J'(\hat{\boldsymbol{\beta}}_z - \boldsymbol{\beta}), \tag{7}$$

*where*

$$J = \left(\theta_{i-j} \vdots -\phi_{i-j} \vdots -\boldsymbol{\Theta}_{i-j} \vdots -\boldsymbol{\Phi}_{i-j}\right) \tag{8}$$

*and the $(i, j)$th entry in each partitioned matrix is indicated, and $\phi_i$, $\theta_i$, $\boldsymbol{\Phi}_i$ and $\boldsymbol{\Theta}_i$ are defined more generally for any integer $i$ as the negative of the coefficient of $B^i$ in their respective polynomials $\phi(B)$, $\theta(B)$, $\boldsymbol{\Phi}(B)$ and $\boldsymbol{\Theta}(B)$.*

McLeod's result has the following implications :

(1) In diagnostic checks:

   (a) $n\,\mathrm{var}\{\hat{r}(1)\} \cong \phi_p^2\theta_q^2$ and $n\,\mathrm{var}\{\hat{r}(s)\} \cong \boldsymbol{\Phi}_{p_s}^2 \cdot \boldsymbol{\Theta}_{q_s}^2$, where $\hat{r}(i)$ is the residual ACF at lag $(i)$.

   (b) The following was shown in Chap. 4 of McLeod's 1977 Ph.D. thesis.

**Theorem 2** *The asymptotic distribution of $\sqrt{n} \cdot \hat{\boldsymbol{r}}$ in the ARMA $(p, q)$ model with coefficients $\phi_1, \ldots, \phi_p, \theta_1, \ldots, \theta_q$ is exactly the same as in the ARMA $(p+q, 0)$ model with coefficients $\phi_1^*, \ldots, \phi_{p+q}^*$ defined by $\phi^*(B) = \phi(B)\theta(B)$.*

The result facilitates the calculation of asymptotic standard errors of $\hat{\boldsymbol{r}}$.

(2) In Model Identification:
Hipel et al. [24] was one of the earliest papers in advocating the use of inverse autocorrelations and inverse partial autocorrelations for time series model identification after Cleveland [6]. The inverse autocorrelations (IACF) are just the autocorrelations of the dual process, and they behave like partial autocorrelations of the primal. If a high order AR($q$) is used to approximate the multiplicative seasonal ARMA model, then the IACF can be estimated using the usual expression for the ACF of the dual MA($q$). The AR estimates are treated as if they are the MA estimates of the dual.

(3) Improved Box–Jenkins estimators:
Calculation of the likelihood function of ARMA models was a challenge in the 70s. For ARMA $(p, q)$ models with $n$ observations the likelihood function can be written

$$L(\phi, \theta, \sigma^2) \propto \sigma^{-n} \left| M_n^{(p,q)}(\phi, \theta) \right|^{1/2} \exp\left\{ -\frac{S(\phi, \theta)}{2\sigma^2} \right\},$$

where $M_n^{p,q}(\phi, \theta)$ is the determinant function of $\sigma^2 V$, where $V$ is the $n \times n$ covariance matrix of the series. Let

$$m_{p,0}(\Phi) = |M_n^{(p,0)}(\Phi)|.$$

McLeod [51] showed that if

$$\phi^*(B) = \phi(B)\theta(B),$$

then as $n \to \infty$, the limit of $M_n^{p,q}(\phi, \theta)$,

$$m_{p,q}(\phi, \theta) = \frac{m_{p,0}^2(\phi) m_{p,0}^2(\theta)}{m_{p+q,0}(\phi^*)}. \tag{9}$$

In other words, the calculation now reduces to that of AR models. Hence, the MLE can be computed much faster. McLeod [54] showed further that for the multiplicative seasonal ARMA model, using notations therein,

$$m(p, q, p_s, q_s) \cong m(p, q) \cdot \{m(p_s, q_s)\}^s.$$

The duality idea has been further exploited in later works: See [63]. Duality for hyperbolic decay time series was studied in [56].

## 5 Water Resources and Intervention Analysis

Since the Seventies, both Ian McLeod and Keith Hipel have contributed enormously to the time series modelling of water resources data. Naturally, many of the results obtained were based on the McLeod–Hipel time series package; furthermore, many of the papers are also closely related to the application of intervention analysis. The 1983 paper by McLeod, Hipel and Camacho suggested a range of statistical tools for explanatory and confirmatory data analysis of water quality time series. This paper won the 1984 American Water Resources Association (AWRA) award for the best paper of 1983 that appeared in the Water Resources Bulletin. Transfer function noise models and periodic autoregressive models were introduced to the water resources literature in some of these papers. See for example [65, 68]. The 1988 paper with Noakes, Hipel Jimenez and Yakowitz ably introduced the idea of fractional Gaussian noise and fractional ARMA models using various geophysical time series including annual river flows.

Intervention Analysis amounts to the dummy variable technique applied to time series analysis. Ian's work on this dated back to the early Seventies. A lot of empirical studies have been made using Canadian river flow data by Ian together with Keith Hipel. An interesting finding obtained by using intervention analysis is that there was a significant drop in flow since 1903 when the Aswan Dam was brought into operation [21]. Apart from many papers on environmental impacts using intervention analysis, power computation for intervention analysis had been long overdue prior to McLeod and Vingilis [61]. The result enables the sample size computation required for detecting an intervention with a prescribed power and level. See also [62].

A recent contribution by Ian is to road safety in Ontario. This appears in the paper, "Road safety impact of Ontario street racing and stunt driving law", by Aizhan Meirambayeva, Evelyn Vingilis, A. Ian McLeod et al., 2014. In this work, the impact of Ontario's street racing and stunt driving legislation was assessed using intervention analysis. Speeding-related collision casualties were examined separately in age and gender groups, and a covariate adjustment using non-speeding casualties was utilized. Some interesting findings include (1) The legal intervention was associated with significantly reduced casualties in young male drivers; (2) No significant change was observed in either young or mature female driver groups; and (3) Results suggest the presence of the deterrent effect of the new street racing law.

Another interesting recent application of intervention analysis is to the increased number of nephrology consults when estimated glomerular filtration rates were reported in addition to serums creatinine [29]. The intervention analysis was done for the whole Ontario population aged 25 or above. It was found that there is an increase of about 23 consults per nephrologist per year. The result would have potential impact in resources allocation and might lead to improved treatment for those with chronic kidney diseases.

There is no one in time series that I know of who has done so much with intervention analysis as Ian.

## 6 Epilogue

Ian's contribution to time series and other areas in statistics are clearly broad and of fundamental importance. Areas not covered in this article include

1. Simulation procedures for ARMA models
2. Trend assessment of water quality time series arising from the 1993 (*Water Resources Bulletin* paper with Hipel and Camacho)
3. Drawing simple random sample (*Applied Statistics*, 1983 with D. Bellhouse)
4. Multivariate contemporaneous ARMA (Works with Hipel and Camacho [58])
5. Kendall's Tau (*Annals of Statistics*, 1995 with Thompson and Valz; *American Statistician*, 1990 with Valz)
6. Subset autoregression (*Journal of Statistical Software* with Y. Zhang)
7. Algorithms in R (with Hao and Krougly)
8. *e*-publications

My apology to Ian and his many co-authors if I have not listed some of their works above.

Finally, I would like to express my personal indebtedness to Ian for his guidance, mentorship, patience and generosity while I was a Ph.D. student at UWO in the late 70s. I have learnt and received so much from Ian. Without Ian as my Ph.D. supervisor, I would not have prospered today. I hope this article will be a small token of my appreciation for what Ian has done for me.

## References


1. Baillie, R. T., Bollerslev, T., & Mikkelsen, H. O. (1996). Fractionally integrated generalized autoregressive conditional heteroscedasticity. *Journal of Econometrics*, *74*, 3–30.
2. Baillie, R. T., Chung, C. F., & Tieslau, M. A. (1995). Analyzing inflation by the fractionally integrated ARFIMA-GARCH Model. *Journal of Applied Economics*, *11*, 23–40.
3. Bollerslev, T. (1986). Generalized autoregressive conditional heteroskedasticity. *Journal of Econometrics*, *31*, 307–327.
4. Box, G. E. P., & Pierce, D. A. (1970). Distribution of the residual autocorrelations in autoregressive integrated moving average time series models. *Journal of the American Statistical Association*, *65*, 1509–1526.
5. Brockwell, P. J., & Davis, R. A. (1991). *Time series: Theory and methods* (2nd ed.). New York: Springer.
6. Cleveland, W. S. (1972). The inverse autocorrelations of a time series and their applications. *Technometrics*, *14*, 277–298.
7. Cryer, J. D., & Chan, K. S. (2008). *Time series analysis with applications in R. Springer texts in statistics* (2nd ed.). New York: Springer.
8. Davidson, J. (2004). Moment and memory properties of linear conditional heteroscedasticity models, and a new model. *Journal of Business & Economic Statistics*, *22*, 16–29.
9. Ding, Z., Granger, C. W. J., & Engle, R. F. (1993). A long memory property of stock market returns and a new model. *Journal of Empirical Finance*, *1*, 83–106.

10. Duchesne, P., & Roy, R. (2003). Robust tests for independence of two time series. *Statistica Sinica*, *13*, 827–852.
11. Dufour, J. M., & Roy, R. (1985). Some robust exact results on sample autocorrelations and test of randomness. *Journal of Econometrics*, *29*, 257–273.
12. Dufour, J. M., & Roy, R. (1986). Generalized portmanteau statistics and tests of randomness. *Communications in Statistics—Theory and Methods*, *15*, 2953–2972.
13. Engle, R. F. (1982). Autoregressive conditional heteroscedasticity with estimates of the variance of United Kingdom inflation. *Econometrica*, *50*, 987–1007.
14. Engle, R. F. (1983). Estimates of the variance of US inflation based on the ARCH model. *Journal of Money, Credit and Banking*, *15*, 286–301.
15. Feller, W. (1951). The asymptotic distribution of the range of sums of independent random variables. *The Annals of Mathematical Statistics*, *22*, 427–432.
16. Giraitis, L., Kokoszka, P., Leipus, R., & Teyssiere, G. (2003). Rescaled variance and related tests for long memory in volatility and levels. *Journal of Econometrics*, *112*, 265–294.
17. Granger, C. W. J., & Andersen, A. P. (1978). *Introduction to bilinear time series models*. Göttingen: Vandenhoeck and Ruprecht.
18. Granger, C. W. J., & Joyeux, R. (1980). An introduction to long memory time series models and fractional differencing. *Journal of Time Series Analysis*, *1*, 15–39.
19. Granger, C. W. J., Spear, S., & Ding, Z. (2000). Stylized facts on the temporal and distributional properties of absolute returns: An update. In W. S. Chan, W. K. Li, & H. Tong (Eds.), *Statistics and finance: An interface*. London: Imperial College Press.
20. Haugh, L. D. (1976). Checking the independence of two covariance-stationary time series: A univariate residual cross-correlation approach. *Journal of the American Statistical Association*, *71*, 378–385.
21. Hipel, K. W., Lennox, W. C., Unny, T. E., & McLeod, A. I. (1975). Intervention analysis in water resources. *Water Resources Research*, *11*, 855–861.
22. Hipel, K. W., & McLeod, A. I. (1978). Preservation of the rescaled adjusted range. 2. Simulation studies using Box–Jenkins models. *Water Resources Research*, *14*, 509–516.
23. Hipel, K. W., & McLeod, A. I. (1994). *Time series modelling of water resources and environmental systems*. Amsterdam: Elsevier Science.
24. Hipel, K. W., McLeod, A. I., & Lennox, W. C. (1977). Advances in Box–Jenkins modelling. Part 1. Theory. *Water Resources Research*, *13*, 567–575.
25. Hosking, J. R. M. (1980). The multivariate portmanteau statistic. *Journal of the American Statistical Association*, *75*, 602–607.
26. Hosking, J. R. M. (1981). Fractional differencing. *Biometrika*, *68*, 165–176.
27. Hurst, H. E. (1951). Long-term storage capacity of reservoirs. *The American Society of Civil Engineers*, *116*, 770–808.
28. Hurst, H. E. (1956). Methods of using long-term storage in reservoirs. *Proceedings of the Institution of Civil Engineers*, *1*, 519–543.
29. Jain, A. K., McLeod, A. I., Huo, C., Cuerden, M. S., Akbari, A., Tonelli, M., et al. (2009). When laboratories report estimated glomerular filtration rates in addition to serum creatinines, nephrology consults increase. *Kidney International*, *76*, 318–323.
30. Kheoh, T. S., & McLeod, A. I. (1992). Comparison of two modified portmanteau tests for model adequacy. *Computational Statistics & Data Analysis*, *14*, 99–106.
31. Kwan, W., Li, W. K., & Li, G. (2012). On the estimation and diagnostics checking of the ARFIMA-HYGARCH Model. *Computational Statistics & Data Analysis*, *56*, 3632–3644.
32. Li, G., & Li, W. K. (2005). Diagnostic checking for time series models with conditional heteroscedasticity estimated by the least absolute deviation approach. *Biometrika*, *92*, 691–701.
33. Li, G., & Li, W. K. (2008). Least absolute deviation estimation for fractionally integrated autoregressive moving average time series model with conditional heteroscedasticity. *Biometrika*, *95*, 399–414.
34. Li, M. Y., Li, W. K., & Li, G. (2013). On mixture memory models. *Journal of Time Series Analysis*, *34*, 606–624.

35. Li, M. Y., Li, W. K., & Li, G. (2015). A new hyperbolic GARCH model. *Journal of Econometrics*, *189*, 428–436.
36. Li, W. K. (1981). Unpublished Ph.D. thesis. University of Western Ontario, London, Canada
37. Li, W. K., & Hui, Y. V. (1994). Robust residual cross correlation tests for lagged relations in time series. *Journal of Statistical Computation and Simulation*, *49*, 103–109.
38. Li, W. K., & Mak, T. K. (1994). On the squared residual autocorrelations in non-linear time series with conditional heteroskedasticity. *Journal of Time Series Analysis*, *15*, 627–636.
39. Li, W. K., & McLeod, A. I. (1981). Distribution of the residual autocorrelations in multivariate ARMA time series models. *Journal of the Royal Statistical Society: Series B*, *43*, 231–239.
40. Li, W. K., & McLeod, A. I. (1986). Fractional time series modeling. *Biometrika*, *73*, 217–221.
41. Lin, J. W., & McLeod, A. I. (2006). Improved Peña–Rodriguez portmanteau test. *Computational Statistics & Data Analysis*, *51*, 1731–1738.
42. Lin, J. W., & McLeod, A. I. (2008). Portmanteau tests for ARMA models with infinite variance. *Journal of Time Series Analysis*, *29*, 600–617.
43. Ling, S., & Li, W. K. (1997a). On fractionally integrated autoregressive moving-average time series models with conditional heteroscedasticity. *Journal of the American Statistical Association*, *92*, 1184–1194.
44. Ling, S., & Li, W. K. (1997b). Diagnostic checking of nonlinear multivariate time series with multivariate ARCH errors. *Journal of Time Series Analysis*, *18*, 447–464.
45. Ljung, G. M., & Box, G. E. P. (1978). On a measure of lack of fit in time series models. *Biometrika*, *65*, 297–303.
46. Luukkonen, R., Saikkonen, P., & Teräsvirta, T. (1988). Testing linearity against smooth transition autoregressive models. *Biometrika*, *75*, 491–499.
47. Mahdi, E., & McLeod, A. I. (2012). Improved multivariate portmanteau diagnostic test. *Journal of Time Series Analysis*, *33*, 211–222.
48. Mandelbrot, B. B. (1965). Une classe de processus stochastiques homothétiques a soi: application à la loi climatologique de H. E. Hurst. *Comptes Rendus de l'Académie des Sciences*, *260*, 3274–3276.
49. Mandelbrot, B. B., & Wallis, J. R. (1968). Noah, Joseph and operational hydrology. *Water Resources Research*, *4*, 909–918.
50. McLeod, A. I. (1975). Derivation of the theoretical autocovariance function of autoregressive-moving average time series. *Applied Statistics*, *24*, 255–256.
51. McLeod, A. I. (1977). Improved Box–Jenkins estimators. *Biometrika*, *64*, 531–534.
52. McLeod, A. I. (1978). On the distribution and applications of residual autocorrelations in Box–Jenkins modelling. *Journal of the Royal Statistical Society: Series B*, *40*, 296–302.
53. McLeod, A. I. (1979). Distribution of the residual cross correlations in univariate ARMA time series models. *Journal of the American Statistical Association*, *74*, 849–855.
54. McLeod, A. I. (1984). Duality and other properties of multiplicative autoregressive-moving average models. *Biometrika*, *71*, 207–211.
55. McLeod, A. I. (1994). Diagnostic checking of periodic autoregression models with application. *Journal of Time Series Analysis15*, 221–233, Addendum. *Journal of Time Series Analysis*, *16*, 647–648.
56. McLeod, A. I. (1998). Hyperbolic decay time series. *Journal of Time Series Analysis*, *19*, 473–484.
57. McLeod, A. I., & Hipel, K. W. (1978). Preservation of the rescaled adjusted range, Part 1, A reassessment of the Hurst phenomenon. *Water Resources Research*, *14*, 491–508.
58. McLeod, A. I., Hipel, K. W., & Camacho, F. (1983). Trend assessment of water quality time series. *Water Resources Bulletin*, *19*, 537–547.
59. McLeod, A. I., Hipel, K. W., & Lennox, W. C. (1977). Advances in Box–Jenkins modelling. Part 2. Applications. *Water Resources Research*, *13*, 576–586.
60. McLeod, A. I., & Li, W. K. (1983). Diagnostic checking ARMA time series models using squared-residual autocorrelations. *Journal of Time Series Analysis*, *4*, 269–273.
61. McLeod, A. I., & Vingilis, E. R. (2005). Power computations for intervention analysis. *Technometrics*, *47*, 174–180.

62. McLeod, A. I., & Vingilis, E. R. (2008). Power computations in time series analyses for traffic safety interventions. *Accident Analysis & Prevention*, *40*, 1244–1248.
63. McLeod, A. I., & Zhang, Y. (2008). Faster ARMA maximum likelihood estimation. *Computational Statistics & Data Analysis*, *52*, 2166–2176.
64. Noakes, D. J., Hipel, K. W., McLeod, A. I., Jimenez, C. J., & Yakowitz, S. (1988). Forecasting annual geophysical time series. *International Journal of Forecasting*, *4*, 103–115.
65. Noakes, D. J., McLeod, A. I., & Hipel, K. W. (1985). Forecasting monthly riverflow time series. *International Journal of Forecasting*, *1*, 179–190.
66. Pierce, D. A. (1970). A duality between autoregressive and moving average processes concerning their least squares parameter estimates. *The Annals of Mathematical Statistics*, *41*, 722–726.
67. Robinson, P. M., & Zaffaroni, P. (2006). Pseudo-maximum likelihood estimation of ARCH($\infty$) models. *Annals of Statistics*, *34*, 1049–1074.
68. Thompstone, R. M., Hipel, K. W., & McLeod, A. I. (1985). Forecasting quarter-monthly riverflow. *Water Resource Bulletin*, *21*, 731–741.
69. Tong, H., & Lim, K. S. (1980). Threshold autoregression, limit cycles and cyclical data (with discussion), 575–586. *Journal of the Royal Statistical Society: Series B*, *42*, 245–292.
70. Wong, H., & Li, W. K. (1995). Portmanteau test for conditional heteroscedasticity, using ranks of squared residuals. *Journal of Applied Statistics*, *22*, 121–134.
71. Wong, H., & Li, W. K. (2002). Detecting and diagnostic checking multivariate conditional heteroscedastic time series models. *Annals of the Institute of Statistical Mathematics*, *54*, 45–59.

# The Doubly Adaptive LASSO for Vector Autoregressive Models

**Zi Zhen Liu, Reg Kulperger and Hao Yu**

**Abstract** The LASSO (Tibshirani, J R Stat Soc Ser B 58(1):267–288, 1996, [30]) and the adaptive LASSO (Zou, J Am Stat Assoc 101:1418–1429, 2006, [37]) are popular in regression analysis for their advantage of simultaneous variable selection and parameter estimation, and also have been applied to autoregressive time series models. We propose the doubly adaptive LASSO (daLASSO), or PLAC-weighted adaptive LASSO, for modelling stationary vector autoregressive processes. The procedure is doubly adaptive in the sense that its adaptive weights are formulated as functions of the norms of the partial lag autocorrelation matrix function (Heyse, 1985, [17]) and Yule–Walker or ordinary least squares estimates of a vector time series. The existing papers ignore the partial lag autocorrelation information inherent in a VAR process. The procedure shows promising results for VAR models. The procedure excels in terms of VAR lag order identification.

**Keywords** Adaptive LASSO · Asymptotic normality · Estimation consistency · LASSO · Oracle property · Doubly adaptive LASSO · PLAC-weighted adaptive LASSO · Selection consistency · VAR · VAR time series · Vector auroregressive processes · Teacher–Student dual

**Mathematics Subject Classifications (2010)** 62E20 · 62F10 · 62F12 · 62H12 · 62J07 · 62M10

Z.Z. Liu
Department of Mathematics, Trent University,
1600 W Bank Dr, Peterborough, ON K9J 7B8, Canada

R. Kulperger (✉) · H. Yu
Department of Statistical and Actuarial Sciences, University of Western Ontario,
1151 Richmond Street, London, ON N6A 5B7, Canada
e-mail: rjk@stats.uwo.ca

H. Yu
e-mail: hyu@stats.uwo.ca

 
W.K. Li et al. (eds.), *Advances in Time Series Methods and Applications*,
Fields Institute Communications 78, DOI 10.1007/978-1-4939-6568-7_2

# 1 Introduction

Consider the linear regression model $\boldsymbol{y} = \boldsymbol{X}\boldsymbol{\beta} + \boldsymbol{\varepsilon}$, where $\boldsymbol{X}$ is the design matrix, $\boldsymbol{\beta} = (\beta_1, \ldots, \beta_p)^T$ the coefficient vector, and $\boldsymbol{\varepsilon}$ the random error vector with zero mean vector and identity covariance matrix. The estimator for $\boldsymbol{\beta}$, $\hat{\boldsymbol{\beta}}^L$, defined by

$$\hat{\boldsymbol{\beta}}^L = \arg\min_{\boldsymbol{\beta}} \left\{ (\boldsymbol{y} - \boldsymbol{X}\boldsymbol{\beta})^T (\boldsymbol{y} - \boldsymbol{X}\boldsymbol{\beta}) + \lambda \sum_{j=1}^{p} |\beta_j| \right\},$$

where $\lambda$ is some positive tuning parameter, is called the LASSO estimator, which was proposed by Tibshirani [30]. Knight and Fu [19] investigated the asymptotic behaviour of the LASSO in linear regression models. They established asymptotic normality of the estimators for the non-zero components of the parameter vector and showed that the LASSO estimator sets some parameters exactly to 0 with a positive probability, which means that the estimators perform model selection and parameter estimation simultaneously. Zhao and Yu [36] and Zou [37] showed that the LASSO would correctly identify the active set only if the irrelevant covariates are roughly orthogonal to the relevant ones, as quantified through the so called *irrepresentable condition*. If the irrepresentable condition are not satisfied the LASSO fails to achieve variable selection consistency. As a remedy, Zou [37] proposed the adaptive LASSO (aLASSO), which is defined as

$$\hat{\boldsymbol{\beta}}^{aL} = \arg\min_{\boldsymbol{\beta}} \left\{ (\boldsymbol{y} - \boldsymbol{X}\boldsymbol{\beta})^T (\boldsymbol{y} - \boldsymbol{X}\boldsymbol{\beta}) + \lambda \sum_{j=1}^{p} w_j |\beta_j| \right\},$$

where $\lambda$ is a positive tuning parameter, and $w_j = 1/|\tilde{\beta}_j|^\gamma$ for $j = 1, \ldots, p$ are adaptive weights with $\tilde{\beta}_j$ being a $\sqrt{n}$-consistent estimate for $\beta_j$ and $\gamma > 0$ a fixed constant. The idea underlying the adaptive LASSO is that smaller penalties are put on important covariates and larger penalties on less important ones. The advantage of the adaptive LASSO over the LASSO is that when the irrepresentable condition fails the adaptive LASSO can still do proper variable selection asymptotically whereas the LASSO cannot [6]. Zou [37] showed that if $\lambda$ and $\gamma$ are properly chosen, the adaptive LASSO enjoys the oracle properties [11], namely, the variable selection consistency and estimation normality.

The (adaptive) LASSO methodology has been widely applied in the statistical literature including time series analysis. Lots of application cases exist for univariate autoregressive (AR) models (For example, [5–7, 20, 22, 24–26, 32, 35]). A bunch of application cases also exist for vector autoregressive (VAR) models. Valdés-Sosa et al. [31] used sparse VAR(1) models to estimate brain functional connectivity where the LASSO is applied to achieve sparsity of VAR(1) models. Fujita et al. [12] applied sparse VAR model to estimate gene regulatory networks based on gene expression profiles obtained from time-series microarray experiments where sparsity was reported to have been achieved by LASSO. Hsu et al. [18] applied the LASSO to achieve subset selection for VAR models of high order. In their methodology, the first step is the determination of the optimal lag order $p_{aic}$ or $p_{bic}$ via Akaike information criterion (AIC) or the Bayesian information criterion (BIC) criterion, respectively. Then they proposed the top-down, bottom-up and hybrid strategies to reduce the full VAR($p_{aic}$) or VAR($p_{aic}$) models to sparse models. Haufe et al. [16] applied the grouped LASSO to VAR models. Ren and Zhang [27] applied the adaptive LASSO to achieve subset selection for VAR models with higher lag order. Similar to Hsu et al. [18], the first step is to use AIC or Hannan and Quinn (HQ) criterion to determine the optimal lag order and then the adaptive LASSO was applied to reduce the full VAR models to sparse ones. Song and Bickel [29] proposed an integrated approach for large VAR processes that yields three types of estimators: the adaptive LASSO with (i) universal grouping, (ii) no grouping, and (iii) segmented grouping. Kock and Callot [21] investigated oracle efficient estimation and forecasting of the adaptive LASSO and the adaptive group LASSO for VAR models.

We proposed a systematic approach called the doubly adaptive LASSO (daLASSO) tailored to several time series models, which incorporates the information of partial autocorrelation embedded in time series data into adaptive LASSO weights [22]. Liu [22] discusses construction of these weights for AR processes and this will also appear in a companion paper. In this paper we focus on issues related to VAR models. In particular, for VAR models, we formulate adaptive weights as functions of the norms of the sample partial lag autocorrelation (PLAC) matrix function [17] and ordinary least squares (OLS) or Yule–Walker estimates of a VAR model. The method may also be called the PLAC-weighted adaptive LASSO, which achieves identification, selection and estimation all in one go. We prove that the PLAC-weighted adaptive LASSO possesses oracle properties. Simulation experiments suggest that our approach shows promising results for VAR models, especially in the identification of VAR lag order.

Section 2 gives a brief review of some basic concepts including the notion of partial lag autocorrelation (PLAC) matrix function [17] and classic methods for building VAR(p) models. Section 3 proposes the doubly adaptive LASSO for VAR models with the lag order unknown a priori, as is the usual case. Section 4 presents the asymptotic properties of the doubly adaptive LASSO estimators. Section 5 implements the algorithm. Section 6 summarizes results from numerical experiments. Section 7 offers a real data analysis example. Section 8 gives some concluding remarks. Proofs are put in Appendix.

## 2 The VAR(p) Process and Standard Modelling Procedure

**Definition 1** (*The VAR(p) process*) The $K$-variate time series $\{\boldsymbol{y}_t\}$, $t \in \mathbb{Z} = \{0, \pm 1, \pm 2, \ldots\}$ is said to be a VAR(p) process if it is stationary, and it is the solution of the specification

$$\boldsymbol{y}_t = \boldsymbol{\Phi}_1 \boldsymbol{y}_{t-1} + \cdots + \boldsymbol{\Phi}_p \boldsymbol{y}_{t-p} + \boldsymbol{\varepsilon}_t,\ t \in \mathbb{Z},$$

where $\boldsymbol{\Phi}_i$'s are fixed $K \times K$ coefficient matrices, and the innovation process $\boldsymbol{\varepsilon}_t \sim \text{WN}_K(\mathbf{0},\ \Sigma_\varepsilon)$. We say that $\{\boldsymbol{y}_t\}$ is an VAR(p) process with mean $\boldsymbol{\mu}$ if $\{\boldsymbol{y}_t - \boldsymbol{\mu}\}$ is a VAR(p) process.

For convenience and without loss of generality, we deal with only the demeaned VAR(p) process in this paper.

**Proposition 1** (The condition for the ergodic stationarity) *The VAR(p) process specified by* (1) *is ergodic stationary if and only if the corresponding characteristic equation satisfies the stability condition, namely,*

$$\det(I - \boldsymbol{\Phi}_1 z - \cdots - \boldsymbol{\Phi}_p z^p) \neq 0$$

*for* $|z| \leq 1$.

See Lütkepohl [23, pp. 14–16] for proof.

### Estimation of the VAR(p) Model

Given the VAR order p there are a variety of approaches to estimating the parameters (see, for example, Lütkepohl [23, pp. 69–102]). If the distribution of the innovation process is known, we can get MLE by maximizing the log-likelihood function. Through the Yule–Walker equations we can obtain the method-of-moments estimator. Maximizing the Gaussian quasi-likelihood yields QMLE if the normal distribution is used as a proxy for the unknown innovation distribution. A further possibility is to treat $\boldsymbol{y}_t = \boldsymbol{\Phi}_1 \boldsymbol{y}_{t-1} + \cdots + \boldsymbol{\Phi}_p \boldsymbol{y}_{t-p} + \boldsymbol{\varepsilon}_t,\ t = 1, \ldots, T$ as multivariate regression equation and employ the ordinary least squares (OLS) method for estimation. Hannan [14] shows that the OLS estimator has nice asymptotic properties such as consistency and asymptotic normality under some regularity conditions.

### Identification Via Information Criteria

A sequence of VAR models are estimated with successively increasing orders $1, 2, \ldots, h$ with $h$ sufficiently large. Then the model that minimizes some criterion is chosen. Some frequently used criteria include the final prediction error (FPE) [1], the AIC [2, 3], the BIC [28], and the HQ [15].

## The Partial Lag Autocorrelation Matrix

We may employ the Box–Jenkins methodology, starting with identification of the lag order. Then parameter estimation follows after the lag order identification. In extending the partial autocorrelation concept to vector time series, Heyse [17] introduced the notion of the partial lag autocorrelation (PLAC) matrix function. The PLAC is the autocorrelation matrix between the elements of $\boldsymbol{y}_t$ and $\boldsymbol{y}_{t+s}$, after removing the linear dependence of each on the intervening vectors $\boldsymbol{y}_{t+1}, \ldots, \boldsymbol{y}_{t+s-1}$, which is defined as the ordinary correlation between the elements of residuals,

$$\boldsymbol{u}_{s-1,t+s} = \boldsymbol{y}_{t+s} - \left(\Psi_{s-1,1}\boldsymbol{y}_{t+s-1} + \cdots + \Psi_{s-1,s-1}\boldsymbol{y}_{t+1}\right),$$

and

$$\boldsymbol{v}_{s-1,t} = \boldsymbol{y}_t - \left(\Theta_{s-1,1}\boldsymbol{y}_{t+1} + \cdots + \Theta_{s-1,s-1}\boldsymbol{y}_{t+s-1}\right)$$

where $\Psi_{s-1,j}$ and $\Theta_{s-1,j}$, $j = 1, \ldots, s-1$ are multivariate linear regression coefficients that minimizes $E[|\boldsymbol{u}_{s-1,t+s}|^2]$ and $E[|\boldsymbol{v}_{s-1,t}|^2]$, respectively.

**Definition 2** (*Partial lag autocorrelation matrix* [17]) The partial lag autocorrelation matrix function of lag $s$ is defined as

$$\boldsymbol{P}(s) = D_{\boldsymbol{v}}(s)^{-1/2}\boldsymbol{V}_{\boldsymbol{vu}}(s)D_{\boldsymbol{u}}(s)^{-1/2},$$

where

$$\begin{aligned}\boldsymbol{V}_{\boldsymbol{u}}(s) &= \mathrm{Var}[\boldsymbol{u}_{s-1,t+s}],\\ \boldsymbol{V}_{\boldsymbol{v}}(s) &= \mathrm{Var}[\boldsymbol{v}_{s-1,t}],\\ \boldsymbol{V}_{\boldsymbol{vu}}(s) &= \mathrm{Cov}(\boldsymbol{v}_{s-1,t}, \boldsymbol{u}_{s-1,t+s}),\end{aligned}$$

and $D_{\boldsymbol{v}}(s)$ and $D_{\boldsymbol{u}}(s)$ are the diagonal matrices of $\boldsymbol{V}_{\boldsymbol{v}}(s)$ and $\boldsymbol{V}_{\boldsymbol{u}}(s)$, respectively.

The $K \times K$ matrix function of the lag $s$, $\boldsymbol{P}(s)$, is a vector extension of the partial autocorrelation function in the same manner as the autocorrelation matrix function is a vector extension of the autocorrelation function. It can be shown that for $s \geq 2$, we have

$$\begin{aligned}\boldsymbol{V}_{\boldsymbol{u}}(s) &= \Gamma(0) - \sum_{k=1}^{s-1} \Psi_{s-1,k}\Gamma(k),\\ \boldsymbol{V}_{\boldsymbol{v}}(s) &= \Gamma(0) - \sum_{k=1}^{s-1} \Theta_{s-1,k}\Gamma'(k),\\ \boldsymbol{V}_{\boldsymbol{vu}}(s) &= \Gamma(s) - \sum_{k=1}^{s-1} \Gamma(s-k)\Psi'_{s-1,k},\end{aligned}$$

where $\Gamma(s)$ is the $K \times K$ lag-s autocovariance matrix, that is, $\Gamma(s) = \mathrm{Cov}(\boldsymbol{y}_t, \boldsymbol{y}_{t+s})$. Note that $\Gamma(s)$ is not symmetric; instead, $\Gamma(s)' = \Gamma(-s)$.

For the case $s = 1$ since there are no intervening vectors between $\boldsymbol{y}_t$ and $\boldsymbol{y}_{t+s}$ we have

$$\begin{aligned} \boldsymbol{V}_{\boldsymbol{u}}(1) &= \mathrm{VAR}(\boldsymbol{y}_{t+1}) = \Gamma(0), \\ \boldsymbol{V}_{\boldsymbol{v}}(1) &= \mathrm{VAR}(\boldsymbol{y}_t) = \Gamma(0), \\ \boldsymbol{V}_{\boldsymbol{vu}}(1) &= \mathrm{Cov}(\boldsymbol{y}_t, \boldsymbol{y}_{t+1}) = \Gamma(1), \end{aligned}$$

and

$$\boldsymbol{P}(1) = D^{-1/2}\Gamma(1)D^{-1/2} = \boldsymbol{\rho}(1),$$

where $D$ is the diagonal matrix of $\Gamma(0)$, and $\boldsymbol{\rho}(1)$ the regular autocorrelation matrix at lag 1.

It can be shown that for $K = 1$ the partial lag autocorrelation matrix function $\boldsymbol{P}(s)$ reduces to the partial autocorrelation function of a univariate autoregressive process.

Analogous to the partial autocorrelation function for the univariate case the partial lag autocorrelation matrix, $\boldsymbol{P}(s)$ has the cut-off property for vector autoregressive processes. So if $\{\boldsymbol{y}_t\}$ is a VAR(p) then $\boldsymbol{P}(s)$ will be nonzero for $s = p$ and will equal 0 for $s > p$. This property makes $\boldsymbol{P}(s)$ a useful tool for identifying VAR processes.

Heyse [17] also proposed a recursive procedure for computing $\boldsymbol{P}(s)$, which is a vector generalization of Durbin's [9] recursive computational procedure for univariate partial autocorrelations. The algorithm requires that we first estimate the sample cross-covariance matrices. Given a realization an K-dimensional vector time serie $\boldsymbol{y}_1, \boldsymbol{y}_2, \ldots, \boldsymbol{y}_T$, the sample autocovariance matrix at lag s is computed by

$$\widehat{\Gamma}(s) = \frac{1}{T}\sum_{t=1}^{T-s}(\boldsymbol{y}_t - \bar{\boldsymbol{y}})(\boldsymbol{y}_{t+s} - \bar{\boldsymbol{y}})',$$

where $\bar{\boldsymbol{y}}$ is the vector of sample mean. The sample partial lag autocorrelation matrix, $\widehat{\boldsymbol{P}}(s)$, can be obtained by using $\widehat{\Gamma}(r)$ of $\Gamma(r)$ for $r = 0, \ldots, s-1$ in the recursive algorithm. For computation details, see Heyse [17], Wei [33, pp. 408–412], or Liu [22].

## VAR Order Identification Via Sample PLAC Matrix

Under the null hypothesis that $\{\boldsymbol{y}_t\}$ is a VAR(s-1) process, the two series of residuals $\{\boldsymbol{u}_{s-1,t+s}\}$ and $\{\boldsymbol{v}_{s-1,t}\}$ are uncorrelated, and each consists of K independent white noise series. Heyes [17] showed that the elements of $\widehat{\boldsymbol{P}}(s)$, denoted by $\widehat{P}_{ij}(s)$, are asymptotically $N(0, 1/T)$ distributed. In addition, $T\left(\widehat{P}_{ij}(s)\right)^2 \sim \chi^2(1)$ asymptotically, which implies that asymptotically $X(s) = T\sum_{i=1}^{K}\sum_{i=1}^{K}\left(\widehat{P}_{ij}(s)\right)^2 \sim \chi^2(K^2)$. $X(s)$ provides a diagnostic aid for determining the order of a vector autoregressive model.

# 3 The PLAC-Weighted Adaptive LASSO

In this section, we use the LASSO methodology to model the VAR(p) process. There are two situations: VAR order is known in advance versus VAR order is unknown in advance.

## 3.1 The Adaptive LASSO

If the VAR order is known in advance, the adaptive LASSO of Zou [37] can be used to build a sparse VAR model. However, when the order is unknown in advance, the adaptive LASSO fails to identify the order properly. For illustration, we generate 1000 data sets of sample size $T = 500$ using R function of `mAr.sim` (R package `mAR`, Barbosa, 2009) from a bivariate VAR(5) process defined by (18) and (19) in Sect. 6.1. The aLASSO was applied to fit 1000 bivariate VAR models, one fit for each replicate. Pretending that we do not know the true lag order ($p = 5$) of the underlying bivariate VAR process, we set the maximum order $h$ to be 10 for each replicate. We use grid-search method and the BIC criteria to find an approximately optimal value of $\gamma$ for each replicates. Specifically, let $\gamma = [2.0, 4.0]_{\Delta=0.25}$, where the subscript $\Delta$ specifies the increment of the sequence. Table 1 shows some empirical statistics such as Bias, MSE, and MAD (See Sect. 6 for definitions of these statistics) of the aLASSO estimates for the VAR order. Table 2 shows the distribution of the aLASSO estimates for the VAR order. From the tables we see clearly that the aLASSO identifies the VAR order as 10 (i.e. VAR(10) models) most frequently and most of time (83 %).

To overcome the issue, we may employ a two-step procedure: First, use the OLS procedure plus the BIC criteria or the PLAC to identify the VAR order; second, apply the aLASSO to get a sparse model. This two-step procedure would work very well. Alternatively, we propose the doubly adaptive LASSO (daLASSO), or partial lag autocorrelation or PLAC-weighted adaptive LASSO. By employing the daLASSO we want to get order identification, subset selection and parameter estimation prop-

**Table 1** Empirical statistics of the aLASSO estimates for the bivariate AR order based on 1000 replicates, each of size T = 500, of the bivariate AR(5) model (18) with coefficients (19)

| True | Min | Max | Mean | Median | Mode | SE | Bias | MSE | MAD |
|---|---|---|---|---|---|---|---|---|---|
| 5 | 7 | 10 | 9.794 | 10 | 10 | 0.488 | 4.794 | 23.22 | 4.794 |

For each replicate, set h = 10 and use the BIC to choose $\lambda_T$ and $\gamma$

**Table 2** Empirical distribution of the aLASSO estimates for the bivariate AR order based on 1000 replicates, each of size T = 500, of the bivariate AR(5) model (18) with coefficients (19)

| Lag order | 5 | 6 | 7 | 8 | 9 | 10 |
|---|---|---|---|---|---|---|
| Percentage | 0 | 0 | 0.3 | 2.8 | 14.5 | 82.8 |

For each replicate, set h = 10 and use the BIC to choose $\lambda_T$ and $\gamma$

erly done in one go. Simulation results as shown in Tables 3 and 4 in Sect. 6.1 suggest that the daLASSO performs much better than the aLASSO in terms of VAR order identification.

## 3.2 The Doubly Adaptive LASSO

Suppose that we observe a time series $\boldsymbol{y}_1, \boldsymbol{y}_2, \ldots, \boldsymbol{y}_T$, which is a realization of a stationary $K$-variate VAR(p) process with the true order $p$ and true parameter matrix $\boldsymbol{\Phi}^o = (\boldsymbol{\Phi}_1^o, \ldots, \boldsymbol{\Phi}_p^o)$ unknown. Because the true lag order $p$ is not known a priori, we set the order to be $h$, which is sufficiently large such that $h > p$. Since the initial values $\boldsymbol{y}_0, \ldots, \boldsymbol{y}_{-h+1}$ are not available, we may use $\boldsymbol{y}_1, \ldots, \boldsymbol{y}_h$ as a presample. This will reduce the effective sample size from $T$ to $T - h$. Now, having the data, we formulate the following VAR(h) model

$$\boldsymbol{y}_t = \boldsymbol{\Phi}_1 \boldsymbol{y}_{t-1} + \cdots + \boldsymbol{\Phi}_h \boldsymbol{y}_{t-h} + \boldsymbol{\varepsilon}_t, \; t = h+1, \ldots T. \tag{1}$$

Let

$$\boldsymbol{\Phi} = (\boldsymbol{\Phi}_1, \boldsymbol{\Phi}_2, \ldots, \boldsymbol{\Phi}_h)_{K \times (hK)}, \tag{2}$$

$$\boldsymbol{x}_t = \left(\boldsymbol{y}_t', \boldsymbol{y}_{t-1}', \ldots, \boldsymbol{y}_{t-h+1}'\right)'_{(hK)\times 1}. \tag{3}$$

Then the model (1) can be written as

$$\boldsymbol{y}_t = \boldsymbol{\Phi} \boldsymbol{x}_{t-1}, \quad t = h+1, \ldots T.$$

To estimate the model via the OLS method, we define

$$\boldsymbol{Y} = \left(\boldsymbol{y}_{h+1}, \; \boldsymbol{y}_{h+2}, \ldots, \boldsymbol{y}_T\right)_{K \times (T-h)}, \tag{4}$$

$$\begin{aligned} \boldsymbol{X} &= (\boldsymbol{x}_h, \; \boldsymbol{x}_{h+1}, \ldots, \boldsymbol{x}_{T-1})_{(hK)\times(T-h)}, \\ &= \begin{pmatrix} \boldsymbol{y}_h & \boldsymbol{y}_{h+1} & \cdots & \boldsymbol{y}_{T-1} \\ \boldsymbol{y}_{h-1} & \boldsymbol{y}_h & \cdots & \boldsymbol{y}_{T-2} \\ \vdots & \vdots & \vdots & \vdots \\ \boldsymbol{y}_1 & \boldsymbol{y}_2 & \cdots & \boldsymbol{y}_{T-h} \end{pmatrix}_{(hK)\times(T-h)}, \end{aligned} \tag{5}$$

$$\boldsymbol{E} = (\boldsymbol{\varepsilon}_{h+1}, \; \boldsymbol{\varepsilon}_{h+2}, \ldots, \boldsymbol{\varepsilon}_T)_{K \times (T-h)},$$

and formulate compactly the multivariate-regression-type equations as

$$\boldsymbol{Y} = \boldsymbol{\Phi} \boldsymbol{X} + \boldsymbol{E}.$$

Equivalently, using *vec* operator, which transforms an $m \times n$ matrix into an $mn \times 1$ vector by stacking the columns, and Kronecker product operator $\otimes$, which for $m \times n$ matrix $A = (a_{ij})$ and $p \times q$ matrix $B = (b_{ij})$ generates an $mp \times nq$ matrix defined by

$$A \otimes B = \begin{pmatrix} a_{11}B & \cdots & a_{1n}B \\ \vdots & & \vdots \\ a_{m1}B & \cdots & a_{mn}B \end{pmatrix},$$

we may formulate the univariate-regression-type equations as

$$\boldsymbol{y} = \left(\boldsymbol{X}' \otimes \boldsymbol{I}_K\right) \boldsymbol{\phi} + \boldsymbol{\varepsilon}, \tag{6}$$

where $\boldsymbol{y}$ and $\boldsymbol{\varepsilon}$ are $K(T-h) \times 1$ vectors defined as

$$\boldsymbol{y} = vec(\boldsymbol{Y}) = \left(\boldsymbol{y}'_{h+1},\ \boldsymbol{y}'_{h+2},\ \ldots, \boldsymbol{y}'_T\right)', \tag{7}$$

$$\boldsymbol{e} = vec(\boldsymbol{E}) = \left(\boldsymbol{\varepsilon}'_{h+1},\ \boldsymbol{\varepsilon}'_{h+2},\ \ldots, \boldsymbol{\varepsilon}'_T\right)', \tag{8}$$

and $\boldsymbol{\phi}$ is a $(hK^2) \times 1$ vector defined as

$$\boldsymbol{\phi} = (\phi_1, \ldots, \phi_l, \ldots, \phi_{hK^2})' \tag{9}$$

$$= vec(\boldsymbol{\Phi}) = \left(vec(\boldsymbol{\Phi}_1)',\ vec(\boldsymbol{\Phi}_2)',\ \ldots,\ vec(\boldsymbol{\Phi}_h)'\right)'$$

$$= \left(\phi_{11,1}, \ldots, \phi_{KK,1}, \phi_{11,2}, \ldots, \phi_{KK,2}, \ldots, phi_{ij,k}, \ldots, \phi_{11,h}, \ldots, \phi_{KK,h}\right)'. \tag{10}$$

Note that the index $l$ in (9) corresponds to the $l$-th element of the vector $\boldsymbol{\phi}$, and the index $(ij, k)$ in (10) corresponds to the $(i, j)$-th element of the matrix $\boldsymbol{\Phi}_k$. The relation between $(i, j, k)$ and $l$ is bijective and defined by

$$l = f(i, j, k) = (k-1)K^2 + (j-1)K + i \tag{11}$$

where $l = 1, 2, \ldots, (hK^2)$, $i, j = 1, 2, \ldots, K$, and $k = 1, 2, \ldots, h$.

The true parameter matrix is $\boldsymbol{\Phi}^o = (\boldsymbol{\Phi}_1^o, \ldots, \boldsymbol{\Phi}_p^o)$, and the parameters vector is

$$\boldsymbol{\phi}^o \equiv \left(\phi_1^o,\ \phi_2^o, \ldots, \phi_{pK^2}^o\right) = vec(\boldsymbol{\Phi}^o) \tag{12}$$

$$= \left(\phi_{11,1}^o, \ldots, \phi_{KK,1}^o, \ldots, \phi_{11,p}^o, \ldots, \phi_{KK,p}^o\right)'.$$

In the context of the doubly adaptive LASSO procedure, we actually estimate the extended true parameter matrix $\boldsymbol{\Phi}^*$ or the extended true parameter vector $\boldsymbol{\phi}^*$ defined by

$$\boldsymbol{\Phi}^* = \left(\boldsymbol{\Phi}_1^*, \ldots, \boldsymbol{\Phi}_p^*, \boldsymbol{\Phi}_{p+1}^*, \ldots, \boldsymbol{\Phi}_h^*\right)',$$

where

$$\boldsymbol{\Phi}_j^* = \begin{cases} \boldsymbol{\Phi}_j^o & \text{if } j \le p \\ \mathbf{0} & \text{if } p < j \le h \end{cases},$$

or

$$\begin{aligned} \boldsymbol{\phi}^* &\equiv \left(\phi_1^*, \phi_2^*, \ldots, \phi_{hK^2}^*\right) \\ &= vec(\boldsymbol{\Phi}^*) = \left(vec(\boldsymbol{\Phi}_1^*)', vec(\boldsymbol{\Phi}_2^*)', \ldots, vec(\boldsymbol{\Phi}_h^*)'\right)' \\ &= \left(\phi_{11,1}^*, \ldots, \phi_{KK,1}^*, \ldots, \phi_{11,p}^*, \ldots, \phi_{KK,p}^*, \ldots, \phi_{11,h}^*, \ldots, \phi_{KK,h}^*\right)' \\ &= \left(\phi_{11,1}^o, \ldots, \phi_{KK,1}^o, \ldots, \phi_{11,p}^o, \ldots, \phi_{KK,p}^o, 0, \ldots, 0\right)'. \end{aligned} \tag{13}$$

It is clear that under appropriate assumptions on the initial values for the VAR(p) and VAR(h) processes, the VAR(p) with the fixed true parameters $\boldsymbol{\Phi}^o$,

$$\boldsymbol{y}_t = \sum_{j=1}^{p} \boldsymbol{\Phi}_j^o \boldsymbol{y}_{t-j} + \boldsymbol{a}_t, \quad t = 1, \ldots, T,$$

and the AR(h) with the fixed extended true parameters $\boldsymbol{\Phi}^*$,

$$\boldsymbol{y}_t = \sum_{j=1}^{h} \boldsymbol{\Phi}_j^* \boldsymbol{y}_{t-j} + \boldsymbol{a}_t, \quad t = 1, \ldots, T$$

are equivalent.

For an $m \times n$ matrix $A$, its entrywise $p$-norm, denoted as $\|A\|_p$, is defined as

$$\|A\|_p = \|\text{vec}(A)\|_p = \left(\sum_{i=1}^{m} \sum_{j=1}^{n} |a_{ij}|^p\right)^{1/p}.$$

The Frobenius norm, which is the special case $p = 2$, is defined as

$$\|A\|_F = \sqrt{\sum_{i=1}^{m} \sum_{j=1}^{n} |a_{ij}|^2}$$

**Definition 3** (*The PLAC-weighted adaptive LASSO*) The PLAC-weighted adaptive LASSO or doubly adaptive LASSO (daLASSO) estimator $\hat{\boldsymbol{\phi}}_T^{daL}$ for $\boldsymbol{\phi}^*$ is defined as

$$\hat{\boldsymbol{\phi}}^{daL} = \arg\min_{\boldsymbol{\phi}} \left\{ \left|\left| \boldsymbol{y} - \left(\boldsymbol{X}' \otimes \boldsymbol{I}_K\right) \boldsymbol{\phi} \right|\right|^2 + \lambda_T \sum_{k=1}^{h} \sum_{i=1}^{K} \sum_{j=1}^{K} \hat{w}_{ij,k} \left|\phi_{ij,k}\right| \right\}, \tag{14}$$

where

$$\hat{w}_{ij,k} = \frac{1}{\left|\tilde{\phi}_{ij,k}\right|^{\gamma_1} \left(\sum_{s=k}^{h} \left\|\widehat{\boldsymbol{P}}(s)\right\|_{\gamma_0}^{\gamma_0}\right)^{\gamma_2}} = \frac{1}{\left|\tilde{\phi}_{ij,k}\right|^{\gamma_1} A_k^{\gamma_2}}, \tag{15}$$

$$A_k = \sum_{s=k}^{h} \left\|\widehat{\boldsymbol{P}}(s)\right\|_{\gamma_0}^{\gamma_0}, \tag{16}$$

$\tilde{\phi}_{ij,k}$ is the ordinary least squares estimate or any other consistent estimate for $\phi_{ij,k}$, $\left\|\widehat{\boldsymbol{P}}(s)\right\|_{\gamma_0} = \left(\sum_{i=1}^{K} \sum_{j=1}^{K} |\widehat{P}_{ij}(s)|^{\gamma_0}\right)^{1/\gamma_0}$ is the entrywise $\gamma_0$-norm of the sample partial lag autocorrelation matrix $\widehat{\boldsymbol{P}}(s)$ at lag $s$, and $\gamma_0 > 0$, $\gamma_1 \geq 0$, and $\gamma_2 \geq 0$ are some fixed constants, and $h$ is the fixed maximum lag set initially.

**Remarks**:

1. Both the LASSO [30] and the adaptive LASSO [37] are special cases of the doubly adaptive LASSO. In former case, $\gamma_1 = \gamma_2 = 0$, and in latter case, $\gamma_2 = 0$.
2. In the daLASSO procedure the PLAC information and the Y-W or OLS estimates of the VAR(h) model work in tandem to perform subset selection and parameter estimation simultaneously. The basic idea can be elucidated from the following points:

   a. Firstly, note that $A_1 \geq \cdots \geq A_p \geq \cdots \geq A_h$. So monotonically increasing penalties are imposed on $\boldsymbol{\Phi}_k$ as $k$ increases from 1 to h. Consequently, a VAR term with smaller lag is more likely to be included in the model compared to one with larger lag.
   b. Secondly, due to the cutoff property of the PLAC, namely, the values of $\|\widehat{\boldsymbol{P}}(s)\|$ for $s = p+1, p+2 \ldots, h$ being relatively very small, if $k$ goes from $h$ backwards to $p$, $A_k$ will exhibit a sharp jump at $k = p$. Consequently, VAR terms with lags greater than $p$ get much more penalties than those with $k \leq p$, and therefore are more likely to be excluded from the model. The true VAR order is thus automatically identified.
   c. Finally, $|\tilde{\phi}_{ij,k}|^{\gamma_1}$ imposes larger penalty on $\phi_{ij,k}$ if the corresponding VAR term is not significant. If a VAR term is not significant, a consistent estimate for the corresponding coefficient is close to zero, and the penalty is therefore close to $\infty$. Consequently, an insignificant VAR term gets more penalty and is more likely to be excluded from the model compared to a significant term.

3. To see the mathematical reason why the daLASSO is superior to aLASSO in identifying the VAR order, observe that $\tilde{\phi}_{ij,k>p} = O_p(T^{-1/2})$ and $A_{k>p} = O_p(T^{-\gamma_0/2})$ so that the aLASSO weights $\hat{w}_{ij,k>p}^{aL} = O_p(T^{\gamma_1/2})$ and the daLASSO weights $\hat{w}_{ij,k>p}^{daL} = O_p(T^{(\gamma_1+\gamma_0\gamma_2)/2})$. Therefore, the daLASSO put more penalties on those parameters with lags that go beyond the true order $p$ compared to the aLASSO.

# 4 The Asymptotic Properties of the PLAC-Weighted Adaptive LASSO

The adaptive LASSO and the doubly adaptive LASSO methods yield biased estimators. In this section, however, we show that with properly chosen values for $\gamma_0$, $\gamma_1$, and $\gamma_2$ in (15), together with a proper choice of $\lambda_T$, the doubly adaptive LASSO enjoys desirable asymptotic properties. We actually study the asymptotic properties of the doubly adaptive LASSO estimator for the extended true parameter vector $\boldsymbol{\phi}^*$ in (13) instead of $\boldsymbol{\phi}^o$ in (12).

First, we clarify notations. Let $\tilde{\phi}_l$ be any consistent estimate for the true $\phi_l^*$, say, the OLS or Yule–Walker estimate. Let $\hat{\phi}_{T,l}^{daL}$ be the doubly adaptive LASSO estimate for $\phi_l^*$. Let $\mathbb{S}$ be the set of the true nonzero coefficient, i.e. $\mathbb{S} = \{l : \phi_l^* \neq 0\} = \text{supp}(\boldsymbol{\phi}^*) \subset \{1, 2, \ldots, hK^2\}$ with $h$ being set large enough such that $h > p$. Let $\mathbb{S}^c = \{1, 2, \ldots, hK^2\} \setminus \mathbb{S}$. Let $s = |\mathbb{S}|$ be the cardinality of the set $\mathbb{S}$. The assumption of the model sparsity implies that $s < pK^2$. Let $\hat{\mathbb{S}}_T = \{l : \hat{\phi}_{T,l}^{daL} \neq 0\}$ and $\hat{\mathbb{S}}_T^c = \{1, 2, \ldots, hK^2\} \setminus \hat{\mathbb{S}}_T$. Let $\boldsymbol{\phi}_{\mathbb{S}}^*$ be the $s$-dimensional vector for true underlying nonzero parameters, and $\boldsymbol{\phi}_{\mathbb{S}^c}^*$ be the vector for true underlying null parameters, i.e. $\boldsymbol{\phi}_{\mathbb{S}}^* = \{\phi_l^* : l \in \mathbb{S}\}$ and $\boldsymbol{\phi}_{\mathbb{S}^c}^* = \{\phi_l^* : l \in \mathbb{S}^c\}$. Let $\hat{\boldsymbol{\phi}}_{T,\mathbb{S}}^{daL}$ be the vector for the daLASSO estimate for $\boldsymbol{\phi}_{\mathbb{S}}^*$ and $\hat{\boldsymbol{\phi}}_{T,\mathbb{S}^c}^{daL}$ the vector for the daLASSO estimate for the zero vector $\boldsymbol{\phi}_{\mathbb{S}^c}^*$, i.e. $\hat{\boldsymbol{\phi}}_{T,\mathbb{S}}^{daL} = \{\hat{\phi}_{T,l}^{daL} : l \in \mathbb{S}\}$ and $\hat{\boldsymbol{\phi}}_{T,\mathbb{S}^c}^{daL} = \{\hat{\phi}_{T,l}^{daL} : l \in \mathbb{S}^c\}$. Let $\hat{\boldsymbol{\phi}}_{\hat{\mathbb{S}}_T}^{daL}$ be the vector for nonzero daLASSO estimates and $\hat{\boldsymbol{\phi}}_{\hat{\mathbb{S}}_T^c}^{daL}$ the vector for zero daLASSO estimates, i.e. $\hat{\boldsymbol{\phi}}_{\hat{\mathbb{S}}_T}^{daL} = \{\hat{\phi}_{T,l}^{daL} : l \in \hat{\mathbb{S}}_T\}$ and $\hat{\boldsymbol{\phi}}_{\hat{\mathbb{S}}_T^c}^{daL} = \{\hat{\phi}_{T,l}^{daL} : l \in \hat{\mathbb{S}}_T^c\}$. Note that $\hat{\boldsymbol{\phi}}_T^{daL} = \hat{\boldsymbol{\phi}}_{\hat{\mathbb{S}}_T}^{daL} \cup \hat{\boldsymbol{\phi}}_{\hat{\mathbb{S}}_T^c}^{daL}$ with the subscript indices being in the same order as those of $\boldsymbol{\phi}^*$.

Let $\boldsymbol{\Gamma}$ be the covariance matrix of $\boldsymbol{x}_t$ in (3), namely,

$$\boldsymbol{\Gamma} = E[\boldsymbol{x}_t \boldsymbol{x}_t'] = \begin{pmatrix} \Gamma(0) & \Gamma(-1) & \cdots & \Gamma(-h+1) \\ \Gamma(1) & \Gamma(0) & \cdots & \Gamma(-h+2) \\ \vdots & \vdots & & \vdots \\ \Gamma(h-1) & \Gamma(h-2) & \cdots & \Gamma(0) \end{pmatrix}_{(hK)\times(hK)},$$

where $\Gamma(s)$ is the lag-s autocovariance matrix of $\boldsymbol{y}_t$. Note that $\boldsymbol{\Gamma}$ is symmetric although $\Gamma(s)$ is not symmetric. We can partition $\boldsymbol{\Gamma}$ as follows

$$\boldsymbol{\Gamma} = \begin{pmatrix} \boldsymbol{\Gamma}_{\mathbb{S}\mathbb{S}} & \boldsymbol{\Gamma}_{\mathbb{S}\mathbb{S}^c} \\ \boldsymbol{\Gamma}_{\mathbb{S}^c\mathbb{S}} & \boldsymbol{\Gamma}_{\mathbb{S}^c\mathbb{S}^c} \end{pmatrix},$$

where we retain the ordering according to the lag index of $\boldsymbol{x}_t$ within each partition.

**Assumptions**:

**A0**: The coefficients matrix $\boldsymbol{\Phi}$ defined in (2) belongs to a compact set.

**A1**: For all $\boldsymbol{\Phi}$, $\det(I - \boldsymbol{\Phi}_1 z - \cdots - \boldsymbol{\Phi}_h z^h) \neq 0$ for $|z| \leq 1$.

**A2**: $\boldsymbol{\varepsilon}_t = (\varepsilon_1, \ldots, \varepsilon_K)'$ is a strong white noise process, i.e. $E[\boldsymbol{\varepsilon}_t] = \mathbf{0}$, $E\left[\boldsymbol{\varepsilon}_t \boldsymbol{\varepsilon}_t'\right] = \Sigma_\varepsilon$ is positive definite, $\varepsilon_t$ and $\varepsilon_s$ are independent for $s \neq t$, and $E|\varepsilon_{it}\varepsilon_{jt}\varepsilon_{kt}\varepsilon_{lt}| < M < \infty$ for $i,\ j,\ k,\ l = 1, \ldots, K$.

**A3**: The submatrix $\boldsymbol{\Gamma}_{\mathbb{SS}}$ is not singular and therefore invertible.

**Remarks on assumptions**:

(1) **A0** is always assumed.
(2) **A1** ensures that $\{\boldsymbol{y}_t\}$ is ergodic stationary
(3) **A2** requires the existence of finite fourth moments of $\{\boldsymbol{y}_t\}$.
(4) **A2** guarantees the existence of the covariance matrix $\boldsymbol{\Gamma}$.

The doubly adaptive LASSO estimator $\hat{\boldsymbol{\phi}}_T^{daL}$ is said to be *consistent* for $\boldsymbol{\phi}^*$ if

$$\| \hat{\boldsymbol{\phi}}_T^{daL} - \boldsymbol{\phi}^* \| \xrightarrow{P} 0 \text{ as } T \to \infty.$$

**Theorem 1** (Estimation Consistency of $\hat{\boldsymbol{\phi}}_T^{daL}$). *Let* $a_T = \sqrt{T} \min_{l \in \mathbb{S}} \left( |\tilde{\phi}_l|^{\gamma_1} A_l^{\gamma_2} \right)$, *where* $\left( |\tilde{\phi}_l|^{\gamma_1} A_l^{\gamma_2} \right)$ *corresponds to* $\left( |\tilde{\phi}_{(ij,k)}|^{\gamma_1} A_k^{\gamma_2} \right)$ *by the bijective function* (11). *If* $\lambda_T = o_p(a_T)$, *then under* ***A0–A2*** *we have:*

$$\left\| \hat{\boldsymbol{\phi}}_T^{daL} - \boldsymbol{\phi}^* \right\| \xrightarrow{P} 0 \textit{ as } T \to \infty,$$

*as* $T \to \infty$.

**Proposition 2** *Let* $a_T = \sqrt{T} \min_{l \in \mathbb{S}} \left( |\tilde{\phi}_l|^{\gamma_1} A_l^{\gamma_2} \right)$, *and* $b_T = \sqrt{T} \max_{l \in \mathbb{S}^c} \left( |\tilde{\phi}_l|^{\gamma_1} A_l^{\gamma_2} \right)$, *where* $\left( |\tilde{\phi}_l|^{\gamma_1} A_l^{\gamma_2} \right)$ *corresponds to* $\left( |\tilde{\phi}_{(ij,k)}|^{\gamma_1} A_k^{\gamma_2} \right)$ *by the bijective function* (11). *If* $\lambda_T = o_p(a_T)$ *and* $\lambda_T / b_T \xrightarrow{P} \infty$, *then under* ***A0–A3***, *we have*

$$\begin{cases} \sqrt{T} \left( \hat{\boldsymbol{\phi}}_{T,\mathbb{S}}^{daL} - \boldsymbol{\phi}_{\mathbb{S}}^* \right) \xrightarrow{D} N\left(\mathbf{0},\ (\boldsymbol{\Gamma}_{\mathbb{SS}})^{-1} \otimes \Sigma_\varepsilon\right), \\ \sqrt{T} \left( \hat{\boldsymbol{\phi}}_{T,\mathbb{S}^c}^{daL} - \boldsymbol{\phi}_{\mathbb{S}^c}^* \right) \xrightarrow{P} \mathbf{0}. \end{cases}$$

Proposition 2 regards the asymptotic property of $\hat{\boldsymbol{\phi}}_{T,\mathbb{S}}^{daL}$, not $\hat{\boldsymbol{\phi}}_{\hat{\mathbb{S}}_T}^{daL}$. To understand the difference between $\hat{\boldsymbol{\phi}}_{T,\mathbb{S}}^{daL}$ and $\hat{\boldsymbol{\phi}}_{\hat{\mathbb{S}}_T}^{daL}$, imagine a Teacher–Student dual in which Teacher is the data generator and Student is the data analyst. Teacher generates, say, 1000 data sets from a sparse VAR(p) model he knows apriori, and Student fits sparse VAR(p) models for Teacher. Teacher will give Student a good mark if Student could statistically identify the sparsity structure and estimate the coefficients with $\sqrt{T}$-consistency. Student does not know $\hat{\boldsymbol{\phi}}_{T,\mathbb{S}}^{daL}$ since he does not know the set $\mathbb{S}$. What

Student knows is $\hat{\boldsymbol{\phi}}^{daL}_{\hat{\mathbb{S}}_T}$ only. Teacher knows everything including $\mathbb{S}$, $\hat{\boldsymbol{\phi}}^{daL}_{T,\mathbb{S}}$, and $\hat{\boldsymbol{\phi}}^{daL}_{\hat{\mathbb{S}}_T}$. Proposition 2 is therefore useful for Teacher to assess analysis results from Student, but of little use for Student.

**Corollary 1** *Let* $a_T = \sqrt{T}\min_{l\in\mathbb{S}}\left(|\tilde{\phi}_l|^{\gamma_1}A_l^{\gamma_2}\right)$, *and* $b_T = \sqrt{T}\max_{l\in\mathbb{S}^c}\left(|\tilde{\phi}_l|^{\gamma_1}A_l^{\gamma_2}\right)$, *where* $\left(|\tilde{\phi}_l|^{\gamma_1}A_l^{\gamma_2}\right)$ *corresponds to* $\left(|\tilde{\phi}_{(ij,k)}|^{\gamma_1}A_k^{\gamma_2}\right)$ *by the bijective function* (11). *If* $\lambda_T = o_p(a_T)$ *and* $\lambda_T/b_T \xrightarrow{P} \infty$, *then under* ***A0–A3****, we have*

$$\mathbb{P}\left(l \in \hat{\mathbb{S}}_T\right) \to 1 \ \text{if} \ l \in \mathbb{S},$$

*as* $T \to \infty$.

This is clear because the $\sqrt{T}$-normality of $\hat{\boldsymbol{\phi}}^{daL}_{T,\mathbb{S}}$ in Proposition 2 implies that $\|\hat{\boldsymbol{\phi}}^{daL}_{T,\mathbb{S}} - \boldsymbol{\phi}^*_{\mathbb{S}}\| = O_p\left(1/\sqrt{T}\right)$. Thus, $\hat{\boldsymbol{\phi}}^{daL}_{T,\mathbb{S}} \xrightarrow{P} \boldsymbol{\phi}^*_{\mathbb{S}}$, which implies that $\forall l \in \mathbb{S}$, we have $\mathbb{P}\left(l \in \hat{\mathbb{S}}_T\right) \to 1$, as $T \to \infty$.

Fan and Li [11] specified the oracle properties of a sparse estimator in the language of Donoho et al. [8]. Heuristically, an oracle procedure can perform as well asymptotically as if the true submodel were known in advance. We extend the notion of the oracle properties of an estimator to the context of VAR times series models. The doubly adaptive positive LASSO estimator $\hat{\boldsymbol{\phi}}^{daL}_T$ for $\boldsymbol{\phi}^*$ is said to have the *oracle properties* if, with probability tending to 1, it could (i) identify the true sparsity pattern, i.e. $\lim P(\hat{\mathbb{S}}_T = \mathbb{S}) = 1$, (ii) identify the true lag order of the VAR process, i,e, $\lim P(\hat{p}^{daL}_T = p) = 1$, and (iii) have an optimal estimation rate of the coefficients as $T \to \infty$.

The following theorem says that the doubly adaptive LASSO procedure is an oracle procedure.

**Theorem 2** (Oracle properties of $\hat{\boldsymbol{\phi}}^{daL}_T$) *Let* $a_T = \sqrt{T}\min_{l\in\mathbb{S}}\left(|\tilde{\phi}_l|^{\gamma_1}A_l^{\gamma_2}\right)$ *and* $b_T = \sqrt{T}\max_{l\in\mathbb{S}^c}\left(|\tilde{\phi}_l|^{\gamma_1}A_l^{\gamma_2}\right)$, *where* $\left(|\tilde{\phi}_l|^{\gamma_1}A_l^{\gamma_2}\right)$ *corresponds to* $\left(|\tilde{\phi}_{(ij,k)}|^{\gamma_1}A_k^{\gamma_2}\right)$ *by the bijective function* (11). *If* $\lambda_T = o_p(a_T)$ *and* $\lambda_T/b_T \xrightarrow{P} \infty$, *then under* ***A0–A3****,* $\hat{\boldsymbol{\phi}}^{daL}_{T'}$ *must satisfy:*

*(i) Selection Consistency:* $\mathbb{P}\left(\hat{\mathbb{S}}_T = \mathbb{S}\right) \to 1$,
*(ii) Identification consistency:* $\mathbb{P}\left(\hat{p}^{daL}_T = p\right) \to 1$, *and*
*(iii) Asymptotic Normality:* $\sqrt{T}\left(\hat{\boldsymbol{\phi}}^{daL}_{\hat{\mathbb{S}}_T} - \boldsymbol{\phi}^*_{\mathbb{S}}\right) \xrightarrow{D} N\left(\mathbf{0},\ (\boldsymbol{\Gamma}_{\mathbb{SS}})^{-1} \otimes \Sigma_\varepsilon\right)$, *as* $T \to \infty$.

Theorem 2 regards the asymptotic properties of $\hat{\boldsymbol{\phi}}_{\hat{\mathbb{S}}_T}^{daL}$, the nonzero subvector of $\hat{\boldsymbol{\phi}}_T^{daL}$. In the Teacher–Student dual, Student knows $\hat{\mathbb{S}}_T$ and $\hat{\boldsymbol{\phi}}_{\hat{\mathbb{S}}_T}^{daL}$ as well. Theorem 2 assures that using the daLASSO Student is able to statistically identify the sparsity structure as if he knew $\mathbb{S}$ and estimate the coefficients with $\sqrt{T}$-consistency. Theorem 2 is therefore useful for Student, the data analyst, to assess the VAR models fitted via the daLASSO.

**Remarks:**

1. Although the asymptotic distributions of $\hat{\boldsymbol{\phi}}_{T,\mathbb{S}}^{daL}$ and $\hat{\boldsymbol{\phi}}_{\hat{\mathbb{S}}_T}^{daL}$ are identical, $\hat{\boldsymbol{\phi}}_{T,\mathbb{S}}^{daL}$ and $\hat{\boldsymbol{\phi}}_{\hat{\mathbb{S}}_T}^{daL}$ represent different identities; $\hat{\boldsymbol{\phi}}_{T,\mathbb{S}}^{daL}$ is the daLASSO estimator for the true non-zero parameter vector unknown in advance whereas $\hat{\boldsymbol{\phi}}_{\hat{\mathbb{S}}_T}^{daL}$ is the vector for non-zeros estimated by the daLASSO.
2. The oracle properties we discuss here concern $\hat{\boldsymbol{\phi}}_{\hat{\mathbb{S}}_T}^{daL}$ rather than $\hat{\boldsymbol{\phi}}_{T,\mathbb{S}}^{daL}$.
3. Proposition 2 concerns $\hat{\boldsymbol{\phi}}_{T,\mathbb{S}}^{daL}$, the daLASSO estimators for the true non-zero parameters, which are unknown in advance whereas Theorem 2 concerns $\hat{\boldsymbol{\phi}}_{\hat{\mathbb{S}}_T}^{daL}$, the non-zeros estimated by the daLASSO.
4. Estimation consistency is necessary for oracle properties whereas oracle properties are sufficient for the former.
5. The LASSO, the aLASSO and the daLASSO all have estimation consistency property under the same asymptotic condition for tuning parameter $\lambda_T$ (and other regularity conditions).
6. The LASSO, the aLASSO and the daLASSO estimators might behaviour quite differently when finite samples are used. We need to investigate and compare their finite sample properties.

## 5 Computation Algorithm for the Doubly Adaptive LASSO

Given values of $\gamma_0$, $\gamma_1$, and $\gamma_2$, the daLASSO procedure is implemented via the *lars* developed by Efron et al. [10]. The *lars* algorithm is very efficient, requiring the same order of computational cost as that of a single least squares fit. The LASSO methodology yields a path of possible solutions defined by the continuum over tuning and weighting parameters. The choice of these parameters determines the tradeoff between model fit and model sparsity. We use the BIC criteria to select the optimal value for $\Lambda_T = (\lambda_T, \gamma_0, \gamma_1, \gamma_2)$. The BIC is defined as

$$BIC = \log(\det \hat{\Sigma}_\varepsilon) + |\hat{\mathbb{S}}_T| \log(T - h), \tag{17}$$

where

$$\hat{\Sigma}_\varepsilon = \frac{1}{T-h}(Y - \hat{\boldsymbol{\Phi}}^{daL} X)(Y - \hat{\boldsymbol{\Phi}}^{daL} X)',$$

$|\hat{\mathbb{S}}_T|$ is the cardinality of the set $\hat{\mathbb{S}}_T$, $\hat{\boldsymbol{\Phi}}$ being the estimates for (2), $\boldsymbol{Y}$ is (4), and $\boldsymbol{X}$ is (5). Algorithm 1 is the detailed computational procedure for the doubly adaptive LASSO given the value of the triple $(\gamma_0, \gamma_1, \gamma_2)$. Algorithm 5 shows the complete computation steps.

---

**Algorithm 1**: The *lars* algorithm for the daLASSO given $(\gamma_0, \gamma_1, \gamma_2)$

---

**Input**: Data $\boldsymbol{y}_t, t = 1, \ldots, T$, and a specific value for $(\gamma_0, \gamma_1, \gamma_2)$ and fixed $h$.

**Output**: $\widehat{\boldsymbol{\Phi}}_T^{daL}$ for the specific $(\gamma_0, \gamma_1, \gamma_2)$.

START

Compute $\hat{w}_{ij,k}$ defined by (15) and transform to $\hat{w}_{T,l}$ according to (11).

Compute $\boldsymbol{X}^* = \boldsymbol{X}\widehat{\boldsymbol{W}}^{-1}$, where $\widehat{\boldsymbol{W}} = \text{diag}[\hat{w}_1, \ldots, \hat{w}_{hK^2}]$, i.e.
$\boldsymbol{x}_l^* = \boldsymbol{x}_l/\hat{w}_l, l = 1, \ldots, hK^2$.

Apply *lars* to obtain the path solution
$\hat{\boldsymbol{\phi}}(\lambda_T|(\gamma_0, \gamma_1, \gamma_2)) = \text{argmin}_{\boldsymbol{\phi}} \left\{(\boldsymbol{y} - \boldsymbol{X}^*\boldsymbol{\phi})^T(\boldsymbol{y} - \boldsymbol{X}^*\boldsymbol{\phi}) + \lambda_T \sum_{j=1}^{hK^2} |\phi_j|\right\}$.

Compute $\hat{\boldsymbol{\phi}}_T^{daL}(\lambda_T|(\gamma_0, \gamma_1, \gamma_2)) = \widehat{\boldsymbol{W}}^{-1}\hat{\boldsymbol{\phi}}$.

Compute $\text{BIC}(\lambda_T|(\gamma_0, \gamma_1, \gamma_2))$ according to (17) for the whole path.

Select $\lambda_T^*$ such that $\text{BIC}(\lambda_T^*|(\gamma_0, \gamma_1, \gamma_2)) \leq \text{BIC}(\lambda_T|(\gamma_0, \gamma_1, \gamma_2))$.

Output $\widehat{\boldsymbol{\Phi}}_T^{daL}(\lambda_T^*|(\gamma_0, \gamma_1, \gamma_2))$.

END

---

**Algorithm 2**: Complete algorithm for the daLASSO

---

**Input**: Data: $\boldsymbol{y}_t, t = 1, \ldots, T$ and fixed $h$

**Output**: The daLASSO estimator $\widehat{\boldsymbol{\Phi}}_T^{daL}$

Start: Set up a grid $\mathcal{G} = \gamma_0 \times \gamma_1 \times \gamma_2$ with $G = |\mathcal{G}|$.

**for** $g \leftarrow 1$ *to* $G$ **do**

  Apply Algorithm 1 to get $\widehat{\boldsymbol{\Phi}}_T\left(\lambda_T^{*(g)} \middle| \left(\gamma_0^{(g)}, \gamma_1^{(g)}, \gamma_2^{(g)}\right)\right)$.

  Calculate BIC $\left(\lambda_T^{*(g)}, \gamma_0^{(g)}, \gamma_1^{(g)}, \gamma_2^{(g)}\right)$.

Choose $\Lambda_T^*$ such that
$\text{BIC}(\Lambda_T^*) = \min\left\{\text{BIC}\left(\lambda_T^{*(g)}, \gamma_0^{(g)}, \gamma_1^{(g)}, \gamma_2^{(g)}\right) : \forall g = 1, \ldots, G\right\}$.

Output $\widehat{\boldsymbol{\Phi}}_T^{daL} \leftarrow \widehat{\boldsymbol{\Phi}}_T(\Lambda_T^*)$.

End

---

## 6 Monte Carlo Study

We use Monte Carlo to investigate the sampling properties of the PLAC-weighted adaptive LASSO estimator for VAR models. Specifically, we would like to assess its performance in terms of order identification, the parameter estimation, and subset selection. The empirical statistics such as minimum, maximum, mean, medium, mode (for VAR lag order only), standard error, bias, MSE, MAD, and selection proportion were summarized based on 1000 replications. The definitions of empirical bias, MSE, and MAD are listed below for reference (and the rest omitted):

$$\widehat{Bias}(\hat{p}^{daL}) = \hat{E}[\hat{p}^{daL}] - p = \frac{1}{M}\sum_{m=1}^{M}(\hat{p}^{daL})^{(m)} - p$$

$$\widehat{MSE}(\hat{p}^{daL}) = \hat{E}[\hat{p}^{daL} - p]^2 = \frac{1}{M}\sum_{m=1}^{M}((\hat{p}^{daL})^{(m)} - p)^2$$

$$\widehat{MAD}(\hat{p}^{daL}) = \hat{E}|\hat{p}^{daL} - p| = \frac{1}{M}\sum_{m=1}^{M}|(\hat{p}^{daL})^{(m)} - p|$$

$$\widehat{Bias}(\hat{\phi}_j^{daL}) = \hat{E}[\hat{\phi}_j^{daL}] - \phi_j^* = \frac{1}{M}\sum_{m=1}^{M}\left(\hat{\phi}_j^{daL}\right)^{(m)} - \phi_j^*$$

$$\widehat{MSE}(\hat{\phi}_j^{daL}) = \hat{E}[\hat{\phi}_j^{daL} - \phi_j^*]^2 = \frac{1}{M}\sum_{m=1}^{M}\left(\left(\hat{\phi}_j^{daL}\right)^{(m)} - \phi_j^*\right)^2$$

$$\widehat{MAD}(\hat{\phi}_j^{daL}) = \hat{E}|\hat{\phi}_j^{daL} - \phi_j^*| = \frac{1}{M}\sum_{m=1}^{M}\left|\left(\hat{\phi}_j^{daL}\right)^{(m)} - \phi_j^*\right|$$

where $M$ denotes the total number of MC runs.

## 6.1 A Bivariate VAR(5) Process

We use R function of `mAr.sim` (R package `mAR`, Barbosa, 2009) to replicate 1000 data sets, denoted as $\mathscr{D}^{(m)}, m = 1, \ldots, 1000$, of sample size $T = 500$ from the stationary and stable bivariate VAR(5) process defined by (18) and (19).

$$\boldsymbol{y}_t = \Phi_1 \boldsymbol{y}_{t-1} + \Phi_2 \boldsymbol{y}_{t-2} + \Phi_4 \boldsymbol{y}_{t-4} + \Phi_5 \boldsymbol{y}_{t-5} + \boldsymbol{e}_t, \tag{18}$$

where

$$\Phi_1 = \begin{pmatrix} 0.4 & 1.2 \\ 0.3 & 0.0 \end{pmatrix}, \quad \Phi_2 = \begin{pmatrix} 0.35 & -0.3 \\ 0.0 & -0.5 \end{pmatrix}, \quad \Phi_4 = \begin{pmatrix} 0.0 & -0.5 \\ 0.4 & 0.0 \end{pmatrix}, \quad \Phi_5 = \begin{pmatrix} 0.0 & 0.0 \\ 0.4 & -0.3 \end{pmatrix}, \tag{19}$$

and $\boldsymbol{e}_t$ is a Gaussian white noise with

$$\Sigma = \mathrm{Cov}(\boldsymbol{e}_t) = \begin{pmatrix} 1.0 & -0.6 \\ 0.0 & 2.5 \end{pmatrix}.$$

The daLASSO was applied to fit 1000 bivariate VAR models to $\mathscr{D}^{(m)}, m = 1, \ldots, 1000$. Pretending that the true lag order ($p = 5$) of the underlying bivariate VAR process is unknown, we set the maximum order $h$ to be 10 for each data. To find an approximately optimal combination of $\gamma_0$, $\gamma_1$, and $\gamma_2$, we use grid-search method and the BIC criteria. Specifically, let $\mathscr{G} = \gamma_0 \times \gamma_1 \times \gamma_2 = [2.0, 4.0]_{\Delta=0.25} \times [1.5, 8.0]_{\Delta=0.25} \times [1.5, 8.0]_{\Delta=0.25}$, where the subscript $\Delta$ specifies the increment of the sequence. The same 3-dimensional grid $\mathscr{G}$ is used for all 1000 replicates. Algorithm 3 describes the computational procedure for simulation study.

**Algorithm 3**: Algorithm for Monte Carlo

```
Input: Data 𝒟^(m), m = 1, ..., 1000 = M and Grid 𝒢.
Output: The LASSO estimate Φ̂^{daL(m)}, m = 1, ..., M.
Start
for m ← 1 to M do
 | Apply Algorithm 5 to get Φ̂^{daL(m)}.
Compute empirical statistics.
End
```

Table 3 shows some empirical statistics such as Bias, MSE, and MAD of the VAR order estimates. Table 4 shows the distribution of the VAR order estimates. Table 5 shows empirical statistics for VAR coefficients. We summarize a few observations as follows:

1. VAR lag order identification. Table 3 shows that the mode of 1000 bivariate VAR order estimates is 5, the true lag order. Table 4 shows that almost 72 % the fitted models have the order 5. The last column in Table 5 shows that autoregressors $\boldsymbol{y}_{t-k}$ for $k > 5$ have very slight chance to be included in models. Table 3 shows the mean and median of VAR order estimates are 5.55 and 5, respectively, indicating that the distribution of VAR order estimates is slightly skewed to the right with a right tail in distribution as evident in Table 4. This example confirms that the daLASSO procedure is very excellent in identifying the order of a VAR process.
2. VAR subset selection. The last column in Table 5 shows that the non-zero coefficients were selected into the model almost 100 % of time. On the other hand, some variables that are not included in the true bivariate VAR(5) process are also selected with quite high false inclusion rate. For example, $\Phi_3^* = \mathbf{0}$, but 25–54 % of time it was falsely estimated as non-zero. The variables corresponding to the coefficients $\phi_{22,1}$, $\phi_{21,2}$, and $\phi_{22,4}$ are falsely included in the models 39, 44,

**Table 3** Empirical statistics of the daLASSO estimates for the bivariate AR order based on 1000 replicates, each of size T = 500, of the bivariate AR(5) model (18) with coefficients (19)

| True | Min | Max | Mean | Median | Mode | SE | Bias | MSE | MAD |
|---|---|---|---|---|---|---|---|---|---|
| 5 | 5 | 10 | 5.546 | 5 | 5 | 1.015 | 0.546 | 1.328 | 0.546 |

For each replicate, set h = 10 and use the BIC to choose $\lambda_T$, $\gamma_0$, $\gamma_1$, and $\gamma_2$

**Table 4** Empirical distribution of the daLASSO estimates for the bivariate AR order based on 1000 replicates, each of size T = 500, of the bivariate AR(5) model (18) with coefficients (19)

| Lag order | 5 | 6 | 7 | 8 | 9 | 10 |
|---|---|---|---|---|---|---|
| Percentage | 71.6 | 11.4 | 10.8 | 3.9 | 1.6 | 0.7 |

For each replicate, set h = 10 and use the BIC to choose $\lambda_T$, $\gamma_0$, $\gamma_1$, and $\gamma_2$

and 45 % of time, respectively. This confirms the suggestion that the daLASSO procedure have strong power and be conservative in terms of subset selection.

3. VAR coefficients estimation. The Mean, Median, SE, BIAS, and MSE columns in Table 5 suggest that the parameters are consistently estimated. In addition, the minimum and maximum columns in Table 5 shows that the signs of parameters are identified correctly almost 100 % of times: if the true value of a parameter is positive, the minimum of estimates never falls below 0; if the true value of a parameter is negative, the maximum of estimates never goes beyond 0. This example confirms the suggestion that the daLASSO procedure estimate the parameters consistently.

## 6.2 A Trivariate VAR(5) Process

We also conduct simulation study on a sparse trivariate VAR(5) process. We use R function of `mAr.sim` (R package `mAR`, Barbosa, 2009) to generate 1000 data sets of sample size $T = 500$ from the stationary process defined by (20) and (21). The daLASSO was applied to fit 1000 models. For each data set, we use grid-search method and the BIC criteria to find an approximately optimal combination of $\gamma_0$, $\gamma_1$, and $\gamma_2$. Specifically, let $\mathcal{G} = \gamma_0 \times \gamma_1 \times \gamma_2 = [2.0, 4.0]_{\Delta=0.25} \times [1.5, 8.0]_{\Delta=0.25} \times [1.5, 8.0]_{\Delta=0.25}$. The same 3-dimensional grid $\mathcal{G}$ is used for all 1000 replicates.

$$y_t = \Phi_1 y_{t-1} + \Phi_2 y_{t-2} + \Phi_4 y_{t-4} + \Phi_5 y_{t-5} + e_t, \tag{20}$$

where

**Table 5** Empirical statistics of the daLASSO estimates for the bivariate AR coefficients $\boldsymbol{\Phi}_1 - \boldsymbol{\Phi}_5$ based on 1000 replicates, each of size T = 500, of the bivariate AR(5) model (18) with coefficients (19)

| Coeff | True | Min | Max | Mean | Median | SE | Bias | MSE | MAD | P(Select) |
|---|---|---|---|---|---|---|---|---|---|---|
| $\phi_{11,1}$ | 0.4 | 0.2228 | 0.5071 | 0.4003 | 0.4009 | 0.0352 | 0.0003 | 0.0012 | 0.0269 | 1 |
| $\phi_{21,1}$ | 0.3 | 0 | 0.5177 | 0.3001 | 0.3009 | 0.0629 | 0.0001 | 0.0039 | 0.0491 | 0.999 |
| $\phi_{12,1}$ | 1.2 | 1.1155 | 1.2684 | 1.1985 | 1.1983 | 0.0218 | −0.0015 | 0.0005 | 0.0173 | 1 |
| $\phi_{22,1}$ | 0 | −0.1091 | 0.1333 | 0.0002 | 0 | 0.0349 | 0.0002 | 0.0012 | 0.0191 | 0.389 |
| $\phi_{11,2}$ | 0.35 | 0.2384 | 0.4406 | 0.3475 | 0.3484 | 0.0265 | −0.0025 | 0.0007 | 0.0195 | 1 |
| $\phi_{21,2}$ | 0 | −0.2051 | 0.2316 | 0.0004 | 0 | 0.0615 | 0.0004 | 0.0038 | 0.0353 | 0.442 |
| $\phi_{12,2}$ | −0.3 | −0.4301 | −0.0972 | −0.3007 | −0.3033 | 0.046 | −0.0007 | 0.0021 | 0.0361 | 1 |
| $\phi_{22,2}$ | −0.5 | −0.8041 | −0.0958 | −0.5004 | −0.5003 | 0.0867 | −0.0004 | 0.0075 | 0.0676 | 1 |
| $\phi_{11,3}$ | 0 | −0.1295 | 0.0948 | −0.0017 | 0 | 0.0216 | −0.0017 | 0.0005 | 0.0091 | 0.251 |
| $\phi_{21,3}$ | 0 | −0.2374 | 0.2041 | 0.0004 | 0 | 0.0525 | 0.0004 | 0.0027 | 0.0292 | 0.441 |
| $\phi_{12,3}$ | 0 | −0.1178 | 0.1204 | 0.0009 | 0 | 0.0292 | 0.0009 | 0.0009 | 0.0136 | 0.303 |
| $\phi_{22,3}$ | 0 | −0.2952 | 0.2556 | 0.0025 | 0 | 0.0762 | 0.0025 | 0.0058 | 0.0458 | 0.539 |
| $\phi_{11,4}$ | 0 | −0.118 | 0.0796 | −0.0007 | 0 | 0.017 | −0.0007 | 0.0003 | 0.006 | 0.195 |
| $\phi_{21,4}$ | 0.4 | 0.1667 | 0.6153 | 0.3979 | 0.4004 | 0.0631 | −0.0021 | 0.004 | 0.0486 | 1 |
| $\phi_{12,4}$ | −0.5 | −0.6263 | −0.3478 | −0.4973 | −0.4983 | 0.0323 | 0.0027 | 0.001 | 0.0244 | 1 |
| $\phi_{22,4}$ | 0 | −0.2723 | 0.2849 | −0.001 | 0 | 0.0614 | −0.001 | 0.0038 | 0.0331 | 0.448 |
| $\phi_{11,5}$ | 0 | −0.0825 | 0.0841 | 0.0003 | 0 | 0.0098 | 0.0003 | 0.0001 | 0.0026 | 0.128 |
| $\phi_{21,5}$ | 0.4 | 0.2103 | 0.5691 | 0.3974 | 0.3952 | 0.0418 | −0.0026 | 0.0018 | 0.0313 | 1 |
| $\phi_{12,5}$ | 0 | −0.1346 | 0.1454 | 0.0014 | 0 | 0.0241 | 0.0014 | 0.0006 | 0.0087 | 0.197 |
| $\phi_{22,5}$ | −0.3 | −0.5738 | 0 | −0.2941 | −0.2982 | 0.0844 | 0.0059 | 0.0072 | 0.0617 | 0.975 |
| $\phi_{11,6}$ | 0 | −0.0846 | 0.0311 | −0.0002 | 0 | 0.0038 | −0.0002 | 0 | 0.0003 | 0.007 |

(continued)

**Table 5** (continued)

| Coeff | True | Min | Max | Mean | Median | SE | Bias | MSE | MAD | P(Select) |
|---|---|---|---|---|---|---|---|---|---|---|
| $\phi_{21,6}$ | 0 | −0.1525 | 0.1818 | −0.0002 | 0 | 0.0201 | −0.0002 | 0.0004 | 0.0043 | 0.061 |
| $\phi_{12,6}$ | 0 | −0.099 | 0.124 | 0.0002 | 0 | 0.008 | 0.0002 | 0.0001 | 0.001 | 0.023 |
| $\phi_{22,6}$ | 0 | −0.2432 | 0.2121 | −0.0018 | 0 | 0.0377 | −0.0018 | 0.0014 | 0.0105 | 0.099 |
| $\phi_{11,7}$ | 0 | −0.0318 | 0.0467 | 0.0001 | 0 | 0.0022 | 0.0001 | 0 | 0.0001 | 0.005 |
| $\phi_{21,7}$ | 0 | −0.1472 | 0.1762 | 0 | 0 | 0.0138 | 0 | 0.0002 | 0.0021 | 0.032 |
| $\phi_{12,7}$ | 0 | −0.0628 | 0.0873 | 0.0002 | 0 | 0.0051 | 0.0002 | 0 | 0.0005 | 0.014 |
| $\phi_{22,7}$ | 0 | −0.2132 | 0.1937 | −0.0005 | 0 | 0.0276 | −0.0005 | 0.0008 | 0.0066 | 0.075 |
| $\phi_{11,8}$ | 0 | 0 | 0.0422 | 0 | 0 | 0.0013 | 0 | 0 | 0 | 0.001 |
| $\phi_{21,8}$ | 0 | −0.0635 | 0.109 | 0.0003 | 0 | 0.0065 | 0.0003 | 0 | 0.0005 | 0.009 |
| $\phi_{12,8}$ | 0 | −0.0284 | 0 | 0 | 0 | 0.001 | 0 | 0 | 0 | 0.002 |
| $\phi_{22,8}$ | 0 | −0.2056 | 0.1646 | −0.0006 | 0 | 0.0182 | −0.0006 | 0.0003 | 0.0028 | 0.028 |
| $\phi_{11,9}$ | 0 | 0 | 0 | 0 | 0 | 0 | 0 | 0 | 0 | 0 |
| $\phi_{21,9}$ | 0 | −0.0713 | 0.166 | 0.0004 | 0 | 0.0075 | 0.0004 | 0.0001 | 0.0006 | 0.008 |
| $\phi_{12,9}$ | 0 | 0 | 0 | 0 | 0 | 0 | 0 | 0 | 0 | 0 |
| $\phi_{22,9}$ | 0 | −0.148 | 0.0944 | −0.0004 | 0 | 0.0084 | −0.0004 | 0.0001 | 0.0007 | 0.009 |
| $\phi_{11,10}$ | 0 | 0 | 0 | 0 | 0 | 0 | 0 | 0 | 0 | 0 |
| $\phi_{21,10}$ | 0 | 0 | 0 | 0 | 0 | 0 | 0 | 0 | 0 | 0 |
| $\phi_{12,10}$ | 0 | 0 | 0 | 0 | 0 | 0 | 0 | 0 | 0 | 0 |
| $\phi_{22,10}$ | 0 | −0.1751 | 0.0478 | −0.0003 | 0 | 0.0076 | −0.0003 | 0.0001 | 0.0005 | 0.007 |

For each replicate, set h = 10 and use the BIC to choose $\lambda_T$, $\gamma_0$, $\gamma_1$, and $\gamma_2$

$$\Phi_1 = \begin{pmatrix} 0.3 & 0.2 & 0.3 \\ 0.5 & 0.0 & 0.0 \\ 0.0 & 0.1 & -0.5 \end{pmatrix}, \quad \Phi_2 = \begin{pmatrix} -0.3 & 0.0 & 0.0 \\ 0.0 & 0.1, & -0.5 \\ 0.7 & 0.2 & 0.0 \end{pmatrix},$$
$$\Phi_4 = \begin{pmatrix} 0.0 & 0.4 & -0.2 \\ 0.6 & 0.0 & 0.0 \\ 0.0 & -0.4, & 0.0 \end{pmatrix}, \quad \Phi_5 = \begin{pmatrix} 0.2 & 0.0 & 0.0 \\ 0.0 & 0.0 & 0.4 \\ 0.0 & 0.3 & 0.3 \end{pmatrix}, \qquad (21)$$

and $\boldsymbol{e}_t$ is a Gaussian white noise with

$$\Sigma = \mathrm{Cov}(\boldsymbol{e}_t) = \begin{pmatrix} 1.0 & -0.6 & 0.4 \\ 0.2 & 1.2 & 0.3 \\ -0.5 & 0.1 & 1.1 \end{pmatrix}.$$

The results are consistent with what we got from the bivariate case. Interested readers may obtain the results from authors through emails.

## 7 Real Data Analysis

We use the data of quarterly West German investment, income, and consumption data (1960–1982) from Lütkepohl [23, pp. 77–79]. Using the software Stata function **var** we fit a VAR(2) model on the first differences of logarithms of the data. The estimated coefficients follow with the significant ones being bold-faced:

$$\hat{\Phi}_1 = \begin{pmatrix} \mathbf{-0.273} & \mathbf{0.337} & 0.652 \\ 0.043 & -0.123 & \mathbf{0.305} \\ 0.003 & \mathbf{0.289} & \mathbf{-0.285} \end{pmatrix}, \quad \hat{\Phi}_2 = \begin{pmatrix} -0.134 & 0.183 & 0.598 \\ \mathbf{0.062} & 0.021 & 0.049 \\ \mathbf{0.050} & \mathbf{0.366} & -0.116 \end{pmatrix}.$$

We use the daLASSO to fit a sparse VAR model. We set $h = 4$ and the grid $\mathscr{G} = \gamma_0 \times \gamma_1 \times \gamma_2 = [1.0, 4.0]_{\Delta=0.5} \times [1.0, 4.0]_{\Delta=0.25} \times [1.0, 5.0]_{\Delta=0.25}$. We use the BIC to select the optimal value for tuning and weighting parameters. A VAR(4) sparse model was fitted with estimated coefficients as follows.

$$\hat{\Phi}_1^{daL} = \begin{pmatrix} -0.261 & 0.381 & 0.399 \\ 0.018 & 0 & 0.534 \\ 0 & 0.456 & -0.139 \end{pmatrix}, \quad \hat{\Phi}_2^{daL} = \begin{pmatrix} 0.399 & 0.030 & 0.426 \\ 0.534 & 0 & 0.378 \\ -0.139 & 0.536 & 0 \end{pmatrix},$$

$$\hat{\Phi}_3^{daL} = \hat{\Phi}_4^{daL} = \mathbf{0}.$$

We observe that (i) all coefficient matrices beyond the lag 2 were shrank to zero, (ii) all significant coefficients were included in the model, (iii) all coefficients that were set to 0 are insignificant, and (iv) some insignificant coefficients were included in the model by the doubly adaptive LASSO procedure.

## 8 Conclusion

In this paper, we propose the doubly adaptive LASSO or PLAC-weighted adaptive LASSO for VAR models. The adaptive LASSO alone fails to identify the VAR order. So one has to identify the lag order of a VAR process before applying the aLASSO. The daLASSO incorporates the partial lag autocorrelation into adaptive LASSO weights, thereby getting order identification, subset selection and parameter estimation done in one go, as shown in Monte Carlo examples and real data analysis example. In the future research, we will develop methods for estimating standard errors of daLASSO estimators and constructing forecasting intervals. We will also conduct comparison studies on forecast performance and goodness-of-fit of classic, aLASSO and daLASSO approaches.

**Acknowledgements** We sincerely thank two anonymous referees for their valuable comments and suggestions that we have adopted to improve this manuscript greatly.

## Appendix

These proofs use three basic results on ergodicity, which are stated here for completeness.

**Theorem 3** (Ergodic theorem. See White [34, pp. 39–46]) *Let the K-variate vector process $\{\boldsymbol{y}_t\}$ be ergodic stationary with $E[\boldsymbol{y}_t] = \boldsymbol{\mu}$ where $E[y_{i,t}] = \mu_i$ is finite for all $i = 1, \ldots, K$. Then*

$$\bar{\boldsymbol{y}}_T = \frac{1}{T}\sum_{t=1}^{T} \boldsymbol{y}_t \longrightarrow \boldsymbol{\mu} \ a.s.$$

**Theorem 4** (Ergodic theorem of functions. See White [34, pp. 39–46]) *Let $f$ be a $\mathscr{F}$-measurable function into $\mathbb{R}^k$ and define $z_t = f(\ldots, \boldsymbol{y}_t, \boldsymbol{y}_{t-1}, \ldots)$, where $\boldsymbol{y}_t$ is $q \times 1$ vector. (i) If $\{\boldsymbol{y}_t\}$ is stationary, then $\{z_t\}$ is stationary. (ii) If $\{\boldsymbol{y}_t\}$ is ergodic stationary and $E[z_t]$ is well-defined then $\{z_t\}$ is ergodic stationary.*

**Theorem 5** (Central Limit Theorem for Martingale Differences. Billingsley [4]) *Let $\{\boldsymbol{v}_t\}$ be an ergodic stationary sequence of square integrable martingale difference vectors such that $Var[\boldsymbol{v}_t] \equiv \Sigma_v$ whose all entries exist and finite, Then*

$$\frac{1}{\sqrt{T}}\sum_{t=1}^{T} \boldsymbol{v}_t \xrightarrow{D} N(\boldsymbol{0}, \ \Sigma_v).$$

**Lemma 1** (Lütkepohl [23, p. 73] states this lemma without proof) *Under **A1–A2**, we have*

1. $\frac{1}{T}\boldsymbol{X}\boldsymbol{X}' \xrightarrow{a.s.} \boldsymbol{\Gamma}$,
2. $\frac{1}{T}(\boldsymbol{X} \otimes I_K)\boldsymbol{e} \xrightarrow{a.s.} \boldsymbol{0}$, *and*
3. $\frac{1}{\sqrt{T}}(\boldsymbol{X} \otimes I_K)\boldsymbol{e} \xrightarrow{D} \boldsymbol{w} \sim N(\boldsymbol{0},\ \boldsymbol{\Gamma} \otimes \Sigma_\varepsilon)$,
   *where $\otimes$ denotes the Kronecker product.*

*Proof* (i) It is easy to check that $\boldsymbol{X}\boldsymbol{X}' = \sum_{t=h}^{T-1} \boldsymbol{x}_t\boldsymbol{x}_t'$. By **A1**, $\boldsymbol{x}_t$ is ergodic stationary. By Theorem 4, $\boldsymbol{x}_t\boldsymbol{x}_t'$ is also ergodic stationary. By Theorem 3 we have

$$\frac{1}{T}\boldsymbol{X}\boldsymbol{X}' \xrightarrow{a.s.} E[\boldsymbol{x}_t\boldsymbol{x}_t'] = \boldsymbol{\Gamma}.$$

(ii) It is not very hard to check that $(\boldsymbol{X} \otimes I_K)\boldsymbol{e} = \sum_{t=h+1}^{T}(\boldsymbol{x}_{t-1} \otimes I_K)\boldsymbol{\varepsilon}_t$. Since $\boldsymbol{x}_t$ is ergodic stationary by **A1**, so is $(\boldsymbol{x}_{t-1} \otimes I_K)\boldsymbol{\varepsilon}_t$ by Theorem 4. Also by Theorem 3, we have

$$\frac{1}{T}(\boldsymbol{X} \otimes I_K)\boldsymbol{e} \xrightarrow{a.s.} E[(\boldsymbol{x}_{t-1} \otimes I_K)\boldsymbol{\varepsilon}_t],$$

where $E[(\boldsymbol{x}_{t-1} \otimes I_K)\boldsymbol{\varepsilon}_t] = E\left[[(\boldsymbol{x}_{t-1} \otimes I_K)\boldsymbol{\varepsilon}_t|\mathscr{F}_{t-1}]\right] = (\boldsymbol{x}_{t-1} \otimes I_K)E[\boldsymbol{\varepsilon}_t|\mathscr{F}_{t-1}] = \boldsymbol{0}$.

(iii) Let $\boldsymbol{v}_t = (\boldsymbol{x}_{t-1} \otimes I_K)\boldsymbol{\varepsilon}_t$. Then $\{\boldsymbol{v}_t\}$ is a vector martingale difference because $E[\boldsymbol{v}_t|\mathscr{F}_{t-1}] = \boldsymbol{0}$. By **A1**, **A2**, and Theorem 5 we have

$$\frac{1}{\sqrt{T}}\sum_{t=h+1}^{T} \boldsymbol{v}_t \xrightarrow{D} N(\boldsymbol{0},\ \Sigma_v),$$

where $\Sigma_v = \mathrm{Var}[\boldsymbol{v}_t] = \mathrm{Var}[(\boldsymbol{x}_{t-1} \otimes I_K)\boldsymbol{\varepsilon}_t] = E[(\boldsymbol{x}_{t-1} \otimes I_K)\boldsymbol{\varepsilon}_t\boldsymbol{\varepsilon}_t'(\boldsymbol{x}_{t-1}' \otimes I_K)] = \boldsymbol{\Gamma} \otimes \Sigma_\varepsilon$. □

**Proof of Theorem** 1

Let $\Psi_T(\boldsymbol{\phi})$ be defined as

$$\Psi_T(\boldsymbol{\phi}) = \| \boldsymbol{y} - (\boldsymbol{X}' \otimes I_K)\boldsymbol{\phi}) \|^2 + \lambda_T \sum_{l=1}^{hK^2} \hat{w}_{T,l}|\phi_l|,$$

where $\boldsymbol{X}$ is defined in (5) and $\boldsymbol{y}$ in (7). Following Fan and Li [11], we show that for every $\varepsilon > 0$ there exists a sufficiently large $C$ such that

$$\mathbb{P}\left(\inf_{\|\boldsymbol{u}\| \geq C} \Psi_T\left(\boldsymbol{\phi}^* + \boldsymbol{u}/\sqrt{T}\right) > \Psi_T(\boldsymbol{\phi}^*)\right) \geq 1 - \varepsilon,$$

which implies that with probability at least $1 - \varepsilon$ that there exists a minimum in the ball $\{\boldsymbol{\phi}^* + \boldsymbol{u}/\sqrt{T} : \|\boldsymbol{u}\| \leq C\}$. Hence there exists a local minimizer such that $\| \hat{\boldsymbol{\phi}}_T^{daL} - \boldsymbol{\phi}^* \| = O_p(T^{-1/2})$. Observe that

$$\begin{aligned}
&\Psi_T\left(\boldsymbol{\phi}^* + \boldsymbol{u}/\sqrt{T}\right) - \Psi_T\left(\boldsymbol{\phi}^*\right)\\
&= \left\| \boldsymbol{y} - (\boldsymbol{X}' \otimes I_K)\left(\boldsymbol{\phi}^* + \boldsymbol{u}/\sqrt{T}\right)\right\|^2 - \|\boldsymbol{y} - (\boldsymbol{X}' \otimes I_K)\boldsymbol{\phi}^*\|^2 + \lambda_T \sum_{l=1}^{hK^2} \hat{w}_{T,l}\left(\left|\phi_l^* + \frac{u_l}{\sqrt{T}}\right| - |\phi_l^*|\right)\\
&= \boldsymbol{u}'\left(\frac{1}{T}(\boldsymbol{X}\boldsymbol{X}' \otimes I_K)\right)\boldsymbol{u} - 2\boldsymbol{u}'\left(\frac{1}{\sqrt{T}}(\boldsymbol{X} \otimes I_K)\boldsymbol{e}\right) + \lambda_T \sum_{l=1}^{hK^2} \hat{w}_{T,l}\left(\left|\phi_l^* + \frac{u_l}{\sqrt{T}}\right| - |\phi_l^*|\right)\\
&= \boldsymbol{u}'\left(\frac{1}{T}(\boldsymbol{X}\boldsymbol{X}' \otimes I_K)\right)\boldsymbol{u} - 2\boldsymbol{u}'\left(\frac{1}{\sqrt{T}}(\boldsymbol{X} \otimes I_K)\boldsymbol{e}\right) + \lambda_T \sum_{l\in\mathbb{S}} \hat{w}_{T,l}\left(\left|\phi_l^* + \frac{u_l}{\sqrt{T}}\right| - |\phi_l^*|\right) + \lambda_T \sum_{l\notin\mathbb{S}} \hat{w}_{T,l}\frac{|u_l|}{\sqrt{T}}\\
&\geq \boldsymbol{u}'\left(\frac{1}{T}(\boldsymbol{X}\boldsymbol{X}' \otimes I_K)\right)\boldsymbol{u} - 2\boldsymbol{u}'\left(\frac{1}{\sqrt{T}}(\boldsymbol{X} \otimes I_K)\boldsymbol{e}\right) + \lambda_T \sum_{l\in\mathbb{S}} \hat{w}_{T,l}\left(\left|\phi_l^* + \frac{u_l}{\sqrt{T}}\right| - |\phi_l^*|\right)\\
&\geq \boldsymbol{u}'\left(\frac{1}{T}(\boldsymbol{X}\boldsymbol{X}' \otimes I_K)\right)\boldsymbol{u} - 2\boldsymbol{u}'\left(\frac{1}{\sqrt{T}}(\boldsymbol{X} \otimes I_K)\boldsymbol{e}\right) - \lambda_T \sum_{l\in\mathbb{S}} \hat{w}_{T,l}\frac{|u_l|}{\sqrt{T}}\ .
\end{aligned}$$

First, consider the third term, which can be expressed as

$$\begin{aligned}
\lambda_T \sum_{l=1}^{hK^2} \hat{w}_{T,l}\frac{|u_l|}{\sqrt{T}} &= \frac{\lambda_T}{\sqrt{T}} \sum_{l\in\mathbb{S}} \left|\tilde{\phi}_l\right|^{-\gamma_1} A_l^{-\gamma_2}|u_l|\\
&\leq \frac{\lambda_T}{\sqrt{T}}\left(\min_{l\in\mathbb{S}}\left(|\tilde{\phi}_l|^{\gamma_1} A_l^{\gamma_2}\right)\right)^{-1} \parallel \boldsymbol{u} \parallel\\
&= \frac{\lambda_T}{a_T} \parallel \boldsymbol{u} \parallel = o_p(1) \parallel \boldsymbol{u} \parallel\ .
\end{aligned}$$

For the second term, by Lemma 1(iii), we have

$$\boldsymbol{u}'\left(\frac{1}{\sqrt{T}}\right)(\boldsymbol{X}' \otimes I_K)'\boldsymbol{e} = \boldsymbol{u}' o_P(\boldsymbol{1}) \leq o_p(1) \parallel \boldsymbol{u} \parallel\ .$$

For the first term, by Lemma 1(i), we have

$$\left(\frac{1}{T}(\boldsymbol{X}\boldsymbol{X}' \otimes I_K)\right) \rightarrow (\boldsymbol{\Gamma} \otimes I_K)\ \text{a.s..}$$

So the first term is a quadratic form in $\boldsymbol{u}$.

Then it follows that in probability

$$\Psi_T\left(\boldsymbol{\phi}^* + \boldsymbol{u}/\sqrt{T}\right) - \Psi_T\left(\boldsymbol{\phi}^*\right) \geq \boldsymbol{u}^T(\boldsymbol{\Gamma} \otimes I_K)\boldsymbol{u} - 2o_p(1) \parallel \boldsymbol{u} \parallel,$$

as $T \rightarrow \infty$. Therefore, for any $\varepsilon > 0$, there exists a sufficiently large C such that the term of quadratic term dominates the other terms with probability $\geq 1 - \varepsilon$. □

**Proof of Proposition** 2

We follow the methodology of Knight and Fu [19] and Zou [37].

Let $\boldsymbol{\phi} = \boldsymbol{\phi}^* + \boldsymbol{u}/\sqrt{T}$ and define

$$\Psi_T(\boldsymbol{u}) = \left\| \boldsymbol{y} - (\boldsymbol{X}' \otimes I_K)\left(\boldsymbol{\phi}^* + \frac{\boldsymbol{u}}{\sqrt{T}}\right)\right\|^2 + \lambda_T \sum_{j=1}^{h} \hat{w}_{T,j} \left|\phi_j^* + \frac{u_j}{\sqrt{T}}\right|,$$

where $\boldsymbol{X}$ is defined by (5) and $\boldsymbol{y}$ by (7). Define the reparameterized objective function as

$$V_T(\boldsymbol{u}) = \Psi_T(\boldsymbol{u}) - \Psi_T(\boldsymbol{0}).$$

Then the minimizing objective is equivalent to minimizing $V_T(\boldsymbol{u})$ with respect to $\boldsymbol{u}$. Let $\hat{\boldsymbol{u}}_T = \arg\min V_T(\boldsymbol{u})$, then

$$\hat{\boldsymbol{\phi}}_T^{daL} = \boldsymbol{\phi}^* + \hat{\boldsymbol{u}}_T/\sqrt{T},$$

or

$$\hat{\boldsymbol{u}}_T = \sqrt{T}\left(\hat{\boldsymbol{\phi}}_T^{daL} - \boldsymbol{\phi}^*\right).$$

Observe that

$$V_T(\boldsymbol{u}) = \boldsymbol{u}'\left(\frac{1}{T}(\boldsymbol{X}\boldsymbol{X}' \otimes I_K)\right)\boldsymbol{u} - 2\boldsymbol{u}'\left(\frac{1}{\sqrt{T}}(\boldsymbol{X} \otimes I_K)\boldsymbol{e}\right) + \frac{\lambda_T}{\sqrt{T}} \sum_{l=1}^{hK^2} \hat{w}_{T,l}\sqrt{T}\left(\left|\phi_l^* + \frac{u_l}{\sqrt{T}}\right| - |\phi_l^*|\right).$$

By Lemma 1 we have $\left(\frac{1}{T}(\boldsymbol{X}\boldsymbol{X}' \otimes I_K)\right) \xrightarrow{a.s.} (\boldsymbol{\Gamma} \otimes I_K)$, and $\frac{1}{\sqrt{T}}(\boldsymbol{X} \otimes I_K)\boldsymbol{e} \xrightarrow{D} \boldsymbol{w} \sim N(\boldsymbol{0},\ \boldsymbol{\Gamma} \otimes \Sigma_\varepsilon)$. Consider the limiting behaviour of the third term. First, by the conditions required in the theorem, we have $\lambda_T \hat{w}_{T,l}/\sqrt{T} \le \lambda_T/\left(\sqrt{T}\min_{l\in\mathbb{S}}\left(\left|\tilde{\phi}_l\right|^{\gamma_1} A_l^{\gamma_2}\right)\right) = \lambda_T/a_T \xrightarrow{P} 0$ for $l \in \mathbb{S}$ and $\frac{\lambda_T}{\sqrt{T}} w_{T,l} = \frac{\lambda_T}{\sqrt{T}} |\tilde{\phi}_l|^{-\gamma_1} A_l^{-\gamma_2} \ge \lambda_T/\left(\sqrt{T}\max_{l\notin\mathbb{S}}\left(\left|\tilde{\phi}_l\right|^{\gamma_1} A_l^{\gamma_2}\right)\right) = \lambda_T/b_T \xrightarrow{P} \infty$ for $l \notin \mathbb{S}$. In summary, we have

$$\frac{\lambda_T}{\sqrt{T}} \hat{w}_{T,l} = \frac{\lambda_T}{\sqrt{T}\left|\tilde{\phi}_l\right|^{\gamma_1} A_l^{\gamma_2}} \xrightarrow{P} \begin{cases} 0 & \text{if } l \in \mathbb{S} \\ \infty & \text{if } l \notin \mathbb{S} \end{cases}.$$

Secondly, we have

$$\sqrt{T}\left(\left|\phi_l^* + \frac{u_l}{\sqrt{T}}\right| - \phi_l^*\right) \to \begin{cases} u_l \operatorname{sgn}(\phi_l^*) & \text{if } l \in \mathbb{S}\ (\phi_l^* = 0) \\ |u_l| & \text{if } l \notin \mathbb{S}\ (\phi_l^* \neq 0) \end{cases}.$$

By Slutsky's theorem, we have the following limiting behaviour of the third term

$$\frac{\lambda_T}{\sqrt{T}} \hat{w}_{T,l}\sqrt{T}\left(\left|\phi_l^* + \frac{u_l}{\sqrt{T}}\right| - |\phi_l^*|\right) \xrightarrow{P} \begin{cases} 0 & \text{if } \forall l \in \mathbb{S} \\ 0 & \text{if } u_l = 0,\ \ \forall l \notin \mathbb{S} \\ \infty & \text{otherwise} \end{cases}.$$

Thus, we have $V_T(\boldsymbol{u}) \to V(\boldsymbol{u})$ for every $\boldsymbol{u}$, where

$$\begin{aligned} V(\boldsymbol{u}) &= \left(\boldsymbol{u}_{\mathbb{S}}' \ \boldsymbol{u}_{\mathbb{S}^c}'\right) \begin{pmatrix} (\boldsymbol{\Gamma} \otimes I_K)_{\mathbb{S}\mathbb{S}} & (\boldsymbol{\Gamma} \otimes I_K)_{\mathbb{S}\mathbb{S}^c} \\ (\boldsymbol{\Gamma} \otimes I_K)_{\mathbb{S}^c\mathbb{S}} & (\boldsymbol{\Gamma} \otimes I_K)_{\mathbb{S}^c\mathbb{S}^c} \end{pmatrix} \begin{pmatrix} \boldsymbol{u}_{\mathbb{S}} \\ \boldsymbol{u}_{\mathbb{S}^c} \end{pmatrix} - 2\left(\boldsymbol{u}_{\mathbb{S}}' \ \boldsymbol{u}_{\mathbb{S}^c}'\right) \begin{pmatrix} \boldsymbol{w}_{\mathbb{S}} \\ \boldsymbol{w}_{\mathbb{S}^c} \end{pmatrix} \\ &\quad + \sum_{l \in \mathbb{S}^c} \frac{\lambda_T}{\sqrt{T}} \hat{w}_{T,l} \sqrt{T} \left( \left| \phi_l^* + \frac{u_l}{\sqrt{T}} \right| - |\phi_l^*| \right) \\ &= \begin{cases} \boldsymbol{u}_{\mathbb{S}}'(\boldsymbol{\Gamma}_{\mathbb{S}\mathbb{S}} \otimes I_K)\boldsymbol{u}_{\mathbb{S}} - 2\boldsymbol{u}_{\mathbb{S}}'\boldsymbol{w}_{\mathbb{S}} & \text{if } \boldsymbol{u}_{\mathbb{S}^c} = \boldsymbol{0} \\ \infty & \text{otherwise} \end{cases}. \end{aligned}$$

$V_T(\boldsymbol{u})$ is convex with the unique minimum $\left(((\boldsymbol{\Gamma}_{\mathbb{S}\mathbb{S}})^{-1} \otimes I_K)\boldsymbol{w}_{\mathbb{S}}, \ \ \boldsymbol{0}\right)'$. Following the epi-convergence results of Geyer [13] and Knight and Fu [19], $\operatorname{argmin}_{\boldsymbol{u}} V_T(\boldsymbol{u}) \xrightarrow{D} \operatorname{argmin}_{\boldsymbol{u}} V(\boldsymbol{u})$, we have

$$\begin{cases} \hat{\boldsymbol{u}}_{\mathbb{S}} \xrightarrow{D} \left((\boldsymbol{\Gamma}_{\mathbb{S}\mathbb{S}})^{-1} \otimes I_K\right) \boldsymbol{w}_{\mathbb{S}} \\ \hat{\boldsymbol{u}}_{\mathbb{S}^C} \xrightarrow{P} \boldsymbol{0} \end{cases},$$

or

$$\begin{cases} \sqrt{T}\left(\hat{\boldsymbol{\phi}}_{T,\mathbb{S}}^{daL} - \boldsymbol{\phi}_{\mathbb{S}}^*\right) \xrightarrow{D} N\left(\boldsymbol{0}, \ \ (\boldsymbol{\Gamma}_{\mathbb{S}\mathbb{S}})^{-1} \otimes \Sigma_\varepsilon\right) \\ \sqrt{T}\left(\hat{\boldsymbol{\phi}}_{T,\mathbb{S}^c}^{daL} - \boldsymbol{\phi}_{\mathbb{S}^c}^*\right) \xrightarrow{P} \boldsymbol{0} \end{cases}.$$

□

**Proof of Theorem** 2

(i) In view of Corollary 1, we know that $\forall j \in \mathbb{S}, \ \ P(j \in \hat{\mathbb{S}}_T) \to 1$. So it suffices to show that $\forall m \notin \mathbb{S}$, $P(m \in \hat{\mathbb{S}}_T) \to 0$. Now, we follow the methodology of Zou [37].

Consider the event $\{m \in \hat{\mathbb{S}}_T\}$. The KKT conditions entail that

$$2(\boldsymbol{X} \otimes I_K)_{(m,\cdot)} \left(\boldsymbol{y} - (\boldsymbol{X}' \otimes I_K)\hat{\boldsymbol{\phi}}_T^{daL}\right) = \lambda_T \hat{w}_{T,m} \operatorname{sgn}\left(\hat{\phi}_{T,m}^{daL}\right),$$

where the subscript $(m, \ \cdot)$ denotes the $m$-th row of a matrix, so $(\boldsymbol{X} \otimes I_K)_{(m,\cdot)}$ is the $m$-th row of $(T-h)K \times hK^2$ matrix $(\boldsymbol{X} \otimes I_K)$. If $\lambda_T / b_T \ \xrightarrow{P} \ \infty$, we have

$$\frac{\lambda_T}{\sqrt{T}} \hat{w}_{T,m} = \frac{\lambda_T}{\sqrt{T}} \frac{1}{|\tilde{\phi}_m|^{\gamma_1} A_m^{\gamma_2}} \geq \frac{\lambda_T}{b_T} \xrightarrow{P} \infty,$$

whereas

$$\frac{(\boldsymbol{X} \otimes I_K)_{(m,\cdot)} \left(\boldsymbol{y} - (\boldsymbol{X}' \otimes I_K)\hat{\boldsymbol{\phi}}_T^{daL}\right)}{\sqrt{T}} = \left(\frac{(\boldsymbol{X} \otimes I_K)_{(m,\cdot)}(\boldsymbol{X}' \otimes I_K)}{T}\right) \sqrt{T}\left(\boldsymbol{\phi}^* - \hat{\boldsymbol{\phi}}_T^{daL}\right) + \frac{(\boldsymbol{X} \otimes I_K)_{(m,\cdot)}\boldsymbol{e}}{\sqrt{T}}.$$

Note that $(\boldsymbol{X} \otimes I_K)_{(m,\cdot)}\boldsymbol{e}$ is the $m$-th element of the vector $(\boldsymbol{X} \otimes I_K)\boldsymbol{e}$, denoted by $((\boldsymbol{X} \otimes I_K)\boldsymbol{e})_m$. By Lemma 1, we have

$$\frac{1}{\sqrt{T}}((\boldsymbol{X} \otimes I_K)\boldsymbol{e})_m \xrightarrow{D} N\left(0,\ \ (\boldsymbol{\Gamma} \otimes \Sigma_\varepsilon)_{(m,m)}\right) ,$$

where $(\boldsymbol{\Gamma} \otimes \Sigma_\varepsilon)_{(m,m)}$ is the $m$-th diagonal element of $(\boldsymbol{\Gamma} \otimes \Sigma_\varepsilon)$. Note also that $(\boldsymbol{X} \otimes I_K)_{(m,\cdot)}(\boldsymbol{X}' \otimes I_K)$ is the $m$-th row of the matrix $(\boldsymbol{X}\boldsymbol{X}' \otimes I_K)$, denoted by $(\boldsymbol{X}\boldsymbol{X}' \otimes I_K)_{(m,\cdot)}$. By Lemma 1, we have

$$\frac{1}{T}(\boldsymbol{X}\boldsymbol{X}' \otimes I_K)_{(m,\cdot)} \xrightarrow{a.s.} (\boldsymbol{\Gamma} \otimes I_K)_{(m,\cdot)} .$$

By Slutsky's theorem and the results of (i), we see that

$$\frac{1}{T}(\boldsymbol{X} \otimes I_K)_{(m,\cdot)}(\boldsymbol{X}' \otimes I_K)\sqrt{T}\left(\boldsymbol{\phi}^* - \hat{\boldsymbol{\phi}}_T^{daL}\right) \xrightarrow{D} (\boldsymbol{\Gamma} \otimes I_K)_{(m,\cdot)}\boldsymbol{z} ,$$

where $\boldsymbol{z}$ is a normally-distributed vector, and thus $(\boldsymbol{\Gamma} \otimes I_K)_{(m,\cdot)}\boldsymbol{z}$ a normally-distributed scalar variable. Therefore,

$$P(m \in \hat{\mathbb{S}}_T) \leq P\left(2(\boldsymbol{X} \otimes I_K)_{(m,\cdot)}\left(\boldsymbol{y} - (\boldsymbol{X}' \otimes I_K)\hat{\boldsymbol{\phi}}_T^{daL}\right) = \lambda_T \hat{w}_m \mathrm{sgn}\left(\hat{\phi}_{T,m}^{daL}\right)\right) \to 0.$$

(ii) The VAR order estimated by the doubly adaptive LASSO is

$$\hat{p}_T^{daL} = \min\left\{s : \hat{\phi}_{ij,k}^{daL} = 0, \forall k = s+1, s+2, \ldots, h,\ \text{and}\ i, j = 1, \ldots, K\right\},$$

or equivalently, in light of the bijective function (11),

$$\hat{p}_T^{daL} = \min\left\{s : (k-1)K^2 + (i-1)K + j \in \hat{\mathbb{S}}_T^c, \forall k = s+1, s+2, \ldots, h,\ \text{and}\ i, j = 1, \ldots, K\right\}. \tag{22}$$

The true order $p$ of the VAR model is

$$p = \min\left\{s : (k-1)K^2 + (i-1)K + j \in \mathbb{S}^c, \forall k = s+1, s+2, \ldots, h,\ \text{and}\ i, j = 1, \ldots, K\right\}. \tag{23}$$

We have from (i) that $\hat{\mathbb{S}}_T^c \to \mathbb{S}^c$ in probability, so the RHS of (22) and (23) are equal in probability. Therefore, $\lim P(\hat{p}_T^{daL} = p) = 1$.

(iii) From (i), we have that $\lim \mathbb{P}\left(\hat{\boldsymbol{\phi}}_{\hat{\mathbb{S}}_T}^{daL} = \hat{\boldsymbol{\phi}}_{T,\mathbb{S}}^{daL}\right) \to 1$. Then, from Proposition 2, the asymptotic normality of $\hat{\boldsymbol{\phi}}_{\hat{\mathbb{S}}_T}^{daL}$ follows. □

## References


1. Akaike, H. (1969). Fitting autoregressive models for prediction. *Annals of the Institute of Statistical Mathematics*, *21*, 243–247.
2. Akaike, H. (1974). A new look at the statistical model identification. *IEEE Transactions on Automatic Control*, *AC–19*, 716–723.
3. Akaike, H. (1978). A Bayesian analysis of the minimum AIC procedure. *Annals of the Institute of Statistical Mathematics*, *30*(Part A), 9–14.
4. Billingsley, P. (1961). The Lindeberg-Levy theorem for martingales. *Proceedings of the American Mathematical Society*, *12*, 788–792.
5. Caner, M., & Knight, K. (2013). An alternative to unit root tests: bridge estimators differentiate between nonstationary versus stationary models and select optimal lag. *Journal of Statistical Planning and Inference*, *143*, 691–715.
6. Chand, S. (2011). *Goodness of fit and lasso variable selection in time series analysis*. Ph.D. thesis, University of Nottingham.
7. Chen, K., & Chan, K. (2011). Subset ARMA selection via the adaptive Lasso. *Statistics and Its Interface*, *4*, 197–205.
8. Donoho, D. L., Michael Elad, M., & Temlyakov, V. N. (2006). Stable recovery of sparse overcomplete representations in the presence of noise. *IEEE Transactions on Iinformation Theory*, *52*(1), 6–18.
9. Durbin, J. (1960). The fitting of time series models. *Review of the Institute of International Statistics*, *28*, 233–244.
10. Efron, B., Johnstone, I., & Tibshirani, R. (2004). Least angle regression. *Annals of Statistics*, *32*(2), 407–499.
11. Fan, J., & Li, R. (2001). Variable selection via nonconcave penalized likelihood and its oracle properties. *Journal of the American Statistical Association*, *96*, 1348–1360.
12. Fujita, A., Sato, J. R., Garay-Malpartida, H. M., Yamaguchi, R., Miyano, S., Sogayar, M. C., et al. (2007). Modeling gene expression regulatory networks with the sparse vector autoregressive model. *BMC Systems Biology*, *1*, 39.
13. Geyer, C. (1994). On the asymptotics of constrained M-estimation. *The Annals of Statistics*, *22*, 1993–2010.
14. Hannan, E. J. (1970). *Multiple time series*. New York: Wiley.
15. Hannan, E. J., & Quinn, B. G. (1979). The determination of the order of an autoregression. *Journal of the Royal Statistical Society*, *B41*, 190–195.
16. Haufem, N. K. S., Muller, S. K, Nolte, G., & Kramer. (2008). Sparse causal discovery in multivatiate time series. In *JMLR: Workshop and conference proceedings* (Vol. 1, pp. 1–16).
17. Heyse, J. F.(1985). *Partial lag autocorrelation and partial process autocorrelation for vector time series, with applications*. Ph.D. dissertation, Temple University.
18. Hsu, N., Hung, H., & Chang, Y. (2008). Subset selection for vector autoregressive processes using LASSO. *Computational Statistics and Data Analysis*, *52*, 3645–3657.
19. Knight, K., & Fu, W. (2000). Asymptotics for LASSO-type estimators. *The Annals of Statistics*, *28*, 1356–1378.
20. Kock, A. B. (2012). *On the oracle property of the adaptive lasso in stationary and nonstationary autoregressions*. CREATES research papers 2012-05, Aarhus University.
21. Kock, A. B., & Callot, L. A. F. (2012). *Oracle inequalities for high dimensional vector autoregressions*. CREATES research paper 2012-12, Aarhus University.
22. Liu, Z. Z. (2014). *The doubly adaptive LASSO methods for time series analysis*. University of Western Ontario - Electronic Thesis and Dissertation Repository. Paper 2321.
23. Lütkepohl, H. (2006). *New introduction to multiple time series analysis*. Berlin: Springer.
24. Medeiros, M. C, & Mendes, E. F. (2012). *Estimating high-dimensional time series models*. CREATES research paper 2012-37.
25. Nardi, Y., & Rinaldo, A. (2011). Autoregressive process modeling via the LASSO procedure. *Journal of Multivariate Analysis*, *102*(3), 528–549.

26. Park, H., & Sakaori, F. (2013). Lag weighted lasso for time series model. *Computational Statistics*, *28*, 493–504.
27. Ren, Y., & Zhang, X. (2010). Subset selection for vector autoregressive processes via the adaptive LASSO. *Statistics and Probability Letters*, *80*, 1705–1712.
28. Schwarz, G. (1978). Estimating the dimension of a model. *Annals of Statistics*, *6*(2), 461–464.
29. Song, S., & Bickel, P. J. (2011). *Large vector auto regressions*. arXiv:1106.3915v1 [stat.ML].
30. Tibshirani, R. (1996). Regression shrinkage and selection via the lasso. *Journal of the Royal Statistical Society, Series B*, *58*(1), 267–288.
31. Valdés-Sosa, P. A., Sánchez-Bornot, J. M., Lage-Castellanos, A., Vega-Hernández, M., Bosch-Bayard, J., Melie-Garía, L., et al. (2005). Estimating brain functional connectivity with sparse multivariate autoregression. *Philosophical Transactions Royal Society B*, *360*(1457), 969–981.
32. Wang, H., Li, G., & Tsai, C. (2007). Regression coefficients and autoregressive order shrinkage and selection via the lasso. *Journal of the Royal Statistical Society, Series B*, *69*(1), 63–78.
33. Wei, W. S. (2005). *Time series analysis: Univariate and multivariate methods* (2nd ed.). Reading, MA: Addison-Wesley.
34. White, H. (2001). *Asymptotic theory for econometricians* (Revised ed.). New York: Academic Press.
35. Yoon, Y., Park, C., & Lee, T. (2013). Penalized regression models with autoregressive error terms. *Journal of Statistical Computation and Simulation*, *83*(9), 1756–1772.
36. Zhao, P., & Yu, B. (2006). On model selection consistency of Lasso. *Journal of Machine Learning Research*, *7*, 2541–2563.
37. Zou, H. (2006). The adaptive LASSO and its oracle properties. *Journal of the American Statistical Association*, *101*, 1418–1429.

# On Diagnostic Checking Autoregressive Conditional Duration Models with Wavelet-Based Spectral Density Estimators

**Pierre Duchesne and Yongmiao Hong**

**Abstract** There has been an increasing interest recently in the analysis of financial data that arrives at irregular intervals. An important class of models is the autoregressive Conditional Duration (ACD) model introduced by Engle and Russell (Econometrica 66:1127–1162, 1998, [22]) and its various generalizations. These models have been used to describe duration clustering for financial data such as the arrival times of trades and price changes. However, relatively few evaluation procedures for the adequacy of ACD models are currently available in the literature. Given its simplicity, a commonly used diagnostic test is the Box-Pierce/Ljung-Box statistic adapted to the estimated standardized residuals of ACD models, but its asymptotic distribution is not the standard one due to parameter estimation uncertainty. In this paper we propose a test for duration clustering and a test for the adequacy of ACD models using wavelet methods. The first test exploits the one-sided nature of duration clustering. An ACD process is positively autocorrelated at all lags, resulting in a spectral mode at frequency zero. In particular, it has a spectral peak at zero when duration clustering is persistent or when duration clustering is small at each individual lag but carries over a long distributional lag. As a joint time-frequency decomposition method, wavelets can effectively capture spectral peaks and thus are expected to be powerful. Our second test checks the adequacy of an ACD model by using a wavelet-based spectral density of the estimated standardized residuals over the whole frequency. Unlike the Box-Pierce/Ljung-Box tests, the proposed diagnostic test has a convenient asymptotic "nuisance parameter-free" property—parameter estimation uncertainty has no impact on the asymptotic distribution of the test statistic. Moreover, it can check a wide range of alternatives and is powerful when the spectrum of the standardized duration residuals is nonsmooth, as can arise from neglected persistent duration clustering, seasonality, calender effects and business cycles. For

P. Duchesne (✉)
Département de Mathématiques et de Statistique, Université de Montréal,
CP 6128, Succ Centre-Ville, Montreal, QC H3C 3J7, Canada
e-mail: duchesne@dms.umontreal.ca

Y. Hong
Department of Economics, Cornell University,
492 Uris Hall, Ithaca, NY 14853-7601, USA
e-mail: yhong.cornell@gmail.com

 
W.K. Li et al. (eds.), *Advances in Time Series Methods and Applications*,
Fields Institute Communications 78, DOI 10.1007/978-1-4939-6568-7_3

each of the two new tests, we propose and justify a suitable data-driven method to choose the finest scale—the smoothing parameter in wavelet estimation. This makes the methods fully operational in practice. We present a simulation study, illustrating the merits of the wavelet-based procedures. An application with tick-by-tick trading data of Alcoa stock is presented.

**Keywords** Autoregressive conditional duration · Duration clustering · High frequency financial time series · Model adequacy · Parameter estimation uncertainty · Spectral density · Standardized duration residual · Wavelet

**AMS Mathematics Subject Classifications (2010)** 62M10 · 62M15 · 62E20

## 1 Introduction

There has been an increasing interest recently in modeling high frequency financial data that arrive at irregular time intervals. Conventional time series analysis assumes that the data are collected at a fixed time interval. When this is not the case, it is standard practice to aggregate the data and to analyze them as if they were generated by a fixed time stochastic process (e.g., [36]). This will unavoidably hide useful information when the time interval chosen is too large, or exhibit excessive heteroskedasticity when the time interval chosen is too small. Nowadays, the power and storage capacity of modern computers allow us to have access to virtually every transaction data available on financial markets. Such financial data usually arrive and are recorded at irregular time intervals. This is a situation where we have to our disposal the so-called ultra-high frequency data (cf. [20]). Other examples of irregularly spaced data are credit card purchases or sales of any commodity using scanning devices. Such financial data contain rich information about market microstructure and economic agents' behaviors.

It seems natural to formulate models for irregularly spaced financial data to study duration clustering. The Autoregressive Conditional Duration (ACD) model proposed by [22] is an important contribution toward this direction, since it proposes an alternative methodology to fixed time interval analysis of time series data. The model treats the arrival time intervals between events of interest (e.g., trades, price changes) as a nonnegative stochastic process, and studies the time series properties of the expected duration between events. It assumes that the expectation of duration, conditional on the past information, may be expressed as a function of past durations. It can be used to infer the pattern of duration clustering, to forecast the intensity of arrival times that is closely linked to (e.g.) transaction volume or price volatility, and to test market microstructure theories. The ACD model has an interesting interpretation in the context of time deformation modeling because it is formulated in transaction time and models the frequency and distribution of arrival times between events of interest (e.g., [31, 65, 66]).

There has been a variety of important extensions and applications of ACD modeling. In [22], the author suggests the possibility of nonlinear ACD models analogous to the NGARCH model of [8]. In [4], they propose the logarithmic ACD (log-ACD) model, which allows to introduce additional variables without sign restrictions on the coefficients. In [6] they also propose an asymmetric ACD model, where the duration depends on the state of the price process. The asymmetric ACD model becomes a log-ACD model under certain parameter restrictions. In [73], they introduce the threshold autoregressive conditional duration model, which allows the conditional duration to depend nonlinearly on the past information set. Applications of threshold duration models are considered in [67], who also introduces a bivariate model for the process of price change and the associated duration. Another member in the class of ACD models is the Burr-ACD model of [34]. The model is based on the Burr distribution. It includes the exponential ACD (EACD) and Weibull ACD (WACD) models as special cases. The works [32] and [46] consider a fractionally integrated ACD model to capture persistent duration clustering. In [33], they combine ACD models and GARCH-type effects and propose an ACD-GARCH model. The papers [67, 68] present the basic characteristics of ACD models and a comprehensive survey is given in [59].

Wavelets have been documented to be capable of capturing nonsmooth or singular features such as spectral modes/peaks (e.g., [28, 47, 57, 60, 63, 69, 70]). Applications of wavelets in various time series contexts include [24, 29, 30, 72]. In time series, testing for serial correlation and ARCH effects generated many wavelet-based test statistics (e.g., [15–17, 43, 44, 50, 51]). Spectral peaks often occur in economic and financial time series, due to strong serial dependence, seasonality, business cycles, calendar effects and other forms of periodicities (e.g., [9, 35, 71]). In this paper, new wavelet-based tests for duration clustering and for diagnostic checking ACD models are developed.

Before modeling an ACD process for the arrival time intervals between events of interest, one may like to appreciate whether there exists duration clustering in the arrival times and its nature; that is, whether there do exist ACD effects and which lags seem significant. Market microstructure theories (e.g., [1, 19, 48]) suggest that the frequency of transactions should carry information about the state of the market. They predict the existence of transaction clustering. In the literature, commonly used tests for ACD effects are Box-Pierce/Ljung-Box (BP/LB) tests (e.g., [21, 22]). Like an ARCH process, an ACD process always has nonnegative autocorrelation at any lag, resulting in a spectral mode at frequency zero under (and only under) the alternative hypothesis. BP/LB tests do not exploit such a one-sided nature. We first propose a one-sided consistent test for ACD effects, which exploits the one-sided nature of the alternative. We use a wavelet-based spectral density estimator at frequency zero. In the present context, spectral peaks can arise when ACD effects are strong or persistent or when ACD effects are weak at each individual lag but carry over a very long distributional lag. We thus expect that our one-sided test for ACD effects will be powerful for such alternatives in small and finite samples. We note that [41] proposed a one-sided test for ACD effects using an alternative approach. Our second objective is to propose a wavelet-based diagnostic test for

ACD models. Although various ACD models are available in the literature (e.g., [4, 6, 22, 32–34, 46]), there have been relatively few diagnostic tests for the adequacy of ACD models. A possible diagnostic test, suggested in [21, 22], is the BP/LB tests adapted to the estimated standardized residuals of an ACD model. In the same spirit, the authors in [5, pp. 83–84] consider as a diagnostic statistic the LB test based on the residuals or the squared residuals. The BP/LB tests are conjectured to follow an asymptotic $\chi^2$ distribution with the degrees of freedom equal to the number of lags used in the test. Nevertheless, there is no formal analysis of the statistical properties (e.g., asymptotic distribution) for these tests. Following the reasoning analogous to [53], it could be shown that the asymptotic critical values used in practice are incorrect, because they do not take into account parameter estimation uncertainty. A valid method is given in [54], who derived an asymptotically valid portmanteau test for checking ACD models based on the estimated standardized duration autocorrelations. Lagrange multipliers tests are also investigated in [39, 56]. In [18], spectral tests are constructed for the adequacy of ACD models, based on kernel-based spectral density estimators of the standardized innovation process. Using the truncated uniform kernel, this gives a generalized BP/LB portmanteau test. Generalized spectral derivative tests are derived in [45], which are motivated by the so-called generalized spectral density. As discussed in Sect. 2.1, ACD models admit weak Autoregressive Moving Average (ARMA) representations. See [25]. Thus an ACD model can be written as an ARMA model where the error term is a martingale difference sequence (MDS). To test the adequacy reduces to check if the noise is MDS. A possible method is described in [23], which do not depend on a kernel or a bandwidth. However, the asymptotic null distribution depends on the choice of the data generating process (DGP) and it is no longer standard. Resampling methods as the bootstrap can be used to approximate the asymptotic critical values of the tests. For the weak ARMA model, the spectral test in [74] represents also an alternative approach to test the adequacy of ACD models written as weak ARMA models.

In this paper, we contribute to the literature of diagnostic checking ACD models by proposing an asymptotically valid test for the adequacy of ACD models by examining if there is remaining structure in the standardized residuals of ACD models, using wavelet methods. More precisely, we compare a wavelet-based spectral density estimator with a flat spectrum; the later is implied by the adequacy of an ACD model. Unlike the BP/LB tests, our test has a convenient asymptotic "nuisance parameter free" property—parameter estimation uncertainty has no impact on the limit distribution of the test statistic. Moreover, it can detect a wide range of model inadequacy. In particular it is powerful against misspecifications that result in nonsmooth spectrum for the standardized residuals. Such alternatives may arise from neglected strong dependence, seasonality, calender effects and business cycles. Unlike one-sided testing for ACD effects, a misspecified ACD model generally does not produce a one-sided alternative. Negative autocorrelations in the standardized duration residuals may occur. As a consequence, we have to check the wavelet-based spectral density estimator over all frequencies rather than at frequency zero only.

For each of the two new tests, we propose and justify a suitable data-driven method to select the finest scale—the smoothing parameter in wavelet estimation. This makes the wavelet-based tests entirely operational.

We describe hypotheses of interest in Sect. 2. Section 3 introduces wavelet analysis and proposes a consistent one-sided test for ACD effects. In Sect. 4, we develop a diagnostic test for the adequacy of ACD models. Section 5 presents two sets of Monte Carlo study, examining the finite sample performance of the proposed tests. The wavelet-based tests are compared to the most popular methods currently used in the literature. In Sect. 6, an application with tick-by-tick trading data of Alcoa stock on June 7, 2010 is presented. Section 7 concludes. All proofs are given in the appendix. Unless indicated, all limits are taken as the sample size $n \to \infty$; $A^*$ denotes the complex conjugate of $A$; $A^\top$ is the transpose of the matrix $A$; $C$ a bounded constant that may differ from place to place; and $\mathbb{Z} = \{0, \pm 1, \ldots\}$ the set of integers. An R code to implement the new methodology is available upon request from the authors.

## 2 Framework and Hypotheses

### 2.1 ACD Processes

Let $X_t$ be the interval between two arrival times of a random event (e.g., trades, price changes), which is called a duration. Throughout, we consider the following DGP:

**Assumption 1** The strictly stationary nonnegative duration process $X_t = D_t^0 \varepsilon_t$, where $\{\varepsilon_t\}$ is a nonnegative iid sequence with probability density $p(\cdot)$, mean $E(\varepsilon_t) = 1$, and finite fourth moment, and $D_t^0 \equiv D^0(\mathscr{I}_{t-1})$ is a measurable nonnegative function of $\mathscr{I}_{t-1}$, the information set available at time $t - 1$.

We make no distributional assumption on the innovations $\{\varepsilon_t\}$. Examples include the exponential, Weibull, generalized gamma, log logistic and lognormal distributions, as often considered in the literature. An ACD model with exponential innovations is called an EACD, while the case with Weibull innovations is denoted a WACD (see [22]). The textbook [68, pp. 303–304] writes explicitly the conditional log-likelihood functions of EACD and WACD models (note that the Weibull conditional log-likelihood reduces to the conditional exponential log-likelihood when the shape parameter of the Weibull distribution is equal to one). In [34], they consider the case with the Burr distribution for the innovations, called a Burr-ACD model, which includes as special cases the EACD and WACD models. The existing literature apparently focuses on various specifications of the innovation distribution and pays relatively little attention to the specification for the key ingredient $D_t^0 = E(X_t|\mathscr{I}_{t-1})$, the conditional duration. Our main interest here is the important issue of model specification for $D_t^0$. Specifically, we first check whether there exists duration clustering in $\{X_t\}$, and if so, whether a parametric ACD model fitted to the data is adequate. We use a frequency domain approach in combination of wavelet analysis, a powerful mathematical tool (cf. [12]). Our methodology does not assume and so is robust to any distribution misspecification for the innovations.

Suppose $D_t$ follows a general linear process

$$D_t = \beta_0 + \sum_{h=1}^{\infty} \beta_h X_{t-h}, \tag{1}$$

where $\beta_0 > 0$, $\sum_{h=1}^{\infty} \beta_h < 1$, and $\beta_h \geq 0$ for all $h \geq 1$, which ensures strict positiveness of the conditional duration. The class (1) contains an $m$-memory conditional duration process

$$D_t = \beta_0 + \sum_{h=1}^{m} \beta_h X_{t-h},$$

which only depends on the most recent $m$ durations. It also includes the more general Autoregressive Conditional Duration process

$$D_t = \beta_0 + \sum_{h=1}^{m} \alpha_h X_{t-h} + \sum_{h=1}^{l} \gamma_h D_{t-h}, \tag{2}$$

whose coefficients $\beta_h$'s, which are a function of $\{\alpha_h, \gamma_h\}$, decay to zero exponentially as $h \to \infty$. The process (2) is called an ACD($m$,$l$). The ACD($m$,$l$) has an ARMA[$\max(m, l), m$] representation. The class (1) also contains a fractionally integrated ACD process proposed in [32, 46], where the coefficients $\beta_h \to 0$ as $h \to \infty$ at a slow hyperbolic rate.

## 2.2 Hypothesis of ACD Effects

Our first objective is to develop a consistent one-sided test for the existence of ACD effects. Under class (1), the null hypothesis of interest of no ACD effects is

$$\mathbb{H}_0^{\mathscr{E}} : \beta_h = 0 \text{ for all } h > 0.$$

The alternative hypothesis that ACD effects exist is

$$\mathbb{H}_1^{\mathscr{E}} : \beta_h \geq 0 \text{ for all } h > 0, \ldots \text{with at least one strict inequality.}$$

Under $\mathbb{H}_0^{\mathscr{E}}$, the true theoretical parameters of model (1) are on the boundary of the parameter space. In [3, 26], they consider inference problems in that context. Note that $\mathbb{H}_1^{\mathscr{E}}$ is a one-sided alternative. A conventional approach to testing ACD effects is the LM test or BP/LB tests. The latter are commonly used in the literature (cf. [21, 22]). However, both the LM test and BP/LB tests fail to account for the one-sided nature of the alternative hypothesis $\mathbb{H}_1^{\mathscr{E}}$. Exploration of such a one-sided nature is expected to yield better power in finite samples. Large samples are usually available for financial data, but in the present context the events of interest may be defined with specific characteristics. This may lead to a relatively small sample size. In [22], for example, they examine the price changes of IBM transaction data where the data of interest consists of only those with a price change. This leads to a new sample that is only 3 % of the original IBM transaction data. Furthermore, market participants may be interested by volume durations, which represent the times between trades (defined either on the quote or trade process) such that a volume of $c$ shares, say, is traded. For example, [5] applies ACD models for various stocks, including AWK, Disney, IBM and SKS stocks. In some occasions, the sample sizes for the volume durations were less than $n = 300$. To exploit the one-sided nature of the alternative hypothesis may be highly desirable in such cases.

It seems that [41] is apparently the first to propose a one-sided test for no ACD effects against an ACD($m$) alternative with a pre-specified fixed order $m > 0$. This test is analogous to the locally most mean powerful unbiased based score test of [49] for ARCH effects. The test statistic for an ACD($m, l$) alternative is numerically identical to that for an ACD($m$) alternative. In the neighborhood of the null hypothesis of no ACD effects, the test maximizes the mean curvature of the power function. In his Monte Carlo experiments, [41] shows that in finite sample his test has a reasonable level and is more powerful than the two-sided LM test, suggesting the gain of exploiting the one-sided alternative. Like the LM test or BP/LB tests, the lag order $m$ has to be prespecified in Higgins'test described in [41] and the choice of it may affect power considerably. Obviously, the optimal choice of $m$ should depend on the alternative, which is usually unknown in practice. For a given $m$, Higgins' test in [41] has no power against the alternatives for which ACD effects exist only at higher order lags. In some cases, even if the alternative were known, it may still be difficult to determine an optimal lag to maximize the power. An example is the fractionally integrated ACD process (cf. [32, 46]).

As the first objective of this paper, we propose a consistent one-sided test for ACD effects that complements the test of [41]. We use a frequency domain approach. Let $f_X(\omega)$ be the standardized spectral density of $\{X_t\}$; that is,

$$f_X(\omega) = (2\pi)^{-1} \sum_{h=-\infty}^{\infty} \rho_X(h) e^{-ih\omega}, \quad \omega \in [-\pi, \pi], \; i = \sqrt{-1},$$

where $\rho_X(h)$ is the autocorrelation function of $\{X_t\}$. Note that (1) implies

$$X_t = \beta_0 + \sum_{h=1}^{\infty} \beta_h X_{t-h} + v_t,$$

where $v_t = X_t - D_t$ is a martingale difference sequence with respect to $\mathscr{I}_{t-1}$; that is, $E(v_t|\mathscr{I}_{t-1}) = 0$ almost surely ($a.s.$), where $\mathscr{I}_{t-1}$ is as in Assumption 1. Under $\mathbb{H}_0^{\mathscr{E}}$, $X_t = \beta_0 + v_t$ is a white noise process, so we have $f_X(0) = (2\pi)^{-1}$. Under

$\mathbb{H}_1^{\mathscr{E}}$, we have $\rho_X(h) \geq 0$ for all $h \neq 0$ and there exists at least one $h \neq 0$ such that $\rho_X(h) > 0$. Thus, $f_X(0) > (2\pi)^{-1}$ under $\mathbb{H}_1^{\mathscr{E}}$. This forms a basis for constructing a consistent one-sided test for $\mathbb{H}_0^{\mathscr{E}}$ versus $\mathbb{H}_1^{\mathscr{E}}$. We can use a consistent wavelet-based estimator $\hat{f}_X(0)$ for $f_X(0)$ and check if $\hat{f}_X(0) > (2\pi)^{-1}$ significantly. Such a test is expected to be powerful in the present context because under the alternative $\mathbb{H}_1^{\mathscr{E}}$ there is always a spectral mode at frequency zero due to positive autocorrelations in $\{X_t\}$. In particular, a spectral peak at frequency zero arises when there exist persistent ACD effects. Wavelets are effective in capturing spectral peaks/modes. Of course, conventional methods such as the kernel method could be used as well. However, this may not deliver an equally powerful procedure, because the kernel method often tends to underestimate spectral mode/peaks (cf. [61, pp. 547–556]).

## 2.3 *Adequacy of ACD Models*

Suppose we have evidence of duration clustering and have decided to use an ACD model, say $D(\mathscr{I}_{t-1}, \boldsymbol{\theta})$, to fit the data, where $\boldsymbol{\theta}$ is an unknown finite dimensional parameter. We may like to test if the model fits the data adequately. One approach is to examine whether the standardized duration residual,

$$e_t = X_t / D(\mathscr{I}_{t-1}, \boldsymbol{\theta}_0),$$

contains any remaining duration structure, where $\boldsymbol{\theta}_0 = \text{plim}\, \hat{\boldsymbol{\theta}}$ and $\hat{\boldsymbol{\theta}}$ is an estimator of $\boldsymbol{\theta}$. Suppose $D(\mathscr{I}_{t-1}, \boldsymbol{\theta}_0) = D^0(\mathscr{I}_{t-1})$ almost surely. Then the model $D(\mathscr{I}_{t-1}, \boldsymbol{\theta}_0)$ is adequate for modelling the conditional duration $D_t^0$. In this case we have $e_t = \varepsilon_t$, an iid white noise process with a flat spectrum. Alternatively, suppose that $D(\mathscr{I}_{t-1}, \boldsymbol{\theta})$ is inadequate in modeling $X_t$ in the sense that $P[D(\mathscr{I}_{t-1}, \boldsymbol{\theta}) = D^0(\mathscr{I}_{t-1})] < 1$ for all $\boldsymbol{\theta}$, then $e_t \neq \varepsilon_t$. In this case we will generally have a non-flat spectrum for $e_t$ because $e_t$ contains some remaining dependent structure for the conditional duration. Therefore, one can check the adequacy of an ACD model $D(\mathscr{I}_{t-1}, \boldsymbol{\theta})$ by examining whether the spectrum of $\{e_t\}$ is flat.

Let $f_e(\omega)$ be the standardized spectral density of $\{e_t\}$. Then under the null hypothesis of adequacy of an ACD model, we have

$$\mathbb{H}_0^{\mathscr{A}} : f_e(\omega) = (2\pi)^{-1}, \quad \omega \in [-\pi, \pi].$$

When the model is misspecified, we generally have

$$\mathbb{H}_1^{\mathscr{A}} : f_e(\omega) \neq (2\pi)^{-1}.$$

Under $\mathbb{H}_1^{\mathscr{A}}$, the standardized duration residual $e_t$ could exhibit negative and/or positive autocorrelations, and any departure of $(2\pi)^{-1}$ can be anticipated for $f_e(\omega)$. Thus the alternative hypothesis $\mathbb{H}_1^{\mathscr{A}}$ is not one-sided and we cannot proceed as in testing for an ACD effect. Instead, we have to compare a spectral density estimator of $\{e_t\}$ with a flat spectrum over all frequencies. We note that our second test here complements the methods in [27]. In that work, the authors consider testing the distribution specification for $\{\varepsilon_t\}$, given correct specification of $D_t(\mathscr{I}_{t-1}, \boldsymbol{\theta})$.

In [21, 22], they consider the BP/LB tests as diagnostic tests for ACD models. See also [34] in the modelling of a Burr-ACD model. The BP/LB tests are assumed to have an asymptotic $\chi^2_m$, where $m$ is the number of lags used in the test. However, even in the simpler case of ARMA modeling, it is well-known that the degrees of freedom of the $\chi^2$ asymptotic distribution of the BP/LB statistics need an adjustment due to parameter estimation uncertainty, which depends on the autoregressive and moving-average orders (cf. [10]). In the present context, we expect that a modification for the asymptotic distribution of the test statistics or the test statistics themselves is also needed, because $D(\mathscr{I}_{t-1}, \boldsymbol{\theta})$ is a conditional mean model similar to an ARMA model. Moreover, the adjustment does not seem so simple as in the ARMA case, since an ACD model is highly nonlinear. This may be rather complicated in view of the results of [53] for diagnostic testing for ARCH models. Finally, the choice of $m$ in BP/LB statistics is somewhat arbitrary and there has been no theoretic guidance for the choice of $m$. In practice the user may prefer an automatic method with reasonable power toward many directions. Our tests have such an appealing feature and they should prove useful to complement other methods, such as the methods in [54], the kernel-based methods in [18], or the generalized spectral derivative tests in [45].

## 3 Testing for ACD Effects

### *3.1 Wavelet Analysis*

Throughout, we use multiresolution analysis (cf. [55]). This is an analytic method to decompose a square-integrable function at different scales. The key is a real-valued function called mother wavelet $\psi(\cdot)$. An example is the Haar wavelet:

$$\psi(x) = \begin{cases} 1 & \text{if } 0 \leq x < \frac{1}{2}, \\ -1 & \text{if } \frac{1}{2} \leq x \leq 1, \\ 0 & \text{otherwise.} \end{cases}$$

This wavelet has bounded support, which ensures that $\psi(\cdot)$ is well localized in time (or space) domain. In [12], it is shown that for any nonnegative integer $d$, there exists an orthonormal compact supported mother wavelet whose first $d$ moments vanish. The mother wavelet $\psi(\cdot)$ can also have infinite support, but it must decay to zero sufficiently fast at $\infty$. An example is the Franklin wavelet $\psi(\cdot)$, which is defined via its Fourier transform

$$\hat{\psi}(z) = (2\pi)^{-1/2} e^{iz/2} \frac{\sin^4(z/4)}{(z/4)^2} \left\{ \frac{1 - (2/3)\cos^2(z/4)}{[1 - (2/3)\sin^2(z/2)][1 - (2/3)\sin^2(z/4)]} \right\}^{1/2}. \quad (3)$$

See (e.g.) [40] for more examples.

To represent a standardized spectral density $f(\cdot)$, which is $2\pi$-periodic and thus is not square-integrable on the real line $\mathbb{R} \equiv (-\infty, \infty)$, we must construct a $2\pi$-periodic basis, where

$$\Psi_{jk}(\omega) = (2\pi)^{-1/2} \sum_{m=-\infty}^{\infty} \psi_{jk}\left(\frac{\omega}{2\pi} + m\right), \quad \omega \in \mathbb{R},$$

with

$$\psi_{jk}(x) = 2^{j/2}\psi(2^j x - k). \quad (4)$$

The integers $j$ and $k$ in (4) are called the dilation and translation parameters. Intuitively, $j$ localizes analysis in frequency and $k$ localizes analysis in time (or space). This joint time-frequency decomposition of information is the key feature of wavelet analysis. With these $2\pi$-periodic orthonormal bases, we can represent the spectral density

$$f(\omega) = (2\pi)^{-1} + \sum_{j=0}^{\infty} \sum_{k=1}^{2^j} \alpha_{jk} \Psi_{jk}(\omega), \ \omega \in [-\pi, \pi],$$

where the wavelet coefficient satisfies

$$\alpha_{jk} = \int_{-\pi}^{\pi} f(\omega)\Psi_{jk}(\omega)d\omega = (2\pi)^{-1/2} \sum_{h=-\infty}^{\infty} \rho(h)\hat{\Psi}^*_{jk}(h), \quad (5)$$

and $\{\rho(h), h \in \mathbb{Z}\}$ is the autocorrelation function of the time series and $\hat{\Psi}_{jk}(\cdot)$ denotes the Fourier transform of $\Psi_{jk}(\cdot)$; that is,

$$\hat{\Psi}_{jk}(h) = (2\pi)^{-1/2} \int_{-\pi}^{\pi} \Psi_{jk}(\omega)e^{-ih\omega}d\omega, \ h = 0, \pm 1, \ldots$$

The second equality in (5) is obtained using Parseval's identity. See [44] and [50] for more discussions.

The wavelet coefficient $\alpha_{jk}$ only depends on the local property of $f(\omega)$, because $\Psi_{jk}(\omega)$ is essentially nonzero only in an interval of size $2\pi/2^j$ centered at $k/2^j$. This is fundamentally different from the Fourier representation of $f(\omega)$, where the Fourier coefficient is the autocorrelation, $\rho(h)$, which depends on the global property of $f(\omega)$. This is why wavelets are particularly capable of capturing spectral peaks or nonsmooth features.

Our test statistic for ACD effects is based on a wavelet spectral density estimator at frequency zero of the duration process $\{X_t\}$. When testing for the adequacy of an ACD model, we need to evaluate a wavelet spectral density estimator for the estimated standardized duration residuals over the entire frequency domain $[-\pi, \pi]$. In an unified manner, we consider wavelet spectral density estimators in these two situations.

### 3.2 Test Statistics

We first describe wavelet-based spectral density estimation for the duration process $\{X_t\}$. Suppose we have a sample $\{X_t\}_{t=1}^n$ of size $n$. Define the sample autocorrelation function of $\{X_t\}_{t=1}^n$ as

$$\hat{\rho}_X(h) = \hat{R}_X(h)/\hat{R}_X(0), \quad h = 0, \pm 1, \ldots, \pm(n-1),$$

where the sample autocovariance of $\{X_t\}$ is given by

$$\hat{R}_X(h) = n^{-1} \sum_{t=|h|+1}^{n} (X_t - \bar{X})(X_{t-|h|} - \bar{X}), \tag{6}$$

with $\bar{X} = n^{-1} \sum_{t=1}^{n} X_t$. A wavelet spectral estimator for $f_X(\omega)$ can be given as

$$\hat{f}_X(\omega) = (2\pi)^{-1} + \sum_{j=0}^{J} \sum_{k=1}^{2^j} \hat{\alpha}_{jk} \Psi_{jk}(\omega),$$

where the empirical wavelet coefficient is

$$\hat{\alpha}_{jk} = (2\pi)^{-1/2} \sum_{h=1-n}^{n-1} \hat{\rho}_X(h) \hat{\Psi}_{jk}^*(h),$$

and the truncation parameter $J \equiv J(n)$ is called the finest scale parameter, a smoothing parameter. In this sense it is similar to the bandwidth in kernel-based spectral density estimation. However, a fundamental difference is that $J$ is not a lag truncation parameter, since even when $J = 0$ the empirical wavelet coefficient $\hat{\alpha}_{jk}$ is still a weighted sum of all $n-1$ sample autocorrelations $\{\hat{\rho}_X(h)\}_{h=1}^{n-1}$ provided $\hat{\psi}(\cdot)$ has unbounded support. In contrast, $J$ corresponds to the highest resolution level used in the wavelet approximation. Given each $J$, there are totally $2^{J+1} - 1$ empirical wavelet coefficients in $\hat{f}_X(\omega)$. To reduce the bias of $\hat{f}_X(\omega)$, we must let $J \to \infty$ as $n \to \infty$. On the other hand, to ensure that the variance of $\hat{f}_X(\omega)$ vanishes, $2^{J+1} - 1$ must grow slower than $n$. Thus, we need to choose $J$ properly to balance the bias and variance of $\hat{f}_X(\omega)$. We note that we use linear rather than nonlinear wavelet estimators here, because our test statistics are based on quadratic forms. In the space of square-integrable functions, linear and nonlinear wavelet estimators achieve the same convergence rate (cf. [57]). Among other advantages, the use of linear wavelet estimators allow us to obtain the asymptotic distribution theory relatively easily. In testing for serial correlation, simulation results of [17] and [51] demonstrate that for certain fixed alternatives, linear wavelet estimators may perform better than nonlinear ones. Other simulation studies (e.g., [50]) show that linear wavelet estimators outperform kernel estimators in finite samples when the spectral density is not smooth.

Our test for ACD effects is based on the spectral density estimator $\hat{f}_X(\cdot)$ at the zero frequency. It is defined as

$$\mathscr{E}(J) = [V_n(J)]^{-1/2} n^{1/2} \pi \left[ \hat{f}_X(0) - (2\pi)^{-1} \right], \tag{7}$$

where the asymptotic variance estimator

$$V_n(J) = \sum_{h=1}^{n-1} (1 - h/n) \left[ \sum_{j=0}^{J} \lambda(2\pi h/2^j) \right]^2$$

and

$$\lambda(z) = 2\pi \hat{\psi}^*(z) \sum_{m=-\infty}^{\infty} \hat{\psi}(z + 2\pi m), \quad z \in \mathbb{R}. \tag{8}$$

From [50], $2^{-J} V_n(J) \to V_0$ as $J \to \infty$, $2^J/n \to 0$, where

$$V_0 = \int_0^{2\pi} |\Gamma(z)|^2 dz, \tag{9}$$

with $\Gamma(z) = \sum_m \hat{\psi}(z + 2\pi m)$. The result is also stated in the Appendix as Lemma 2. Note that $V_n(J)$ is nonstochastic and is readily computable given $\hat{\psi}(\cdot)$ and $J$. Because $\hat{f}_X(0)$ is close to $(2\pi)^{-1}$ under $\mathbb{H}_0^{\mathscr{E}}$ and is significantly larger that $(2\pi)^{-1}$ under $\mathbb{H}_1^{\mathscr{E}}$, the test statistic delivers a one-sided consistent testing procedure for $\mathbb{H}_1^{\mathscr{E}}$. How large $\mathscr{E}(J)$ must be in order to be considered as significantly larger than zero is described by the sampling distribution of $\mathscr{E}(J)$.

## 3.3 Asymptotic Distribution

To derive the asymptotic distribution of $\mathscr{E}(J)$, we impose the following regularity conditions.

**Assumption 2** The function $\psi : \mathbb{R} \to \mathbb{R}$ is an orthonormal mother wavelet such that $\int_{-\infty}^{\infty} \psi(x)dx = 0$, $\int_{-\infty}^{\infty} |\psi(x)|dx < \infty$, $\int_{-\infty}^{\infty} \psi^2(x)dx = 1$ and $\int \psi(x)\psi(x - k)dx = 0$ for all integers $k$, $k \neq 0$.

**Assumption 3** The Fourier transform of $\psi(\cdot)$ satisfies $|\hat{\psi}(z)| \leq C \min\{|z|^q, (1+|z|)^{-\tau}\}$ for some $q > 0$ and $\tau > 1$.

**Assumption 4** The function $\lambda : \mathbb{R} \to \mathbb{R}$ is square-integrable, where $\lambda(\cdot)$ is defined in (8).

Assumption 2 is a standard condition for an orthonormal mother wavelet $\psi(\cdot)$. Assumption 3 requires that $\hat{\psi}(\cdot)$ have some regularity (i.e., smoothness) at zero and sufficiently fast decay at $\infty$. The condition $|\hat{\psi}(z)| \leq C|z|^q$ is effective as $z \to 0$, where $q$ governs the degree of smoothness of $\hat{\psi}(\cdot)$ at zero. If $\int_{-\infty}^{\infty}(1+|x|^\nu)|\psi(x)|dx < \infty$ for some $\nu > 0$, then $|\hat{\psi}(z)| \leq C|z|^q$ for $q = \min(\nu, 1)$; see (e.g.) [62]. When $\psi(\cdot)$ has first $d$ vanishing moments (i.e., $\int_{-\infty}^{\infty} x^r \psi(x)dx = 0$ for $r = 0, \ldots, d-1$), we have $|\hat{\psi}(z)| \leq C|z|^d$ as $z \to 0$. On the other hand, the condition $|\hat{\psi}(z)| \leq C(1+|z|)^{-\tau}$ is effective as $z \to \infty$. This holds with $\tau = \infty$ for the so-called band-limited wavelets, whose $\hat{\psi}(\cdot)$'s have compact supports. In Assumption 4, the condition that $\lambda(\cdot)$ is real-valued implies $\lambda(-z) = \lambda(z)$ because $\hat{\psi}^*(z) = \hat{\psi}(-z)$. In addition, Assumptions 2 and 3 ensure that $\lambda(\cdot)$ is continuous almost everywhere in $\mathbb{R}$, $\lambda(0) = 0$ and $|\lambda(z)| \leq C$. Most commonly used wavelets satisfy Assumptions 2–4. Examples are the compactly supported wavelets of a positive order in [12], the Franklin wavelet, the Lemarie-Meyer wavelets, the Littlewood-Paley wavelets, and spline wavelets. Assumption 3 rules out the Haar wavelet, however, because its Fourier transform, $\hat{\psi}(z) = -ie^{iz/2}\sin^2(z/4)/(z/4)$, goes to zero at a rate of $|z|^{-1}$ only.

The asymptotic normality of $\mathscr{E}(J)$ is stated below.

**Theorem 1** *Suppose Assumptions 1–4 hold, and* $2^J/n \to 0$. *Then* $\mathscr{E}(J) \to^d N(0, 1)$ *under* $\mathbb{H}_0^{\mathscr{E}}$.

Both small and large (i.e., fixed and increasing as $n \to \infty$) finest scales $J$ are allowed here. Thus, the choice of $J$ has no impact on the null limit distribution of $\mathscr{E}(J)$, as long as $2^J$ grows slower than the sample size $n$. Of course, it may have impact on the finite sample distribution of $\mathscr{E}(J)$. Note that because $\mathscr{E}(J)$ is a one-sided test, it is appropriate to use upper-tailed $N(0, 1)$ critical values. The critical value at the 5 % level, for example, is 1.645.

Consider the class of local alternatives:

$$\mathbb{H}_{1n}^{\mathscr{E}}(a_n) : D_t = D_0 \left[1 + a_n \sum_{j=1}^{\infty} \beta_j(\varepsilon_{t-j} - 1)\right],$$

where $\beta_j \geq 0$ with at least one strict inequality and $\sum_{j=1}^{\infty} \beta_j < \infty$. Following a reasoning similar to [44, Theorem 2], if $J$ is fixed, then under $\mathbb{H}_{1n}^{\mathscr{E}}(n^{-1/2})$, it is possible to show that $\mathscr{E}(J) \to^d N(\mu(J), 1)$, where $\mu(J) = [V_0(J)]^{-1/2} \sum_{h=1}^{\infty} d_J(h)\beta_h$, $V_0(J) = \sum_{h=1}^{\infty} d_J^2(h)$ and $d_J(h) = \sum_{j=0}^{J} \lambda(2\pi h/2^j)$. When $J \to \infty$, with the more restrictive rate $2^{2J}/n \to \infty$, we can obtain the asymptotic distribution of $\mathscr{E}(J)$ under $\mathbb{H}_{1n}^{\mathscr{E}}(2^{J/2}/n^{1/2})$, which is now $\mathscr{E}(J) \to^d N(\mu, 1)$, where $\mu = V_0^{-1/2}\sum_{j=1}^{\infty}\beta_j$, with $V_0$ defined as (9). Thus, under a fixed finest scale $J$, $\mathscr{E}(J)$ has nontrivial power against alternatives in $\mathbb{H}_{1n}^{\mathscr{E}}(n^{-1/2})$. However, if $J \to \infty$ such that $2^{2J}/n \to 0$, $\mathscr{E}(J)$ offers nontrivial power against alternatives only in $\mathbb{H}_{1n}^{\mathscr{E}}(2^{J/2}/n^{1/2})$. The discussion in [44, pp. 1060–1061] also applies.

### 3.4 Adaptive Choice of the Finest Scale

Although the choice of $J$ has no impact on the null limit distribution of $\mathscr{E}(J)$, it may significantly affect the power of $\mathscr{E}(J)$ in finite samples. It is not easy to choose an optimal $J$ to maximize power, especially in light of the facts that $J$ is not a lag order and that usually no prior information on the alternative is available. Therefore, it is desirable to choose $J$ via suitable data-driven methods, which adapt to unknown alternatives and are more objective than any arbitrary choice of $J$ or any simple "rule-of-thumb". To allow for this possibility, we consider using a data-dependent finest scale $\hat{J}$. We simply plug $\hat{J}$ in $\mathscr{E}(\cdot)$, and then obtain the statistic $\mathscr{E}(\hat{J})$. Such a test has the appealing advantage of being totally operational in practice. Before we discuss specific methods to choose $\hat{J}$, we first justify the use of $\hat{J}$ in our test $\mathscr{E}(\hat{J})$.

**Theorem 2** *Suppose Assumptions 1–4 hold, and $\hat{J}$ is a data-dependent finest scale such that $2^{\hat{J}}/2^J = 1 + o_P(2^{-J/2})$ for some nonstochastic $J$ satisfying $2^{2J}/n \to 0$. Then $\mathscr{E}(\hat{J}) - \mathscr{E}(J) \to^p 0$ and $\mathscr{E}(\hat{J}) \to^d N(0, 1)$ under $\mathbb{H}_0^{\mathscr{E}}$.*

Theorem 2 implies that the effect of sampling randomness in $\hat{J}$ has negligible impact on the limit distribution of $\mathscr{E}(\hat{J})$ as long as $\hat{J}$ converges to $J$ sufficiently fast. The conditions on $\hat{J}$ are weak. When $J$ is fixed ($J = 0$ say), as may occur under $\mathbb{H}_0^{\mathscr{E}}$ for sensible data-driven methods, $2^{\hat{J}}/2^J = 1 + o_P(2^{-J/2})$ becomes $2^{\hat{J}}/2^J \to^p 1$; no rate condition on $\hat{J}$ is required. Often, $\hat{J}$ and $J$ have the forms of $2^{\hat{J}+1} = \hat{c}n^{\nu}$ and $2^{J+1} = cn^{\nu}$, where $c \in (0, \infty)$ is a tuning constant and $\hat{c}$ is its estimator. For the parametric plug-in method considered below, we generally have $\hat{c}/c = 1 + O_P(n^{-1/2})$, thus satisfying the condition on $\hat{J}$ for wavelets with $q > \frac{1}{2}$.

We now consider a specific data-driven method to select $J$ for $\mathscr{E}(J)$. We impose the following additional conditions:

**Assumption 5** For $\psi(\cdot)$, there exists a largest number $q \in [1, \infty)$ such that $0 < \lambda_q < \infty$, where $\lambda_q \equiv \frac{(2\pi)^q}{1-2^{-q}} \lim_{z\to 0} \frac{\lambda(z)}{|z|^q}$.

**Assumption 6** The duration process $\{X_t\}$ is fourth order stationary with the conditions (i) $\sum_{h=-\infty}^{\infty} R_X^2(h) \le C$; (ii) $\sum_{j=-\infty}^{\infty}\sum_{k=-\infty}^{\infty}\sum_{l=-\infty}^{\infty} |\kappa_X(j,k,l)| \le C$, where the quantity $\kappa_X(j,k,l)$ denotes the fourth order cumulant of the joint distribution of the random variables $\{X_t - \mu_X, X_{t+j} - \mu_X, X_{t+k} - \mu_X, X_{t+l} - \mu_X\}$, and $E(X_t) = \mu_X$; and (iii) $\sum_{h=-\infty}^{\infty} |h|^q |R_X(h)| < \infty$, where $q \in [1, \infty)$ is as in Assumption 5.

Obviously, the smoother is $\lambda(\cdot)$ at 0, the larger is the value of $q$ for which $\lambda_q$ is nonzero and finite. If $q$ is an even positive integer, then $\lambda_q = \frac{(2\pi)^q}{1-2^{-q}} \frac{1}{q!} d^q \lambda(0)/dz^q$, and $\lambda_q < \infty$ if and only if $\lambda(\cdot)$ is $q$-time differentiable at zero. For the Franklin wavelet (3), $q = 2$. In this case $\lambda_q = 0$ for $q < 2$, $\lambda_2 = (8/3) \sum_{l=0}^{\infty} (2l+1)^{-2}$ and $\lambda_q = \infty$ for $q > 2$. In general, if the mother wavelet $\psi(\cdot)$ has and only has first $\upsilon$ vanishing moments, then $\lambda_q = 0$ for $q < \upsilon$, $\lambda_q = \infty$ for $q > \upsilon$, and $\lambda_\upsilon \neq 0$.

Because $\mathscr{E}(J)$ is based on $\hat{f}_X(0)$, it is appropriate to adapt $\hat{J}$ to the unknown $f_X(\cdot)$ at zero rather than over the entire frequency domain $[-\pi, \pi]$. Thus, we consider a choice of $\hat{J}$ minimizing the asymptotic mean squared error (MSE) of $\hat{f}_X(0)$. From a theoretical point of view, the choice of $\hat{J}$ based on the MSE criterion may not maximize the power of the test. A more sensible alternative is to develop a data-driven $\hat{J}$ using a suitable power criterion, or a criterion that trades off level distortion and power loss. This, however, would necessitate higher order asymptotic analysis and is far beyond the scope of this paper. Here, we are content with the MSE criterion, which delivers reasonable power (see the simulation below).

To derive the MSE, we need to obtain the bias and the variance of $\hat{f}_X(0)$. To characterize the bias of $\hat{f}_X(0)$, we define the generalized derivative of $f_X(\cdot)$ at 0:

$$f_X^{(q)}(0) = \frac{1}{2\pi} \sum_{h=-\infty}^{\infty} |h|^q R_X(h).$$

If $q$ is an even integer and $f(\cdot)$ is $q$-time differentiable at 0, then $f^{(q)}(0) = (-1)^{\frac{q}{2}} d^q f(0)/d^q \omega$. However, there is no simple relationship between the two for a general $q$.

To characterize the variance of $\hat{f}_X(0)$, we define the integral

$$D_\psi = \int_0^{2\pi} \left| \sum_{m=-\infty}^{\infty} \hat{\psi}(z + 2m\pi) \right|^2 dz,$$

which exists and is finite given Assumptions 2 and 3.

We now state the MSE for $\hat{f}_X(0)$. For simplicity we assume here that the spectral density for establishing the MSE is unnormalized.

**Theorem 3** *Suppose Assumptions 2–6 hold, and* $2^{J+1}/n^{1/(2q+1)} \to c \in (0, \infty)$. *Then*

$$\lim_{n\to\infty} n^{\frac{2q}{2q+1}} MSE\left\{\hat{f}_X(0, J), f_X(0)\right\} = 2cD_\psi f_X^2(0) + \lambda_q^2 c^{-2q} \left[f_X^{(q)}(0)\right]^2.$$

The MSE criterion provides a basis to choose an optimal $J$. The optimal convergence rate for the MSE can be attained by setting the derivative of the MSE with respect to the tuning constant $c$ to zero. This yields the optimal finest scale $J^0$:

$$2^{J^0+1} = \left[q\lambda_q^2 \alpha(q) n / D_\psi\right]^{\frac{1}{2q+1}},$$

where $\alpha(q) = \left[f_X^{(q)}(0)/f_X(0)\right]^2$. This optimal $J^0$ is infeasible because $\alpha(q)$ is unknown. Nevertheless, we can plug-in an estimator $\hat{\alpha}(q)$ for $\alpha(q)$. This gives a "plug-in" data-driven $\hat{J}$:

$$2^{\hat{J}+1} = \left[q\lambda_q^2\hat{\alpha}(q)n/D_\psi\right]^{\frac{1}{2q+1}}. \tag{10}$$

Because $\hat{J}$ must be a nonnegative integer for each $n \geq 1$, we use

$$\hat{J} = \max\{\lfloor(2q+1)^{-1}\log_2\left(q\lambda_q^2\hat{\alpha}(q)n/D_\psi\right) - 1\rfloor, 0\},$$

where $\lfloor\cdot\rfloor$ denotes the integer part.

**Corollary 1** *Suppose Assumptions 2–6 hold, $\hat{J}$ is given as in (10), and $\hat{\alpha}(q) = \alpha_\xi + o_P(n^{-\delta})$, where $\delta = 1/(2(2q+1))$ if $\alpha_\xi \in (0,\infty)$ and $\delta = 1/(2q+1)$ if $\alpha_\xi = 0$. Then $\mathscr{E}(\hat{J}) - \mathscr{E}(J) \to^p 0$ and $\mathscr{E}(\hat{J}) \to^d N(0,1)$ under $\mathbb{H}_0^{\mathscr{E}}$.*

In Corollary 1, we require $\hat{\alpha}(q) \to^p \alpha_\xi$ at a rate faster than $n^{-\frac{1}{2(2q+1)}}$ when $\alpha_\xi > 0$ and the more stringent rate $n^{-\frac{1}{2q+1}}$ when $\alpha_\xi = 0$. This ensures that the use of $\hat{\alpha}(q)$ rather than $\alpha_\xi$ has no impact on $\mathscr{E}(J)$ asymptotically.

Plug-in methods can be parametric or nonparametric (see, e.g., [2], [11], and [58], for kernel-based spectral density estimation). Parametric plug-in methods use a parsimonious approximating model for $\hat{\alpha}(q)$. It yields a less variable smoothing parameter, but when the approximating model is misspecified, it will not attain the minimum asymptotic MSE, although this has no impact on the consistency of $\mathscr{E}(\hat{J})$ against $\mathbb{H}_1^{\mathscr{E}}$. On the other hand, nonparametric plug-in methods estimate $\hat{\alpha}(q)$ nonparametrically. It attains the minimum MSE asymptotically but still involves the choice of a preliminary smoothing parameter.

Both parametric and nonparametric plug-in methods can be used here. Below, we consider a parametric "plug-in" method in spirit similar to [2], who considers data-driven methods to choose a bandwidth in kernel-based spectral density estimation at frequency zero. As an approximate parametric model for the duration process $\{X_t\}$, we consider an AR($p$) model

$$X_t = \beta_0 + \sum_{h=1}^{p}\beta_h X_{t-h} + v_t. \tag{11}$$

The spectral density under the model (11) is given by

$$g(\omega;\boldsymbol{\beta}) = \frac{1}{2\pi}\left|1 - \sum_{h=1}^{p}\beta_h\exp(-ih\omega)\right|^{-2},$$

where $\boldsymbol{\beta} = (\beta_0, \beta_1, \ldots, \beta_p)^\top$. Here we have normalized the variance $\sigma_v^2 = 1$. This has no impact on $\hat{\alpha}(q)$ because the variance $\sigma_v^2$ cancel in the numerator and denominator of $\hat{\alpha}(q)$. We can estimate $\boldsymbol{\beta}$ with the ordinary least-squares method. In practice, the autoregressive order $p$ can be chosen by the Akaike's Information Criterion (AIC) or Bayesian Information Criterion (BIC). We then obtain, for $q = 2$,

$$\hat{\alpha}(2) = \left[ \frac{g''(0; \hat{\boldsymbol{\beta}})}{g(0; \hat{\boldsymbol{\beta}})} \right]^2.$$

Although OLS may deliver negative parameter estimate $\hat{\boldsymbol{\beta}}$ for the non-negative duration process $\{X_t\}$, this causes no problem here because $\hat{\alpha}(2) \geq 0$. It could be shown that under proper conditions, $\hat{\alpha}(2) = \alpha_\xi + O_P(n^{-1/2})$ where $\alpha_\xi = p\lim_{n\to\infty} \hat{\alpha}(2)$, thus satisfying the conditions in Corollary 1. We note that the method developed here can be adapted to the ARCH test in [44].

To summarize, our test procedure for ACD effects can be described as follows: (i) Find $\hat{J}$ by the autoregression model (11); (ii) compute the test statistic $\mathscr{E}(\hat{J})$ in (7); (iii) reject the null hypothesis of no ACD effects if $\mathscr{E}(\hat{J}) > z_{1-\alpha}$, where $z_{1-\alpha}$ is the $1 - \alpha$ quantile of the standard normal distribution $N(0, 1)$.

## 4 Diagnostic Testing for ACD Models

### 4.1 *Test Statistic*

We now turn to test the adequacy of an ACD model $D(\mathscr{I}_{t-1}, \boldsymbol{\theta})$. Suppose $\hat{\boldsymbol{\theta}}$ is a parameter estimator. Then the standardized duration residual is given by

$$\hat{e}_t = X_t / \hat{D}_t = X_t / D(\mathscr{I}_{t-1}, \hat{\boldsymbol{\theta}}).$$

Let $\hat{R}_e(h)$ be the sample autocovariance function of $\{\hat{e}_t\}$ defined in the same way as $\hat{R}_X(h)$ in (6). Throughout, we impose the following regularity conditions on the model $D(\mathscr{I}_{t-1}, \boldsymbol{\theta})$ and the estimator $\hat{\boldsymbol{\theta}}$, and an additional condition on the mother wavelet $\psi(\cdot)$.

**Assumption 7** *(i) For each $\boldsymbol{\theta} \in \Theta$, $D(\cdot, \boldsymbol{\theta})$ is a measurable nonnegative function of $\mathscr{I}_{t-1}$; (ii) with probability one, $D(\mathscr{I}_{t-1}, \cdot)$ is twice continuously differentiable with respect to $\boldsymbol{\theta}$ in a neighborhood of $\Theta_0$ of $\boldsymbol{\theta}_0 = plim\, \hat{\boldsymbol{\theta}}$, with*

$$\lim_{n\to\infty} n^{-1} \sum_{t=1}^{n} E \sup_{\boldsymbol{\theta}\in\Theta_0} \left\| \frac{\partial}{\partial \boldsymbol{\theta}} D(\mathscr{I}_{t-1}, \boldsymbol{\theta}) \right\|^2 < \infty$$

*and* $\lim_{n\to\infty} n^{-1} \sum_{t=1}^{n} E \sup_{\boldsymbol{\theta}\in\Theta_0} \left\| \frac{\partial^2}{\partial \boldsymbol{\theta} \partial \boldsymbol{\theta}^\top} D(\mathscr{I}_{t-1}, \boldsymbol{\theta}) \right\| < \infty.$

**Assumption 8** $\hat{\theta} - \theta_0 = O_P(n^{-1/2})$.

**Assumption 9** $\hat{\psi}(z) = e^{iz/2}b(z)$ or $\hat{\psi}(z) = -ie^{iz/2}b(z)$, where $b(\cdot)$ is a real-valued function.

In Assumption 8, we allow any $n^{1/2}$-consistent estimator $\hat{\theta}$, such as the QMLE considered in [22]. We also allow the semi-parametric estimators considered in [13, 14], which attain an efficiency bound.

The Franklin wavelet (3) satisfies Assumption 9. In fact, most commonly used wavelets satisfy this assumption. For example, the Lemarie-Meyer family is of the form $\hat{\psi}(z) = e^{iz/2}b(z)$, where $b(\cdot)$ is a real-valued and symmetric. Another family is the spline wavelets of order $m > 0$. For odd $m$, this family is of the form $\hat{\psi}(z) = e^{iz/2}b(z)$, where $b(\cdot)$ is real-valued and symmetric; for even $m$, it is of the form $\hat{\psi}(z) = -ie^{iz/2}b(z)$, where $b(\cdot)$ is real-valued and odd function. See [40] for more discussion.

To test the adequacy of an ACD model, we need a spectral density estimator of $\{\hat{e}_t\}$ over all frequencies $\omega \in [-\pi, \pi]$. A feasible wavelet estimator of $f_e(\omega)$ can be given as

$$\hat{f}_e(\omega) = (2\pi)^{-1} + \sum_{j=0}^{J} \sum_{k=1}^{2^j} \hat{\alpha}_{ejk} \Psi_{jk}(\omega),$$

where the empirical wavelet coefficient is

$$\hat{\alpha}_{ejk} = (2\pi)^{-1/2} \sum_{h=1-n}^{n-1} \hat{\rho}_e(h) \hat{\Psi}^*_{jk}(h),$$

and $\hat{\rho}_e(h)$ is the sample autocorrelation function of $\{\hat{e}_t\}$.

We consider a test by comparing $\hat{f}_e(\cdot)$ and the flat spectrum. We use the quadratic form,

$$Q(\hat{f}_e, f_e^0) = \int_{-\pi}^{\pi} [\hat{f}_e(\omega) - (2\pi)^{-1}]^2 d\omega = \sum_{j=0}^{J} \sum_{k=1}^{2^j} \hat{\alpha}_{ejk}^2,$$

where the second equality follows by Parseval's identity and the orthonormality of wavelet bases $\{\Psi_{jk}(\cdot)\}$. Because $\{\hat{e}_t\}$ could exhibit positive and/or negative autocorrelation, the quadratic metric is appropriate. Other divergence measures such as the Hellinger metric or the Kullback-Leibler Information Criteria could be used. However, $Q(\hat{f}_e, f_e^0)$ is convenient to compute, since there is no need to calculate the numerical integration over $\omega \in [-\pi, \pi]$.

Our diagnostic test statistic for ACD models is a normalized version of $Q(\hat{f}_e, f_e^0)$:

$$\mathscr{A}(J) = \left[ 2\pi n \sum_{j=0}^{J} \sum_{k=1}^{2^j} \hat{\alpha}_{ejk}^2 - (2^{J+1} - 1) \right] / [4(2^{J+1} - 1)]^{1/2}. \tag{12}$$

When the ACD model is adequate, $\hat{f}_e(\omega)$ is close to a flat spectrum and so $Q(\hat{f}_e, f_e^0)$ is close to zero. Suppose the ACD model is inappropriate such that $\hat{f}_e(\omega)$ is not flat, $\mathscr{A}(\hat{J})$ will diverge to $+\infty$ as $n \to \infty$ because $Q(\hat{f}_e, f_e^0) \geq C > 0$ as $n \to \infty$. How large $\mathscr{A}(J)$ must be in order to be viewed as significantly larger that zero is determined by the sampling distribution of $\mathscr{A}(J)$.

## 4.2 Asymptotic Distribution

We now state the asymptotic normality of $\mathscr{A}(J)$.

**Theorem 4** *Suppose Assumptions 1–3 and 7–9 hold,* $2^{2J}/n \to 0$ *such that* $J \to \infty$. *If* $D(\mathscr{I}_{t-1}; \boldsymbol{\theta}_0) = D_t^0$ *a.s., then* $\mathscr{A}(J) \xrightarrow{d} N(0, 1)$.

Although the alternative $\mathbb{H}_1^{\mathscr{A}}$ is not a one-sided alternative, it is appropriate to use one-sided critical values for $\mathscr{A}(J)$, since $\mathscr{A}(J)$ is positive for $n$ sufficiently large under $\mathbb{H}_1^{\mathscr{A}}$. Unlike the test $\mathscr{E}(J)$ for ACD effects in Sect. 3, we require here that the finest scale $J \to \infty$ as $n \to \infty$ in order to obtain asymptotic normality for $\mathscr{A}(J)$. An interesting feature of the test $\mathscr{A}(J)$ is that parameter estimation uncertainty in $\hat{\boldsymbol{\theta}}$ has no impact on the limit distribution of $\mathscr{A}(J)$. This delivers a convenient procedure in practice and is in sharp contrast to the BP/LB statistics as considered in [21, 22]. These statistics, adapted to the estimated standardized residuals $\{\hat{e}_t\}$, are assumed to follow an asymptotic $\chi_m^2$ distribution, where $m$ is the number of lags used in the tests. However, the validity of such asymptotic distribution is not established in the literature. In the case of goodness-of-fit tests for ARMA models, it is well-known that the degrees of freedom need to be adjusted (cf. [10]). The correct adjustment is not known for BP/LB goodness-of-fit tests for ACD models and this seems rather complicated in light of the results of [53] for testing ARCH models.

## 4.3 Adaptive Choice of Finest Scale

Like the test $\mathscr{E}(J)$ for ACD effects, the choice of $J$ may have significant impact on the power of $\mathscr{A}(J)$. It will be highly desirable to choose $J$ via suitable data-driven methods, which let data speak for proper finest scales. The data-driven method developed for the test $\mathscr{E}(J)$ of ACD effects could be used for $\mathscr{A}(J)$, but it may not deliver good power because it exploits the information of $f(\cdot)$ at frequency zero only. A more suitable data-driven method for the $\mathscr{A}(J)$ test should exploit the information of $f(\cdot)$ for all frequencies $\omega \in [-\pi, \pi]$.

Before discussing a specific method, we first justify the use of a data-driven finest scale, noted also $\hat{J}$.

**Theorem 5** *Suppose Assumptions 1–3 and 7–9 hold, and $\hat{J}$ is a data-driven finest scale with $2^{\hat{J}}/2^J = 1 + o_P(2^{-J/2})$, where $J$ is a nonstochastic finest scale such that $J \to \infty$ and $2^J/n \to 0$ as $n \to \infty$. If $D(\mathscr{I}_{t-1}, \boldsymbol{\theta}_0) = D_t^0$ a.s., then $\mathscr{A}(\hat{J}) - \mathscr{A}(J) \to^p 0$ and $\mathscr{A}(\hat{J}) \to^d N(0, 1)$.*

Thus, the use of $\hat{J}$ rather than $J$ has no impact on the limit distribution of $\mathscr{A}(\hat{J})$ provided that $\hat{J}$ converges to $J$ sufficiently fast. The condition $2^{\hat{J}}/2^J - 1 = o_P(2^{-J/2})$ is mild. If $2^J \propto n^{1/5}$, for example, we require $2^{\hat{J}}/2^J = 1 + o_P(n^{-1/10})$.

We now develop a data-driven method to choose $\hat{J}$ that will satisfy the conditions of Theorem 5. For this purpose, we first derive the formula for the asymptotic integrated MSE (IMSE) of $\hat{f}_e(\cdot)$ over the entire frequency domain $[-\pi, \pi]$. This differs from the MSE of $\hat{f}_X(0)$. We impose the following additional conditions on $\{e_t\}$.

**Assumption 10** The stochastic process $\{e_t\}$ is a fourth order stationary process with

$$(i) \quad \sum_{h=-\infty}^{\infty} R_e^2(h) \leq C;$$

$$(ii) \quad \sum_{j=-\infty}^{\infty} \sum_{k=-\infty}^{\infty} \sum_{l=-\infty}^{\infty} |\kappa_e(j,k,l)| \leq C,$$

where $\kappa_e(j,k,l)$ denotes the fourth order cumulant of the joint distribution of $\{e_t - \mu_e, e_{t+j} - \mu_e, e_{t+k} - \mu_e, e_{t+l} - \mu_e\}$, where $E(e_t) = \mu_e$.

**Assumption 11** $\sum_{h=-\infty}^{\infty} |h|^q |R_e(h)| \leq C$, where $q \in [1, \infty)$ is as in Assumption 5.

We can study, in passing, the consistency of the test statistic $\mathscr{A}(J)$ under the two previous assumptions. Consider a fixed alternative satisfying Assumptions 10 and 11. Note that under the alternative hypothesis, $e_t \neq \varepsilon_t$, and $D(\mathscr{I}_{t-1}, \boldsymbol{\theta}_0)$ is not equal almost surely to $D^0(\mathscr{I}_{t-1})$. Define the spectral density $f_e(\cdot)$ using the unobservable error. Following reasoning analogous to (but also more tedious than) that of [50, Theorem 2] or [16, Theorem 2], it can be shown that if $2^{2J}/n \to 0$, $J \to \infty$ and $Q(f_e, (2\pi)^{-1}) > 0$, then $\mathscr{A}(J) = O_P(n/2^{J/2})$, and thus the test is consistent. More details can be obtained by communicating with the authors.

To state the next result, we define a pseudo spectral density estimator $\tilde{f}_e(\cdot)$ for $f_e(\cdot)$, using the error series $\{e_t\}_{t=1}^n$; namely,

$$\tilde{f}_e(\omega) \equiv (2\pi)^{-1} \tilde{R}_e(0) + \sum_{j=0}^{J} \sum_{k=1}^{2^j} \tilde{\alpha}_{ejk} \Psi_{jk}(\omega), \quad \omega \in [-\pi, \pi],$$

where $\tilde{R}_e(h) \equiv n^{-1} \sum_{t=|h|+1}^{n} (e_t - \mu_e)(e_{t-|h|} - \mu_e)$ and

$$\tilde{\alpha}_{ejk} \equiv (2\pi)^{1/2} \sum_{h=1-n}^{n-1} \tilde{R}_e(h) \hat{\Psi}_{jk}^*(h).$$

Also define the generalized derivative of $f_e(\omega)$,

$$f_e^{(q)}(\omega) = \frac{1}{2\pi} \sum_{h=-\infty}^{\infty} |h|^q R_e(h) e^{-ih\omega},$$

and put $\vartheta_q = \frac{(2\pi)^{2q+1}}{1-2^{-2q}} \lim_{z\to 0} \frac{|\hat{\psi}(z)|^2}{|z|^{2q}}$. The constant $\vartheta_q$ can be interpreted similarly to $\lambda_q^2$. For the Franklin's wavelet, we have $\vartheta_2 = \pi^4/45$.

**Theorem 6** *Suppose Assumptions 2–3, 5 and 7–11 hold, $J \to \infty$ and $2^J/n \to 0$ as $n \to \infty$. Then $Q(\hat{f}_e, f_e) = Q(\tilde{f}_e, f_e) + o_P(2^J/n + 2^{-2qJ})$, and*

$$E[Q(\tilde{f}_e, f_e)] = \frac{2^{J+1}}{n} \int_{-\pi}^{\pi} f_e^2(\omega) d\omega + 2^{-2q(J+1)} \vartheta_q \int_{-\pi}^{\pi} [f_e^{(q)}(\omega)]^2 d\omega + o(2^J/n + 2^{-2qJ}).$$

Theorem 6 shows that the parameter estimation uncertainty in $\hat{\boldsymbol{\theta}}$ has no impact on the optimal convergence rate of $Q(\hat{f}_e, f_e)$, which is the same as those of $Q(\tilde{f}_e, f_e)$. Note that $E[Q(\tilde{f}_e, f_e)]$ is the IMSE of $\tilde{f}_e(\cdot)$.

To obtain the optimal finest scale that minimizes the asymptotic IMSE of $\tilde{f}_e(\cdot)$, we differentiate the asymptotic IMSE of $\tilde{f}_e(\cdot)$ with respect to $J$ and set the derivative equal to zero. This yields

$$2^{J_0+1} = \left[2q\vartheta_q \xi_0(q) n\right]^{1/(2q+1)}, \tag{13}$$

where

$$\xi_0(q) \equiv \int_{-\pi}^{\pi} [f_e^{(q)}(\omega)]^2 d\omega \Big/ \int_{-\pi}^{\pi} f_e^2(\omega) d\omega. \tag{14}$$

This optimal $J_0$ is infeasible because $\xi_0(q)$ is unknown under $\mathbb{H}_1^{\mathscr{A}}$. However, we can use some estimator $\hat{\xi}_0(q)$ for $\xi_0(q)$. This gives a data-driven finest scale $\hat{J}_0$ :

$$2^{\hat{J}_0+1} \equiv \left[2q\vartheta_q \hat{\xi}_0(q) n\right]^{1/(2q+1)}.$$

Because $\hat{J}_0$ is a nonnegative integer for each $n \geq 1$, we should use

$$\hat{J}_0 \equiv \max\left\{ \left\lfloor \frac{1}{2q+1} \log_2 \left(2q\vartheta_q \hat{\xi}_0(q) n\right) - 1 \right\rfloor, 0 \right\}, \tag{15}$$

where $\lfloor \cdot \rfloor$ denotes the integer part.

**Corollary 2** *Suppose that Assumptions 2–3, 5 and 7–11 hold, and $\hat{J}_0$ is given as in (15), where $\hat{\xi}_0(q) - \zeta_0(q) = o_P(n^{-1/(2q+1)})$, and $\zeta_0(q) \in (0, \infty)$. If $D(\mathscr{I}_{t-1}, \boldsymbol{\theta}_0) = D_t^0$ a.s., then $\mathscr{A}(\hat{J}_0) \xrightarrow{d} N(0, 1)$.*

The condition on $\hat{\xi}_0(q)$ is mild. We do not require the probability limit $p \lim \hat{\xi}_0(2) \equiv \zeta_0(q) = \xi_0(q)$, where $\xi_0(q)$ is as in (14). When (and only when) $\zeta_0(q) = \xi_0(q)$, $\hat{J}_0$ in (15) will converge to the optimal $J_0$ in (13).

For the estimator $\hat{\xi}_0(q)$, we use a parametric AR($p$) model for $\{\hat{e}_t\}$:

$$\hat{e}_t = \gamma_0 + \sum_{h=1}^{p} \gamma_h \hat{e}_{t-h} + v_t, \quad t = 1, \ldots, n, \tag{16}$$

where we set the initial values $\hat{e}_t \equiv 0$ if $t \leq 0$. In practice, we can use the AIC/BIC methods to determine $p$. Suppose $\hat{\boldsymbol{\gamma}} \equiv (\hat{\gamma}_0, \hat{\gamma}_1, \ldots, \hat{\gamma}_p)^\top$ is the OLS estimator in (16). For concreteness, we consider $q = 2$ here. We have

$$\hat{\xi}_0(2) \equiv \int_{-\pi}^{\pi} \left[ \frac{d^2}{d\omega^2} \hat{f}_e(\omega) \right]^2 d\omega \Big/ \int_{-\pi}^{\pi} \hat{f}_e^2(\omega) d\omega, \tag{17}$$

where $\hat{f}_e(\omega) \equiv (2\pi)^{-1} |1 - \sum_{h=1}^{p} \hat{\gamma}_h e^{-ih\omega}|^{-2}$. We have set $\sigma_v^2 = 1$ but this has no impact on $\hat{\xi}_0(q)$. We also have used the fact that the generalized spectral derivative $\hat{f}_e^{(2)}(\omega) = -\frac{d^2}{d\omega^2} \hat{f}_e(\omega)$ for $q = 2$. The estimator $\hat{\xi}_0(2)$ incorporates the information of $f_e(\cdot)$ over $[-\pi, \pi]$ rather than at zero only. This is analogous to [7] and [64], who consider data-driven choices of bandwidths in kernel-based spectral density estimation over the entire frequency domain $[-\pi, \pi]$. We can use one-dimensional numerical integrations to compute $\hat{\xi}_0(2)$. Note that $\hat{\xi}_0(2)$ satisfies the conditions of Corollary 2 with $q = 2$ because for parametric AR($p$) approximations, $\hat{\xi}_0(2) - \zeta_0(2) = O_P(n^{-1/2})$.

To summarize, our diagnostic procedure for the adequacy of ACD models can be described as follows: (i) Find $\hat{J}$ by the autoregression in (16); (ii) compute the test statistic $\mathscr{A}(\hat{J})$ in (12). (iii) reject the null hypothesis of adequacy of the ACD model if the test statistic $\mathscr{A}(\hat{J}) > z_{1-\alpha}$, where $z_{1-\alpha}$ is the $1 - \alpha$ quantile of the standard normal distribution $N(0, 1)$.

## 5 Finite Sample Performance

We now study the finite sample performance of tests $\mathscr{E}(\cdot)$ and $\mathscr{A}(\cdot)$. We use the Franklin wavelet in (3). To examine the impact of the choice of finest scale $J$, we consider $J = 0, 1, 2, 3, 4$ for each sample size $n$, which correspond to using $2^{J+1} - 1 = 1, 3, 7, 15, 31$ empirical wavelet coefficients. We also consider the data-driven methods proposed for $\mathscr{E}(\cdot)$ and $\mathscr{A}(\cdot)$ respectively.

The simulation experiments for testing ACD effects and for testing the adequacy of ACD models are described in the next two sections.

## 5.1 Testing for ACD Effects

We first consider $\mathscr{E}(J)$ for $J$ from 0 to 4 and the data-dependent $\hat{J}$ described in Sect. 3.4. We consider the parametric plug-in autoregressive method, where the order of the autoregression is chosen by the AIC and BIC methods.

We compare $\mathscr{E}(J)$ with two test statistics: one-sided locally most mean powerful test in [41] and the BP/LB two-sided tests. The Higgins' (2000) test statistic in [41] is given by

$$H(m) = \hat{V}_n^{-1/2} \sum_{t=m+1}^{n} \left(X_t/\hat{\mu} - 1\right) \sum_{h=1}^{m} X_{t-h},$$

where $\hat{\mu} = n^{-1} \sum_{t=1}^{n} X_t$ and

$$\hat{V}_n = \{(n-m)^{-1} \sum_{t=m+1}^{n} (X_t/\hat{\mu} - 1)^2\}\{ \sum_{t=m+1}^{n} (\sum_{h=1}^{m} X_{t-h})^2 - ( \sum_{t=m+1}^{n} \sum_{h=1}^{m} X_{t-h})^2/(n-m)\}.$$

The test statistic $H(m)$ is analogous to the test in [49] for ARCH($m$). It has an one-sided asymptotic $N(0,1)$ distribution under $\mathbb{H}_0^{\mathscr{E}}$. The BP/LB test statistics are given as

$$BP(m) = n \sum_{h=1}^{m} \hat{\rho}_X^2(h),$$

$$LB(m) = n^2 \sum_{h=1}^{m} (n-h)^{-1} \hat{\rho}_X^2(h).$$

Both of them have a valid asymptotic $\chi_m^2$ distribution under $\mathbb{H}_0^{\mathscr{E}}$, because they are based on the observed raw data $\{X_t\}_{t=1}^n$, rather than the estimated standardized duration residuals. For $H(m)$, $BP(m)$ and $LB(m)$, the lag order $m$ has to be chosen at priori. These tests will attain their maximal powers when using the optimal lag order, which depends on the true alternative. If the alternative is unknown, as often occurs in practice, these tests may suffer from power losses when a suboptimal lag order is used. There has been no theoretical guidance on the choice of $m$ for these tests. To examine the effect of the choice of $m$, we use $m = 1, 4, 12$.

We consider the following DGP: $X_t = D_t \varepsilon_t$, $t = 1, \ldots, n$, where $\{\varepsilon_t\} \sim$ iid *EXP*(1). We consider $n = 128, 256$. The initial values for $D_t$, $t \leq 0$ are set equal to the unconditional mean of $X_t$. To reduce the effect of the initial values, we generate $2n + 1$ observations and then discard the first $n + 1$ ones. For each experiment, we generate 1000 iterations using the R software on a personal computer.

We study the level of the tests by setting $D_t \equiv 1$. Table 1 reports the empirical level at the 10 and 5 % nominal levels using asymptotic critical values. Based on 1000 iterations, acceptable values at the 5 % level are (3.6 %, 6.4 %) and at the 10 % level are (8.1 %, 11.9 %). We first look at the wavelet test $\mathscr{E}(J)$. For $J = 0, 1$, $\mathscr{E}(J)$ performs reasonably well for both $n = 128, 256$, particularly for $n = 256$ at the 5 %

**Table 1** Size at the 10 and 5 % levels for tests for an ACD effect

| | $n = 128$ | | $n = 256$ | |
|---|---|---|---|---|
| | 10 % | 5 % | 10 % | 5 % |
| $\mathcal{E}(\hat{J})$, AIC | 8.2 | 4.4 | 10.4 | 5.2 |
| $\mathcal{E}(\hat{J})$, BIC | 8.1 | 4.1 | 8.7 | 5.0 |
| $\mathcal{E}(0)$ | 8.0 | 4.2 | 8.9 | 5.6 |
| $\mathcal{E}(1)$ | 7.5 | 4.0 | 9.7 | 4.8 |
| $\mathcal{E}(2)$ | 5.9 | 3.4 | 9.0 | 5.3 |
| $\mathcal{E}(3)$ | 5.1 | 2.6 | 7.8 | 4.2 |
| $\mathcal{E}(4)$ | 3.6 | 1.7 | 6.5 | 3.9 |
| $H(1)$ | 8.8 | 4.3 | 9.1 | 5.4 |
| $BP(1)$ | 8.6 | 3.6 | 11.0 | 6.0 |
| $LB(1)$ | 8.6 | 3.8 | 11.1 | 6.0 |
| $H(4)$ | 7.1 | 3.5 | 8.2 | 4.5 |
| $BP(4)$ | 7.3 | 3.6 | 8.1 | 4.1 |
| $LB(4)$ | 7.7 | 3.7 | 8.4 | 4.3 |
| $H(12)$ | 12.0 | 6.7 | 11.0 | 5.5 |
| $BP(12)$ | 7.3 | 4.5 | 7.5 | 4.8 |
| $LB(12)$ | 9.4 | 5.7 | 8.3 | 5.2 |

(1) DGP: $X_t = \varepsilon_t, \varepsilon_t \sim EXP(1)$. (2) 1000 iterations. (3) The mean and standard deviation (inside the parentheses) of $\hat{J}$ according to AIC method: 0.31 (0.73), 0.30 (0.71) for $n = 128, 256$ respectively; BIC method: 0.02 (0.17), 0.01 (0.14) for $n = 128, 256$ respectively

level. The 5 % level is in general well controlled for $n = 256$ and $J \in \{0, 1, 2, 3\}$. When $J > 1$, $\mathcal{E}(J)$ shows some underrejections, specially for $n = 128$. It seems that given each $n$ the level deteriorates when $J$ becomes larger. For the data-dependent $\hat{J}$, both AIC and BIC give reasonable levels. The AIC method seems to perform very well for $n = 256$ at both the 10 and 5 % levels. We report, in the notes to each table, the mean and standard deviation of the data-dependent $\hat{J}$. The BIC method has a tendency to choose a more parsimonious model than the AIC method, and the resulting $\hat{J}_{BIC}$ is smaller and less variable on the average. On the other hand, the levels of the tests H($m$), BP($m$) and LB($m$) are reasonable. The last H(12) shows some overrejection, specially for $n = 128$.

Next, we investigate the power under the following ACD alternatives:

$$\begin{aligned}
&\text{ACD(1):} && D_t = \beta_0 + \alpha X_{t-1},\\
&\text{ACD(4)}^a\text{:} && D_t = \beta_0 + \alpha \textstyle\sum_{j=1}^{4} X_{t-j},\\
&\text{ACD(4)}^b\text{:} && D_t = \beta_0 + \alpha \textstyle\sum_{j=1}^{4} (1 - j/5) X_{t-1},\\
&\text{ACD(12)}^a\text{:} && D_t = \beta_0 + \alpha \textstyle\sum_{j=1}^{12} X_{t-j},\\
&\text{ACD(12)}^b\text{:} && D_t = \beta_0 + \alpha \textstyle\sum_{j=1}^{12} (1 - j/13) X_{t-j},\\
&\text{ACD(1,1)}^a\text{:} && D_t = \beta_0 + \alpha X_{t-1} + \beta D_{t-1},\\
&\text{ACD(1,1)}^b\text{:} && D_t = \beta_0 + \alpha X_{t-2} + \beta D_{t-2},
\end{aligned}$$

where $\beta_0$ is chosen such that $E(X_t) = 1$. For ACD(1), we set $\alpha = 0.09, 0.36$ respectively. The ACD(1) process has no sharp spectral peak at any frequency. In contrast, ACD(4)$^a$, ACD(4)$^b$, and particularly ACD(12)$^a$ and ACD(12)$^b$ have a relatively long distributional lag, which generates a spectral peak at zero. We set $\alpha = 0.36/4$ for ACD(4)$^a$ and $\alpha = 0.36/\sum_{j=1}^{4}(1 - j/5)$ for ACD(12)$^b$. We set $\alpha = 0.90/12$ for ACD(12)$^a$ and $\alpha = 0.90/\sum_{j=1}^{12}(1 - j/13)$ for ACD(12)$^b$. The ACD(1,1) model is expected to be a workhorse in modelling duration clustering, as is GARCH(1,1) in modelling financial volatility. When $\alpha + \beta < 1$, ACD(1,1) can be expressed as an ACD($\infty$) with exponentially decaying coefficients. We set $(\alpha, \beta) = (0.2, 0.7)$, which displays relatively persistent ACD effects, yielding a spectral peak at zero. We consider the level-corrected power under these alternatives, using the empirical critical values obtained from 1000 replications under $\mathbb{H}_0^{\mathscr{E}}$. This provides comparison among the tests on a fair ground.

Table 2 reports the power against ACD(1). We first consider the wavelet test $\mathscr{E}(J)$. The power of $\mathscr{E}(J)$ is the highest at $J = 0$ and decreases as $J$ increases. The test $\mathscr{E}(0)$ has power similar to H(1), which is optimal for ACD(1). In contrast, BP(1)/LB(1) have slightly smaller power, apparently due to their two-sided nature.

**Table 2** Size-adjusted power against ACD(1) at 10 and 5 % levels for tests of an ACD effect

| | a) $\alpha = 0.09$ | | | | b) $\alpha = 0.36$ | | | |
|---|---|---|---|---|---|---|---|---|
| | $n = 128$ | | $n = 256$ | | $n = 128$ | | $n = 256$ | |
| | 10% | 5% | 10% | 5% | 10% | 5% | 10% | 5% |
| $\mathscr{E}(\hat{J})$, AIC | 30.6 | 18.5 | 40.7 | 29.1 | 88.6 | 80.6 | 96.1 | 94.2 |
| $\mathscr{E}(\hat{J})$, BIC | 36.2 | 22.0 | 49.3 | 34.7 | 94.0 | 86.7 | 98.7 | 97.1 |
| $\mathscr{E}(0)$ | 37.7 | 23.6 | 50.2 | 36.1 | 95.6 | 91.3 | 99.6 | 98.9 |
| $\mathscr{E}(1)$ | 28.0 | 17.9 | 40.9 | 31.9 | 90.5 | 82.7 | 97.8 | 96.4 |
| $\mathscr{E}(2)$ | 20.1 | 12.1 | 28.4 | 16.8 | 67.5 | 56.4 | 84.4 | 76.3 |
| $\mathscr{E}(3)$ | 18.1 | 8.1 | 21.0 | 14.4 | 50.0 | 35.7 | 63.2 | 54.0 |
| $\mathscr{E}(4)$ | 14.3 | 8.6 | 17.2 | 9.8 | 31.0 | 22.7 | 46.1 | 35.6 |
| $H(1)$ | 35.6 | 24.0 | 51.3 | 34.0 | 95.3 | 92.0 | 99.6 | 99.0 |
| $BP(1)$ | 23.7 | 17.4 | 37.8 | 26.7 | 91.8 | 87.4 | 99.0 | 98.3 |
| $LB(1)$ | 23.7 | 17.4 | 37.8 | 26.7 | 91.8 | 87.4 | 99.0 | 98.3 |
| $H(4)$ | 20.2 | 14.0 | 29.9 | 18.4 | 67.4 | 57.6 | 87.3 | 80.7 |
| $BP(4)$ | 19.0 | 11.6 | 30.0 | 21.7 | 82.6 | 75.7 | 97.0 | 95.3 |
| $LB(4)$ | 18.9 | 11.7 | 29.9 | 21.5 | 82.4 | 75.4 | 96.9 | 95.1 |
| $H(12)$ | 14.8 | 7.1 | 18.5 | 12.0 | 31.3 | 21.8 | 50.7 | 39.9 |
| $BP(12)$ | 14.6 | 8.1 | 22.8 | 12.3 | 68.5 | 57.8 | 92.5 | 87.6 |
| $LB(12)$ | 14.7 | 8.1 | 22.4 | 12.1 | 67.0 | 56.4 | 92.5 | 87.3 |

(1) DGP: $X_t = D_t\varepsilon_t$, $D_t = \beta_0 + \alpha X_{t-1}$, $\beta_0 = 1 - \alpha$, where $\varepsilon_t$ follows an $EXP(1)$ distribution. (2) 1000 iterations. (3) The mean and standard deviation (inside the parentheses) of $\hat{J}$ according to AIC method: a) 0.50 (0.79), 0.61 (0.81) for $n = 128, 256$ respectively; b) 1.16 (0.69), 1.33 (0.72) for $n = 128, 256$ respectively; BIC method: a) 0.12 (0.35), 0.21 (0.42) for $n = 128, 256$ respectively. b) 0.92 (0.58), 1.19 (0.52) for $n = 128, 256$ respectively

Appealingly, the wavelet test $\mathscr{E}(\hat{J})$ with the BIC method has power very similar to that of H(1) and $\mathscr{E}(0)$, suggesting the merit of using the BIC method. The tests H($m$) and BP($m$)/LB($m$) have decreasing power as $m$ increases, as is expected. There is no data-driven method to select $m$ for H($m$) and BP($m$)/LB($m$).

Table 3 reports the power under ACD(4). In contrast to ACD(1), the wavelet test $\mathscr{E}(\hat{J})$ now has a ∩-shape power as $J$ increases from 0 to 4, with $J = 2$ giving the maximum power for both $n = 128$ and 256. The power of $\mathscr{E}(2)$ is similar to that of H(4). The BP(4)/LB(4) tests have substantially smaller power, indicating the loss of not exploiting the one-sided nature of the alternative. When $m \neq 4$, H($m$) and BP($m$)/LB($m$) have substantially smaller power than H(4) and BP(4)/LB(4) respectively. Again, there is no rule to select a data-driven $m$. In contrast, the wavelet test $\mathscr{E}(\hat{J})$ with the AIC method has a better power than suboptimal $J$'s. It is less powerful that $\mathscr{E}(2)$, but the difference becomes smaller as $n$ increases.

Table 4 reports the power under ACD(12). Like ACD(4), the wavelet test $\mathscr{E}(J)$ still has a ∩-shape power as $J$ increases, with $J = 3$ having the maximum power. The H($m$) test has an increasing power as $m$ increases under ACD(12)$^a$, but a ∩-shape

**Table 3** Size-adjusted power against ACD(4) at 10 and 5 % levels for tests of an ACD effect

| | *ACD*(4), case a) | | | | *ACD*(4), case b) | | | |
|---|---|---|---|---|---|---|---|---|
| | $n = 128$ | | $n = 256$ | | $n = 128$ | | $n = 256$ | |
| | 10 % | 5 % | 10 % | 5 % | 10 % | 5 % | 10 % | 5 % |
| $\mathscr{E}(\hat{J})$, AIC | 60.1 | 49.2 | 81.6 | 77.1 | 69.8 | 60.5 | 88.6 | 84.7 |
| $\mathscr{E}(\hat{J})$, BIC | 50.4 | 36.7 | 68.3 | 59.2 | 65.8 | 54.3 | 83.9 | 75.6 |
| $\mathscr{E}(0)$ | 48.6 | 34.6 | 65.5 | 53.2 | 65.0 | 52.7 | 82.2 | 72.2 |
| $\mathscr{E}(1)$ | 60.6 | 47.3 | 79.5 | 72.8 | 75.6 | 63.9 | 91.7 | 86.0 |
| $\mathscr{E}(2)$ | 74.6 | 60.3 | 90.9 | 83.5 | 75.6 | 63.0 | 91.7 | 84.0 |
| $\mathscr{E}(3)$ | 59.8 | 46.9 | 75.2 | 68.4 | 58.4 | 45.9 | 73.6 | 66.3 |
| $\mathscr{E}(4)$ | 37.3 | 27.5 | 54.1 | 44.7 | 36.2 | 27.0 | 53.3 | 43.5 |
| $H(1)$ | 42.0 | 31.3 | 61.2 | 46.3 | 60.2 | 49.4 | 80.7 | 66.8 |
| $BP(1)$ | 30.0 | 23.8 | 48.2 | 39.3 | 48.7 | 39.0 | 68.8 | 61.1 |
| $LB(1)$ | 30.0 | 23.8 | 48.2 | 39.3 | 48.7 | 39.0 | 68.8 | 61.1 |
| $H(4)$ | 70.1 | 61.9 | 90.8 | 83.9 | 71.7 | 62.8 | 91.7 | 85.0 |
| $BP(4)$ | 49.8 | 40.0 | 77.9 | 72.7 | 56.2 | 46.6 | 83.0 | 78.1 |
| $LB(4)$ | 49.5 | 40.0 | 77.9 | 72.8 | 55.5 | 46.6 | 82.8 | 77.9 |
| $H(12)$ | 37.6 | 26.5 | 60.5 | 49.3 | 36.9 | 26.4 | 58.4 | 48.6 |
| $BP(12)$ | 37.9 | 27.2 | 67.1 | 57.5 | 43.9 | 32.7 | 73.3 | 63.3 |
| $LB(12)$ | 37.0 | 26.9 | 66.4 | 57.0 | 43.2 | 32.2 | 73.2 | 63.0 |

(1) DGP: $X_t = D_t\varepsilon_t$, $D_t = \beta_0 + \sum_{i=1}^{4} \alpha_i X_{t-i}$, $\beta_0 = 1 - \sum_{i=1}^{4} \alpha_i$, where $\varepsilon_t$ follows an *EXP*(1) distribution. Case a): $\alpha_i = 0.09, i = 1, \ldots, 4$. Case b): $\alpha_i = \alpha(1 - i/5), \alpha = .36/\sum_{i=1}^{4}(1 - i/5)$. (2) 1000 iterations. (3) The mean and standard deviation (inside the parentheses) of $\hat{J}$ according to AIC method: a) 1.10 (1.16), 1.73 (1.13) for $n = 128, 256$ respectively; b) 1.13 (1.02), 1.67 (0.96) for $n = 128, 256$ respectively; BIC method: a) 0.39 (0.79), 0.74 (1.02) for $n = 128, 256$ respectively. b) 0.52 (0.81), 0.94 (0.93) for $n = 128, 256$ respectively

**Table 4** Size-adjusted power against ACD(12) at 10 and 5 % levels for tests of an ACD effect

| | $ACD(12)$, case a) | | | | $ACD(12)$, case b) | | | |
|---|---|---|---|---|---|---|---|---|
| | $n = 128$ | | $n = 256$ | | $n = 128$ | | $n = 256$ | |
| | 10 % | 5 % | 10 % | 5 % | 10 % | 5 % | 10 % | 5 % |
| $\mathscr{E}(\hat{J})$, AIC | 68.0 | 63.4 | 94.3 | 93.4 | 87.1 | 83.7 | 98.5 | 98.2 |
| $\mathscr{E}(\hat{J})$, BIC | 61.9 | 54.0 | 88.1 | 84.5 | 82.2 | 76.1 | 97.2 | 96.0 |
| $\mathscr{E}(0)$ | 60.4 | 51.4 | 86.6 | 81.4 | 81.3 | 73.8 | 97.2 | 94.6 |
| $\mathscr{E}(1)$ | 71.0 | 62.3 | 94.1 | 92.1 | 90.3 | 84.1 | 99.3 | 99.0 |
| $\mathscr{E}(2)$ | 81.1 | 74.1 | 97.4 | 95.8 | 95.0 | 91.3 | 99.7 | 99.4 |
| $\mathscr{E}(3)$ | 87.9 | 83.1 | 99.2 | 98.5 | 96.1 | 92.4 | 99.9 | 99.9 |
| $\mathscr{E}(4)$ | 85.2 | 81.2 | 98.9 | 98.6 | 90.4 | 86.4 | 99.4 | 99.1 |
| $H(1)$ | 53.4 | 45.6 | 83.1 | 73.5 | 74.7 | 66.0 | 95.1 | 89.6 |
| $BP(1)$ | 45.3 | 37.0 | 75.7 | 68.8 | 65.4 | 58.6 | 91.0 | 87.5 |
| $LB(1)$ | 45.3 | 37.0 | 75.7 | 68.8 | 65.4 | 58.6 | 91.0 | 87.5 |
| $H(4)$ | 74.3 | 67.0 | 95.0 | 93.1 | 90.4 | 87.3 | 99.3 | 98.8 |
| $BP(4)$ | 61.3 | 54.9 | 90.9 | 88.6 | 82.8 | 78.9 | 98.5 | 98.1 |
| $LB(4)$ | 61.3 | 55.0 | 90.9 | 88.6 | 82.9 | 78.9 | 98.5 | 98.1 |
| $H(12)$ | 75.9 | 69.4 | 96.9 | 95.8 | 81.1 | 77.2 | 98.0 | 97.6 |
| $BP(12)$ | 70.2 | 64.2 | 97.3 | 95.9 | 83.2 | 78.4 | 99.2 | 98.6 |
| $LB(12)$ | 70.5 | 64.2 | 97.3 | 95.9 | 83.1 | 78.2 | 99.1 | 98.5 |

(1) DGP: $X_t = D_t\varepsilon_t$, $D_t = \beta_0 + \sum_{i=1}^{12} \alpha_i X_{t-i}$, $\beta_0 = 1 - \sum_{i=1}^{12} \alpha_i$, where $\varepsilon_t$ follows an $EXP(1)$ distribution. Case a): $\alpha_i = 0.90/12, i = 1, \ldots, 12$. Case b): $\alpha_i = \alpha(1 - i/13), \alpha = .90/\sum_{i=1}^{12}(1 - i/13)$. (2) 1000 iterations. (3) The mean and standard deviation (inside the parentheses) of $\hat{J}$ according to AIC method: a) 2.39 (2.07), 4.24 (1.50) for $n = 128, 256$ respectively; b) 2.84 (1.67), 4.11 (1.14) for $n = 128, 256$ respectively; BIC method: a) 0.86 (1.44), 2.43 (1.93) for $n = 128, 256$ respectively. b) 1.54 (1.61), 3.08 (1.55) for $n = 128, 256$ respectively

power under ACD(12)$^b$. Interestingly, H(12) is less powerful than H(4), although $m = 12$ is the optimal lag order. The BP($m$)/LB($m$) tests have an increasing power as $m$ increases under both ACD(12)$^a$ and ACD(12)$^b$. Note that the best wavelet test $\mathscr{E}(3)$ is more powerful that H(12) under ACD(12)$^a$ and H(4) under ACD(12)$^b$. For data-driven $\hat{J}$, the test $\mathscr{E}(\hat{J})$ with the AIC method has better power than $\mathscr{E}(\hat{J})$ with the BIC method, and has better power than $\mathscr{E}(J)$ for $J = 0, 1$, but not for $J \geq 2$.

Table 5 reports the power against the ACD(1,1) alternatives. The wavelet test $\mathscr{E}(J)$ has a ∩-shape power as $J$ increases with $J = 1$ having the maximum power under ACD(1,1)$^a$ and $J = 2$ having the maximum power under ACD(1,1)$^b$. The H($m$) and BP($m$)/LB($m$) tests have declining power as $m$ increases under ACD(1,1)$^a$ but have a ∩ -shape power under ACD(1,1)$^b$. Interestingly, although H(1) is theoretically optimal for ACD(1,1), H(4) has better power than H(1) under ACD(1,1)$^b$. Note that under ACD(1,1)$^a$, the best wavelet test $\mathscr{E}(1)$ has better power than H(1), and under ACD(1,1)$^b$, the best wavelet test $\mathscr{E}(2)$ has slightly better power than H(4).

**Table 5** Size-adjusted power against ACD(1,1) at 10 and 5 % levels for tests of an ACD effect

| | $ACD(1,1)$, case a) | | | | $ACD(1,1)$, case b) | | | |
|---|---|---|---|---|---|---|---|---|
| | $n = 128$ | | $n = 256$ | | $n = 128$ | | $n = 256$ | |
| | 10 % | 5 % | 10 % | 5 % | 10 % | 5 % | 10 % | 5 % |
| $\mathscr{E}(\hat{J})$, AIC | 77.4 | 67.1 | 92.6 | 89.2 | 91.0 | 88.1 | 99.4 | 99.2 |
| $\mathscr{E}(\hat{J})$, BIC | 77.2 | 65.8 | 93.2 | 87.8 | 87.9 | 81.4 | 98.7 | 97.9 |
| $\mathscr{E}(0)$ | 77.2 | 65.9 | 93.1 | 86.5 | 87.1 | 79.3 | 98.1 | 96.3 |
| $\mathscr{E}(1)$ | 82.1 | 71.1 | 94.3 | 91.7 | 94.0 | 89.5 | 99.5 | 99.0 |
| $\mathscr{E}(2)$ | 71.9 | 61.3 | 89.4 | 82.5 | 95.0 | 93.0 | 99.7 | 99.6 |
| $\mathscr{E}(3)$ | 56.4 | 45.2 | 72.6 | 65.2 | 94.0 | 89.9 | 99.6 | 99.5 |
| $\mathscr{E}(4)$ | 35.2 | 28.1 | 53.9 | 44.6 | 84.1 | 78.5 | 97.6 | 96.3 |
| $H(1)$ | 75.2 | 65.0 | 91.6 | 84.9 | 83.0 | 74.9 | 96.7 | 94.4 |
| $BP(1)$ | 63.6 | 54.9 | 85.0 | 79.8 | 75.0 | 67.8 | 94.4 | 91.0 |
| $LB(1)$ | 63.6 | 54.9 | 85.0 | 79.8 | 75.0 | 67.8 | 94.4 | 91.0 |
| $H(4)$ | 69.4 | 60.1 | 90.7 | 83.4 | 92.6 | 89.2 | 99.6 | 99.1 |
| $BP(4)$ | 60.6 | 51.6 | 85.8 | 82.1 | 86.2 | 80.7 | 99.0 | 98.7 |
| $LB(4)$ | 60.2 | 51.5 | 85.6 | 82.0 | 86.3 | 80.6 | 99.0 | 98.7 |
| $H(12)$ | 37.4 | 27.6 | 58.0 | 48.5 | 77.1 | 71.6 | 96.7 | 95.0 |
| $BP(12)$ | 49.5 | 38.6 | 77.1 | 69.0 | 82.8 | 76.6 | 99.0 | 98.1 |
| $LB(12)$ | 48.5 | 38.3 | 76.8 | 68.9 | 82.5 | 76.1 | 98.9 | 97.9 |

(1) DGP: $X_t = D_t\varepsilon_t$, $ACD(1,1)$: $D_t = \beta_0 + \alpha X_{t-1} + \beta D_{t-1}$, a) $\alpha = 0.2$, $\beta = 0.45$, $\beta_0 = 1 - \alpha - \beta$, where $\varepsilon_t$ follows an $EXP(1)$ distribution. b) $\alpha = 0.2$, $\beta = 0.75$, $\beta_0 = 1 - \alpha - \beta$, where $\varepsilon_t$ follows an $EXP(1)$ distribution. (2) 1000 iterations. (3) The mean and standard deviation (inside the parentheses) of $\hat{J}$ according to AIC method: a) 1.13 (0.93), 1.55 (0.88) for $n = 128, 256$ respectively; b) 2.36 (1.38), 3.39 (1.10) for $n = 128, 256$ respectively; BIC method: a) 0.60 (0.74), 1.01 (0.77) for $n = 128, 256$ respectively. b) 1.35 (1.30), 2.49 (1.26) for $n = 128, 256$ respectively

The wavelet tests $\mathscr{E}(\hat{J})$ with AIC/BIC perform reasonably well. They have power only marginally smaller than the best wavelet tests with the finite sample optimal $J$.

In summary, we observe: (1) the wavelet test with the finite sample optimal $J$ generally performs better that H($m$) test with the optimal lag. This shows the usefulness and potential of the wavelet method for testing for duration effects; (2) the choice of the finest scale $J$ seems not important for the level (unless $J$ is large and $n$ is small) but affects the power significantly. The data-driven method yields an objective finest scale $\hat{J}$ which delivers reasonable levels for the test; (3) the power of the wavelet test using data-driven $\hat{J}$ was reasonable under all the alternatives considered. It seems that the AIC method performs better for relatively persistent duration processes, while the BIC method performs better for short durations; (4) the tests H($m$) and BP($m$)/LB($m$) generally attain their own maximal powers when using the optimal lag order (except for ACD(12)$^b$), but may suffer from severe power loss using a suboptimal lag.

## 5.2 Testing for the Adequacy of an ACD Model

We now consider $\mathscr{A}(J)$ for $J = 0, 1, 2, 3, 4$ and the data-dependent $\hat{J}$ described in Sect. 4.2. We use the parametric plug-in autoregressive method. The autoregression order is chosen by AIC and a modified AIC method. We modify the AIC method since we need a data-dependent $\hat{J}$ such that $\hat{J} \to \infty$ as $n \to \infty$. This differs from testing for ACD effects where $J$ can be fixed. We take

$$\hat{J}_{modAIC} = \max\left\{\hat{J}_{AIC}, \lfloor 0.2\log_2(32n) - 1 \rfloor\right\},$$

where $\lfloor\cdot\rfloor$ stands for the integer part. The formula $0.2\log_2(32n)$ gives a slow convergence to $\infty$ which corresponds to $2^J = n^{0.2}$. This ensures $\hat{J}_{modAIC} \to \infty$ as $n \to \infty$.

We compare $\mathscr{A}(J)$ with the BP/LB test statistics, which remain very popular procedures. The statistic $\mathscr{A}(J)$ is asymptotically one-sided $N(0, 1)$ under $\mathbb{H}_0^{\mathscr{A}}$. The BP($m$)/LB($m$) statistics were often used in the empirical literature (e.g., [21, 22, 34, 67, 68]). They are assumed to have an asymptotic $\chi^2_m$ distribution under $\mathbb{H}_0^{\mathscr{A}}$. However, our analysis suggests that the test statistics or their limiting distributions need an adjustment, and such an adjustment seems not simple in lights of the results by [53] in diagnostic testing for ARCH models. Therefore, the level of BP($m$)/LB($m$) is troublesome. Nevertheless, we still include BP($m$)/LB($m$) in our simulation comparison. We compare the power of the tests using their empirical critical values, so the power comparison is valid. As was the case for testing for ACD effects, the lag order $m$ has to be chosen at priori. These tests will attain their maximal powers when using the optimal lag order, which depends on the true alternative. If the alternative is unknown, as often occurs in practice, these tests may suffer from power loss when using a suboptimal lag order. To examine the effect of the choice of $m$ for these tests, we use $m = 1, 4, 12$.

We first study the level by fitting an ACD(1,1) model to the following DGP: $X_t = D_t\varepsilon_t$, $t = 1, \ldots, n$, where $\varepsilon_t$ are iid $EXP(1)$ and $D_t$ is an ACD(1,1) with $(\alpha, \beta) = (0.3, 0.65)$. We consider $n = 128, 256$. Estimation is performed by taking the square root of the duration process and setting the mean equal to zero (cf. [22]). Once the ACD(1,1) model is estimated, we can compute the estimated standardized duration residuals $\hat{e}_t = X_t/\hat{D}_t$ and all the relevant test statistics.

Table 6 reports the empirical level at the 10 and 5 % nominal levels using asymptotic critical values. We first look at the wavelet test $\mathscr{A}(J)$. At the 10 % level, the best levels are obtained with $J = 2, 3$. For $J \in \{1, 2, 3, 4\}$, the levels are well controlled at the 5 % level. When $J$ is small, that is, $J = 0, 1$, or when $J$ is too large, say $J = 4$, the wavelet test $\mathscr{A}(J)$ seem to exhibit underrejection. Both AIC and the modified AIC methods give good levels at the 10 % level, but seem to overreject at the 5 % level. The BP/LB tests severally underreject, possibly due to the use of incorrect asymptotic critical values. They seem to have their better level for larger lag orders. It appears that the $\chi^2_m$ distribution is not appropriate, as is expected.

**Table 6** Size at the 10 and 5 % levels for tests for goodness-of-fit of ACD models when the model is ACD(1,1)

| | $n = 128$ | | $n = 256$ | |
|---|---|---|---|---|
| | 10 % | 5 % | 10 % | 5 % |
| $\mathscr{A}(\hat{J})$, AIC | 10.5 | 7.7 | 10.3 | 7.4 |
| $\mathscr{A}(\hat{J})$, modified AIC | 10.7 | 7.7 | 9.9 | 7.4 |
| $\mathscr{A}(0)$ | 4.1 | 2.6 | 4.5 | 2.6 |
| $\mathscr{A}(1)$ | 6.6 | 4.5 | 6.2 | 4.2 |
| $\mathscr{A}(2)$ | 8.1 | 5.2 | 7.0 | 4.4 |
| $\mathscr{A}(3)$ | 7.4 | 4.7 | 7.1 | 4.2 |
| $\mathscr{A}(4)$ | 6.3 | 3.8 | 6.5 | 3.6 |
| $BP(1)$ | 1.5 | 0.5 | 1.7 | 0.4 |
| $LB(1)$ | 1.5 | 0.5 | 1.7 | 0.5 |
| $BP(4)$ | 5.4 | 2.9 | 5.0 | 1.9 |
| $LB(4)$ | 6.1 | 3.2 | 5.2 | 2.1 |
| $BP(12)$ | 5.6 | 3.0 | 5.7 | 2.6 |
| $LB(12)$ | 7.3 | 3.9 | 6.4 | 3.2 |

(1) DGP: $X_t = D_t\varepsilon_t$, $D_t = \beta_0 + \alpha X_{t-1} + \beta D_{t-1}$, $\beta_0 = 1 - \alpha - \beta$, $\alpha = 0.3$, $\beta = 0.65$, where $\varepsilon_t$ is $EXP(1)$. (2) 1000 iterations. (3) The mean and standard deviation (inside the parentheses) of $\hat{J}$ according to AIC method: 0.31 (0.80), 0.24 (0.69) for $n = 128, 256$ respectively; modified AIC method: 1.15 (0.50), 1.11 (0.40) for $n = 128, 256$ respectively

Next, we investigate the power under the following ACD alternatives:

$$\begin{aligned}
&\text{ACD(2,2)}: && D_t = \beta_0 + 0.3X_{t-2} + 0.6D_{t-2},\\
&\text{ACD(4)}: && D_t = \beta_0 + 0.9X_{t-4},\\
&\text{ACD(4,4)}: && D_t = \beta_0 + 0.1X_{t-2} + 0.3X_{t-4} + 0.1D_{t-2} + 0.3D_{t-4},\\
&\text{ACD(12)}: && D_t = \beta_0 + 0.9X_{t-12},
\end{aligned}$$

where $\beta_0$ is chosen such that the unconditional mean of the duration process is 1. We estimate all these DGPs by an ACD(1,1) model. We then consider the level-corrected power under these alternatives, using the empirical critical values from the 1000 replications under $\mathbb{H}_0^{\mathscr{A}}$. The spectral density of the ACD(2,2) process exhibits a single peak at frequency zero. For the ACD(4) alternative, the spectral density has a peak at frequency zero and another peak at a non-zero frequency. The ACD(4,4) alternative has a spectral peak at zero and an other smaller peak at a non-zero frequency. Finally, the ACD(12) process has many spectral peaks at non-zero frequencies. When we fit an ACD(1,1) model to the data, the estimated standardized residuals show remaining dependence in the residuals. Often, if the spectral density of a duration process exhibits a large peak and if we estimate incorrectly an ACD(1,1) model, then $\hat{f}_e(\cdot)$ may still exhibit some peaks (typically of less magnitude, however).

Table 7 reports the power against ACD(2,2) and ACD(4,4). Under ACD(2,2), the choices of $J = 1$ and $J = 2$ give the highest power for the wavelet test for $n = 128$ and $n = 256$, respectively. The best power among the BP($m$)/LB($m$) tests is obtained

**Table 7** Size-adjusted power against ACD($q, q$), $q = 2, 4$ at 10 and 5% Levels for tests of goodness-of-fit when the model is ACD(1,1)

| | ACD(2,2) | | | | ACD(4,4) | | | |
|---|---|---|---|---|---|---|---|---|
| | $n = 128$ | | $n = 256$ | | $n = 128$ | | $n = 256$ | |
| | 10% | 5% | 10% | 5% | 10% | 5% | 10% | 5% |
| $\mathscr{A}(\hat{J})$, AIC | 85.4 | 80.8 | 97.9 | 96.7 | 73.4 | 68.2 | 94.9 | 93.1 |
| $\mathscr{A}(\hat{J})$, modified AIC | 86.5 | 81.5 | 98.0 | 96.5 | 72.5 | 67.8 | 94.2 | 92.8 |
| $\mathscr{A}(0)$ | 79.9 | 73.9 | 96.4 | 95.0 | 58.9 | 51.4 | 85.3 | 80.0 |
| $\mathscr{A}(1)$ | 86.4 | 81.3 | 97.2 | 96.0 | 54.0 | 44.0 | 79.4 | 69.6 |
| $\mathscr{A}(2)$ | 85.9 | 80.9 | 97.9 | 97.0 | 79.1 | 73.8 | 96.2 | 95.1 |
| $\mathscr{A}(3)$ | 83.0 | 78.4 | 97.2 | 96.1 | 77.2 | 70.3 | 95.8 | 94.0 |
| $\mathscr{A}(4)$ | 79.1 | 72.2 | 96.4 | 94.1 | 70.1 | 63.3 | 93.9 | 91.2 |
| $BP(1)$ | 74.0 | 67.9 | 94.6 | 93.1 | 52.0 | 44.1 | 79.6 | 72.8 |
| $LB(1)$ | 74.0 | 67.9 | 94.6 | 93.1 | 52.0 | 44.1 | 79.6 | 72.8 |
| $BP(4)$ | 82.9 | 78.0 | 97.0 | 95.4 | 76.7 | 70.4 | 95.8 | 93.1 |
| $LB(4)$ | 82.7 | 78.0 | 97.0 | 95.4 | 76.7 | 70.5 | 95.8 | 93.1 |
| $BP(12)$ | 80.6 | 75.7 | 96.9 | 94.7 | 73.1 | 65.1 | 94.5 | 92.2 |
| $LB(12)$ | 80.3 | 75.4 | 96.9 | 94.7 | 72.4 | 64.8 | 94.5 | 92.1 |

(1) DGP: $X_t = D_t\varepsilon_t$, a) $D_t = \beta_0 + \alpha X_{t-2} + \beta D_{t-2}, \beta_0 = 1 - \alpha - \beta, \alpha = 0.3, \beta = 0.6$, where $\varepsilon_t$ is $EXP(1)$. b) $D_t = \beta_0 + \alpha_1 X_{t-2} + \alpha_2 X_{t-4} + \beta_1 D_{t-2} + \beta_1 D_{t-4}$, $\beta_0 = 1 - \alpha_1 - \alpha_2 - \beta_1 - \beta_2$, $\alpha_1 = 0.1, \alpha_2 = 0.3, \beta_1 = 0.1, \beta_2 = 0.3$, where $\varepsilon_t$ is $EXP(1)$. (2) 1000 iterations. (3) The mean and standard deviation (inside the parentheses) of $\hat{J}$ according to AIC method: a) 2.22 (1.37), 2.83 (1.09) for $n = 128, 256$ respectively; b) 2.11 (1.48), 2.74 (1.07) for $n = 128, 256$ respectively. Modified AIC method: a) 2.37 (1.16), 2.85 (1.03) for $n = 128, 256$ respectively; b) 2.37 (1.14), 2.81 (0.93) for $n = 128, 256$ respectively

with $m = 4$, but this choice of $m$ is difficult to justify. Under ACD(4,4), the choice $J = 2$ gives the highest power for the wavelet test $\mathscr{A}(J)$, and $m = 4$ gives the higher power for BP($m$)/LB($m$). For both ACD(2,2) and ACD(4,4) the AIC and modified AIC method give a power similar to the optimal $J$, with the advantage that $J$ was data-driven. The test $\mathscr{A}(\hat{J})$ dominates under the ACD(2,2) alternative, and the power of $\mathscr{A}(\hat{J})$ is close to BP(4)/LB(4) under ACD(4,4).

Table 8 reports the power against ACD(4) and ACD(12). Under ACD(4), the choice $J = 2$ gives the highest power for the wavelet test $\mathscr{A}(J)$. The choice $m = 4$ gives the higher power for BP($m$)/LB($m$). Under ACD(12), the choice $J = 4$ gives the highest power for the wavelet test. The best power for BP($m$)/LB($m$) is obtained at $m = 12$. Their power can be very low if $m$ is misspecified: the tests based on $m = 1, 4$ have almost no power. This illustrates the importance of a data-dependent method. On the other hand, the wavelet test based on AIC or modified AIC have a power very similar to the wavelet test $\mathscr{A}(J)$ based on the finite sample optimal $J$, and the power is similar to BP($m$)/LB($m$) with the optimal $m$ .

In summary, (1) when $J$ is fixed, the wavelet test with the finite sample optimal $J$ performs well in power compared to that BP($m$)/LB($m$) tests with the optimal lag. This shows the usefulness of the wavelet method for testing for goodness-of-fit of

**Table 8** Size-adjusted power against ACD($q$), $q = 4$, 12 at 10 and 5 % Levels for tests of goodness-of-fit when the model is ACD(1,1)

| | ACD(4) | | | | ACD(12) | | | |
|---|---|---|---|---|---|---|---|---|
| | $n = 128$ | | $n = 256$ | | $n = 128$ | | $n = 256$ | |
| | 10 % | 5 % | 10 % | 5 % | 10 % | 5 % | 10 % | 5 % |
| $\mathscr{A}(\hat{J})$, AIC | 96.9 | 95.6 | 99.2 | 99.1 | 83.4 | 78.2 | 98.3 | 97.8 |
| $\mathscr{A}(\hat{J})$, modified AIC | 96.9 | 95.9 | 99.2 | 99.1 | 83.2 | 78.2 | 98.3 | 97.8 |
| $\mathscr{A}(0)$ | 64.4 | 53.0 | 90.9 | 86.0 | 24.5 | 15.1 | 32.0 | 21.6 |
| $\mathscr{A}(1)$ | 71.7 | 52.7 | 96.4 | 93.1 | 19.8 | 9.3 | 27.6 | 16.7 |
| $\mathscr{A}(2)$ | 98.6 | 98.1 | 99.7 | 99.6 | 15.5 | 10.2 | 24.3 | 16.8 |
| $\mathscr{A}(3)$ | 97.9 | 97.1 | 99.3 | 99.2 | 20.4 | 14.3 | 26.7 | 20.1 |
| $\mathscr{A}(4)$ | 95.5 | 93.7 | 99.0 | 98.9 | 89.1 | 84.6 | 98.7 | 98.2 |
| $BP(1)$ | 52.9 | 39.1 | 85.3 | 77.1 | 20.2 | 12.6 | 25.3 | 18.0 |
| $LB(1)$ | 52.9 | 39.1 | 85.3 | 77.1 | 20.2 | 12.6 | 25.3 | 18.0 |
| $BP(4)$ | 98.6 | 97.8 | 99.6 | 99.4 | 14.3 | 9.0 | 19.3 | 12.3 |
| $LB(4)$ | 98.6 | 97.8 | 99.6 | 99.4 | 14.3 | 9.1 | 19.3 | 12.3 |
| $BP(12)$ | 96.7 | 94.9 | 99.1 | 99.0 | 93.4 | 91.8 | 99.7 | 99.3 |
| $LB(12)$ | 96.4 | 94.7 | 99.1 | 99.0 | 93.8 | 92.4 | 99.7 | 99.3 |

(1) DGP: $X_t = D_t\varepsilon_t$, a) $D_t = \beta_0 + \alpha X_{t-4}$, $\alpha = 0.9$, where $\varepsilon_t$ is $EXP(1)$. b) $D_t = \beta_0 + \alpha X_{t-12}$, $\beta_0 = 1 - \alpha$, $\alpha = 0.9$, where $\varepsilon_t$ is $EXP(1)$. (2) 1000 iterations. (3) The mean and standard deviation (inside the parentheses) of $\hat{J}$ according to AIC method: a) 3.32 (0.81), 3.49 (0.62) for $n = 128, 256$ respectively; b) 4.37 (1.21), 4.75 (0.51) for $n = 128, 256$ respectively; modified AIC method: a) 3.33 (0.77), 3.49 (0.62) for $n = 128, 256$ respectively. b) 4.43 (1.00), 4.75 (0.48) for $n = 128, 256$ respectively

ACD models. (2) The levels of BP/LB tests do not seem to be those of a $\chi^2_m$ under the null hypothesis, as is expected. In contrast, since we prove rigorously the asymptotic normality of $\mathscr{A}(J)$ and $\mathscr{A}(\hat{J})$, they should be preferred in practice. (2) The choice of the finest scale $J$ seems not important for the level (unless $J$ is large and $n$ is small) but affects the power significantly. The data-driven method yields an objective finest scale $\hat{J}$ that delivers reasonable power against various model misspecifications.

## 6 Application with Tick-by-Tick Trading Data

To assess empirically the ability of the wavelet-based tests $\mathscr{E}(J)$ and $\mathscr{A}(J)$ to detect ACD effects and to check ACD models, we utilize real time series data. The sample consists of the tick-by-tick trading data of Alcoa stock on June 7, 2010. These data are used in [68, p. 324]. The original file includes 78,602 trades. Focusing on the normal trading hours (that is 19h30 to 16h00), we notice 78,413 transactions during that particular day, giving 78,412 time durations between trades. We ignored the zero durations, giving $n = 9{,}596$ nonzero intraday durations. The sample mean and sample variance of these nonzero durations were 2.44 and 6.23, respectively. We focused on the adjusted time duration:

$$X_t = \Delta T_t / f(T_t),$$

where $\Delta T_t = T_t - T_{t-1}$ denotes the $t$th duration and $f(\cdot)$ represents a deterministic function consisting of the cyclical component in the durations $\Delta T_t$. We postulated the following model to explain the cyclical component:

$$\log[f(T_t)] = c_0 + c_1 f_1(T_t) + c_2 f_2(T_t) + c_3 f_3(T_t) + a_t, \tag{18}$$

where

$$\begin{aligned} f_1(T_t) &= (T_t - 43{,}200)/23{,}400, \\ f_2(T_t) &= f_1^2(T_t), \\ f_3(T_t) &= \log(T_t). \end{aligned}$$

Here the random variable $a_t$ plays the role of the error term in the linear regression model (18). In $f_1(\cdot)$, 43,200 denotes the 12h00 noon and 23,400 corresponds to the number of trading hours measured in seconds. An ordinary least-squares fit gave:

$$\tilde{f}(T_t) = \exp[1154.5 + 60.0 f_1(T_t) - 17.2 f_2(T_t) - 108.1 f_3(T_t)].$$

The time series to analyze is thus:

$$\tilde{X}_t = \Delta T_t / \tilde{f}(T_t). \tag{19}$$

See [68, pp. 298–300 and p. 324] for explanations and more details on the cleaning data procedure and the treatment of the diurnal pattern.

We can now apply the test procedures for ACD effects $\mathscr{E}(J)$ defined by (7). We used the Franklin wavelet (3) and computed the test statistic for $J = 0, 1, \ldots, 6$. We also considered $\hat{J}$ using the AIC and BIC methods. We found $\hat{J}_{AIC} = 4$ and $\hat{J}_{BIC} = 3$. The values of the test statistics with their associated $P$ values are given in Table 9. All the results are highly significant, strongly suggesting ACD effects. Having found ACD effects we considered adjusting ACD($m$,$l$) models. In ARMA modelling, the ARMA(1,1) represents the workhorse. Similarly, the ACD(1,1) is a good starting model. In addition, the sample autocorrelation function and sample partial autocorrelation function both suggested $m$ and $l$ being strictly positive. We adjusted EACD(1,1) and WACD(1,1) models. The models were:

$$\begin{aligned} \text{EACD}(1,1) &: \tilde{X}_t = D_t \varepsilon;\ D_t = 0.09 + 0.06\tilde{X}_{t-1} + 0.87 D_{t-1}, \\ \text{WACD}(1,1) &: \tilde{X}_t = D_t \varepsilon;\ D_t = 0.06 + 0.05\tilde{X}_{t-1} + 0.90 D_{t-1}, \end{aligned}$$

and the estimated shape parameter of the Weibull distribution has been found equal to 1.367 (the standard error was 0.098). All the coefficients were found significant.

**Table 9** Test statistics $\mathscr{E}(J)$, $J = 0, 1, \ldots, 6$ for ACD effects defined by (7), applied on the Alcoa stock on June 7, 2010, seasonally adjusted, defined by (19)

| $J$ | $\mathscr{E}(J)$ | $P$ value |
|---|---|---|
| 0 | 10.57 | $0^+$ |
| 1 | 12.24 | $0^+$ |
| 2 | 15.44 | $0^+$ |
| 3 | 16.61 | $0^+$ |
| 4 | 17.31 | $0^+$ |
| 5 | 18.06 | $0^+$ |
| 6 | 19.14 | $0^+$ |

(1) The notation $0^+$ denotes a number smaller than $10^{-4}$. (2) The data-driven $J$ using the AIC and BIC methods were $\hat{J}_{AIC} = 4$ and $\hat{J}_{BIC} = 3$

To check these models, we applied the test procedures $\mathscr{A}(J)$ on the residuals of the adjustment, using again the Franklin wavelet. We also included the popular LB$(q)$ test procedures. Recall that the critical values are not strictly valid, but we follow the literature by examining the values of these test statistics as approximate rules. For example, the Ljung-Box test statistic is advocated in [68], with $q = 10, 20$. These choices of $q$ are somewhat arbitrary, and it may be preferable to include also smaller values of $q$ to appreciate the residual dependence in lower-order lags. For the new wavelet-based test statistic, we include $\mathscr{A}(\hat{J}_{AIC})$ and $\mathscr{A}(\hat{J}_{modAIC})$. Using the residuals of the EACD(1,1) and WACD(1,1), we found $\hat{J}_{AIC} = 0$. The modified rule yielded $\hat{J}_{modAIC} = 4$. Since no automatic choice of $q$ is available for $LB(q)$, we considered $q = 1, 2, 3, 10, 20$. The results for the adjustment of EACD(1,1) and WACD(1,1) are given in Table 10. Based on the values of $q$ recommended in Tsay (2013), one may be tempted to recommend the EACD(1,1) fit. However, the data-driven $J$ based on the AIC method was $\hat{J}_{AIC} = 0$. A low resolution seemed necessary to explain the remaining dependence in the residuals, which is not really surprising: the EACD(1,1)/WACD(1,1) explained a large part of the dependence between the durations. However, these models can be improved, and the null hypothesis of adequacy is strongly rejected using $\mathscr{A}(\hat{J}_{AIC})$. Incidently, the order of magnitude of the test statistic $LB(1) = 4.24$ ($LB(1) = 6.71$) for the EACD(1,1) model (WACD(1,1) model) suggested that the lag-1 residual dependence was still significant, which was hidden by considering larger values of $q$. The test statistics suggested that the EACD(1,1)/WACD(1,1) were not appropriate to explain duration persistence. Thus, we considered EACD(1,2) and WACD(1,2) models. The results were:

$$\text{EACD}(1,2) : \tilde{X}_t = D_t\varepsilon;\ D_t = 0.11 + 0.07\tilde{X}_{t-1} + 0.49D_{t-1} + 0.35D_{t-2},$$
$$\text{WACD}(1,2) : \tilde{X}_t = D_t\varepsilon;\ D_t = 0.07 + 0.07\tilde{X}_{t-1} + 0.47D_{t-1} + 0.41D_{t-2},$$

with an estimated shape parameter given by 1.368 (with a standard error of 0.098). All the coefficients in these models were significant. We applied the test procedures on the residuals of these models. The data-driven $J$ using the AIC and modified

**Table 10** Test statistics $\mathscr{A}(\hat{J}_{AIC}), \mathscr{A}(\hat{J}_{modAIC}), LB(q), q = 1, 2, 3, 10, 20$, applied on the residuals of the EACD(1,1) and WACD(1,1) models, for the Alcoa stock on June 7, 2010

| | | Test Statistic | $P$ value |
|---|---|---|---|
| EACD(1,1) | $\mathscr{A}(0)$ | 3.53 | $2 \times 10^{-4}$ |
| | $\mathscr{A}(4)$ | −0.13 | 0.55 |
| | $LB(1)$ | 4.24 | 0.04 |
| | $LB(2)$ | 4.44 | 0.11 |
| | $LB(3)$ | 4.44 | 0.22 |
| | $LB(10)$ | 11.89 | 0.29 |
| | $LB(20)$ | 20.86 | 0.41 |
| WACD(1,1) | $\mathscr{A}(0)$ | 5.86 | $0^+$ |
| | $\mathscr{A}(4)$ | 2.00 | 0.02 |
| | $LB(1)$ | 6.71 | 0.01 |
| | $LB(2)$ | 6.72 | 0.03 |
| | $LB(3)$ | 6.75 | 0.08 |
| | $LB(10)$ | 18.72 | 0.04 |
| | $LB(20)$ | 30.26 | 0.07 |

(1) The notation $0^+$ denotes a number smaller than $10^{-4}$. (2) The data-driven $J$ using the AIC and modified AIC methods were $\hat{J}_{AIC} = 0$ and $\hat{J}_{modAIC} = 4$

**Table 11** Test statistics $\mathscr{A}(\hat{J}_{AIC}), \mathscr{A}(\hat{J}_{modAIC}), LB(q), q = 1, 2, 3, 10, 20$, applied on the residuals of the EACD(1,2) and WACD(1,2) models, for the Alcoa stock on June 7, 2010

| | | Test Statistic | $P$ value |
|---|---|---|---|
| EACD(1,2) | $\mathscr{A}(0)$ | 0.08 | 0.47 |
| | $\mathscr{A}(4)$ | −0.47 | 0.68 |
| | $LB(1)$ | 0.62 | 0.43 |
| | $LB(2)$ | 1.30 | 0.52 |
| | $LB(3)$ | 1.30 | 0.73 |
| | $LB(10)$ | 9.17 | 0.52 |
| | $LB(20)$ | 18.03 | 0.59 |
| WACD(1,2) | $\mathscr{A}(0)$ | 0.72 | 0.24 |
| | $\mathscr{A}(4)$ | 1.57 | 0.06 |
| | $LB(1)$ | 1.34 | 0.25 |
| | $LB(2)$ | 2.91 | 0.23 |
| | $LB(3)$ | 2.92 | 0.40 |
| | $LB(10)$ | 14.40 | 0.16 |
| | $LB(20)$ | 26.79 | 0.14 |

(1) The data-driven $J$ using the AIC and modified AIC methods were $\hat{J}_{AIC} = 0$ and $\hat{J}_{modAIC} = 4$

AIC methods were again $\hat{J}_{AIC} = 0$ and $\hat{J}_{modAIC} = 4$ for both models. The results are presented in Table 11. They suggest that both models are adequate to describe these data. In applications, the hazard function of WACD models appears more flexible than the one of an EACD model. Furthermore, recall that the shape parameter of the WACD(1,2) based on the Weibull distribution was significantly different from one. Given the large sample size, it may be preferable to include the additional shape parameter in the model and to retain the WACD(1,2) model. See also the discussion in [68, Section 6.5.2].

# 7 Conclusion

We have proposed a consistent one-sided test for duration clustering and a new diagnostic test for ACD models using a wavelet spectral density estimator. The first test exploits the one-sided nature of duration clustering. An ACD process is positively autocorrelated at all lags, resulting in a spectral mode or peak at frequency zero. As a joint time-frequency decomposition method, wavelets can effectively capture spectral peaks. To compare the wavelet-spectral density estimator with the spectral density under the null hypothesis, alternative approaches include using a quadratic norm or a supremum-norm measure. See [44] when testing for serial correlation using wavelets, and also [52, p. 104] when testing for conditional heteroscedasticity using supremum-norm measures. More work is needed to study theoretically these approaches properly adapted to the problem of testing for ACD effects, which should be compared empirically with the methods presented in this paper.

The second test checks the adequacy of an ACD model using a wavelet spectral density of the estimated standardized duration residuals. Unlike the popular BP/LB tests in the ACD literature, our diagnostic test has an asymptotic nuisance free parameter property; that is, parameter estimation uncertainty has no impact on the asymptotic distribution of the test statistic. Moreover, it can check a wide range of alternatives and is powerful when the spectrum of the estimated standardized duration residuals is nonsmooth, which can arise from neglected persistent duration clustering, seasonalities, calendar effect and business cycles. For each of the proposed methods, we developed a suitable data-driven method to choose the finest scale $J$. This makes the proposed methods fully operational.

When testing for duration clustering, the wavelet-based test with the optimal $J$ performed similarly or even better than the existing tests with the optimal lag, suggesting the merits of using wavelets. The data-driven method yielded an objective finest scale $\hat{J}$ which delivers reasonable levels and powers against various alternatives. Similarly, when testing for the adequacy of an ACD model, the wavelet test with the optimal $J$ generally performs better than BP/LB tests with the optimal lag. The BP/LB tests with lag $m$ do not seem to follow a $\chi_m^2$ distribution, since the levels were severally underestimated. In contrast, $\mathscr{A}(J)$ and $\mathscr{A}(\hat{J})$ have a convenient asymptotically valid normal distribution, and the data-driven method yields an objective finest scale $\hat{J}$ which delivers reasonable levels and power against various model misspecifications. The real data analysis suggested the merits of our wavelet-based test statistics. It is hoped that the results presented in this paper will be useful for the practitioner, improving the toolbox of techniques for diagnostic checking ACD models.

**Acknowledgements** The authors would like to thank W. K. Li, David Stanford, Hao Yu, and two referees for constructive suggestions, which led to an improved paper. Funding in partial support of this work was provided by the Natural Science and Engineering Research Council of Canada.

# Appendix

To prove Theorems 1–3, we first state some useful lemmas.

**Lemma 1** *Define* $d_J(h) \equiv \sum_{j=0}^{J} \lambda(2\pi h/2^j)$, $h, J \in \mathbb{Z}$, *where* $\lambda(z)$ *is as in (8). Then*

(i) $d_J(0) = 0$ *and* $d_J(-h) = d_J(h)$ *for all* $h, J \in \mathbb{Z}, J > 0$;
(ii) $|d_J(h)| \le C < \infty$ *uniformly in* $h, J \in \mathbb{Z}, J > 0$;
(iii) *For any given* $h \in \mathbb{Z}, h \neq 0$, $d_J(h) \to 1$ *as* $J \to \infty$;
(iv) *For any given* $r \ge 1$, $\sum_{h=1}^{n-1} |d_J(h)|^r = O(2^J)$ *as* $J, n \to \infty$.

**Lemma 2** *Let* $V_n(J)$ *and* $V_0$ *be defined as in Theorem 1. Suppose* $J \to \infty$, $2^J/n \to 0$. *Then* $2^{-J}V_n(J) \to V_0$, *where* $V_0 = \int_0^{2\pi} |\Gamma(z)|^2 dz$, *with* $\Gamma(z) = \sum_{-\infty}^{\infty} \hat{\psi}(z + 2\pi m)$.

*Proof* For the proofs of Lemmas 1 and 2, see [42, Proof of Lemma A.1] and [44, Proof of Lemma A.2]. □

*Proof* (Theorem 1) The model under the null hypothesis is $D_t \equiv \beta_0$, $X_t = \beta_0\varepsilon_t$, $E(\varepsilon_t) = 1$, $\{\varepsilon_t\}$ an iid process. We write $\bar{R}_X(h) = n^{-1}\sum_{t=|h|+1}^{n}(X_t/\bar{X} - 1)(X_{t-|h|}/\bar{X} - 1)$. Alternatively, $\hat{\rho}_X(h) = \bar{R}_X(h)/\bar{R}_X(0)$. Under the null hypothesis, $\bar{R}_X(h) = \bar{R}_\varepsilon(h)$ where $\bar{R}_\varepsilon(h)$ is defined similarly as $\bar{R}_X(h)$. We define $R_\varepsilon(h) = \mathrm{cov}(\varepsilon_t, \varepsilon_{t-h})$. Let $u_t = \varepsilon_t - 1$. We define $\tilde{R}_\varepsilon(h) = n^{-1}\sum_{t=|h|+1}^{n} u_t u_{t-h}$ and $\tilde{\rho}_\varepsilon(h) = \tilde{R}_\varepsilon(h)/R_\varepsilon(0)$. Define

$$\tilde{f}_\varepsilon(0) \equiv (2\pi)^{-1} + \sum_{j=0}^{J}\sum_{k=1}^{2^j} \tilde{\alpha}_{jk}\Psi_{jk}(0),$$

where $\tilde{\alpha}_{jk} = (2\pi)^{-1/2}\sum_{h=1-n}^{n-1} \tilde{\rho}_\varepsilon(h)\hat{\Psi}_{jk}^*(h)$, $\hat{\Psi}_{jk}(h) \equiv (2\pi)^{-1/2}\int_{-\pi}^{\pi} \Psi_{jk}(\omega) e^{-ih\omega}d\omega$.

Writing $\hat{f}_X(0) - (2\pi)^{-1} = [\hat{f}_X(0) - \tilde{f}_\varepsilon(0)] + [\tilde{f}_\varepsilon(0) - (2\pi)^{-1}]$, we shall prove Theorem 1 by showing Theorems 7–8 below.

**Theorem 7** $[V_n(J)]^{-1/2}n^{1/2}[\hat{f}_X(0) - \tilde{f}_\varepsilon(0)] \to^p 0$.

**Theorem 8** $[V_n(J)]^{-1/2}n^{1/2}\pi[\tilde{f}_\varepsilon(0) - (2\pi)^{-1}] \to^d N(0, 1)$.

*Proof* (Theorem 7) We use the following representations (see [44]):

$$\hat{f}_X(0) = (2\pi)^{-1} + \pi^{-1}\sum_{h=1}^{n-1} d_J(h)\hat{\rho}_X(h), \tag{20}$$

$$\tilde{f}_\varepsilon(0) = (2\pi)^{-1} + \pi^{-1}\sum_{h=1}^{n-1} d_J(h)\tilde{\rho}_\varepsilon(h).$$

We obtain $\pi[\hat{f}_X(0) - \tilde{f}_\varepsilon(0)] = \sum_{h=1}^{n-1} d_J(h)[\hat{\rho}_X(h) - \tilde{\rho}_\varepsilon(h)]$. Because $\bar{R}_X(0) - R_\varepsilon(0) = O_P(n^{-1/2})$ given Assumption 1, it suffices to show

$$[V_n(J)]^{-1/2} n^{1/2} \sum_{h=1}^{n-1} d_J(h)[\bar{R}_X(h) - \tilde{R}_\varepsilon(h)] \to^p 0. \tag{21}$$

We shall show (21) for the case $J \to \infty$, where $2^{-J} V_n(J) \to V_0$ by Lemma 2. The proof for fixed $J$ is similar, with $V_n(J) \to V_0(J)$, where $V_0(J)$ is defined in Lemma 2.

Straightforward algebra yields $\hat{R}_X(h) - \tilde{R}_\varepsilon(h) = (\bar{\varepsilon}^{-1} - 1)\hat{A}_1(h) + (\bar{\varepsilon}^{-1} - 1)\hat{A}_2(h) + (\bar{\varepsilon}^{-1} - 1)^2 \hat{A}_3(h)$, where

$$\hat{A}_1(h) = n^{-1} \sum_{t=|h|+1}^{n} u_t \varepsilon_{t-h}, \; \hat{A}_2(h) = n^{-1} \sum_{t=|h|+1}^{n} u_{t-h} \varepsilon_t, \; \hat{A}_3(h) = n^{-1} \sum_{t=|h|+1}^{n} \varepsilon_t \varepsilon_{t-h}.$$

We first consider $\hat{A}_1(h)$. Note that $E[\hat{A}_1(h)\hat{A}_1(m)] = O(n^{-1}), \forall h, m$. Then expending the square, we show that $E[\sum_{h=1}^{n-1} d_J(h)\hat{A}_1(h)]^2 = O(2^J/n + 2^{2J}/n)$. This shows that $\sum_{h=1}^{n-1} d_J(h)\hat{A}_1(h) = O_P(2^J/n^{1/2})$. Proceeding similarly we show that $\sum_{h=1}^{n-1} d_J(h)\hat{A}_2(h) = O_P(2^J/n^{1/2})$. We show also easily that

$$\sum_{h=1}^{n-1} d_J(h)\hat{A}_3(h) = O_P(2^J).$$

This completes the proof for Theorem 7. □

*Proof* (Theorem 8) Put $\hat{W} \equiv \sum_{h=1}^{n-1} d_J(h)\tilde{R}_\varepsilon(h)/R_\varepsilon(0)$. Write $\hat{W} = n^{-1} \sum_{t=2}^{n} W_t$, where

$$W_t \equiv R_\varepsilon^{-1}(0) u_t \sum_{h=1}^{t-1} d_J(h) u_{t-h}.$$

Observe that $\{W_t, \mathscr{F}_{t-1}\}$ is an adapted martingale difference sequence, where $\mathscr{F}_t$ is the sigma field consisting of all $u_s, s \le t$. Thus, we obtain

$$\mathrm{var}(n^{1/2}\hat{W}) = n^{-1} \sum_{t=2}^{n} E[W_t^2] = n^{-1} \sum_{t=2}^{n} \sum_{h=1}^{t-1} d_J^2(h) = \sum_{h=1}^{n} (1 - h/n) d_J^2(h) = V_n(J).$$

By the martingale central limit theorem in [37, pp.10–11], $[V_n(J)]^{-1/2} n^{1/2} \hat{W} \to^d N(0, 1)$ if we can show

$$[V_n(J)]^{-1}n^{-1}\sum_{t=2}^{n} E\left\{W_t^2 \mathbf{1}[|W_t| > \eta n^{1/2}\{V_n(J)\}^{1/2}]\right\} \rightarrow 0 \text{ for any } \eta > 0, \quad (22)$$

$$[V_n(J)]^{-1}n^{-1}\sum_{t=2}^{n} E(W_t^2|\mathscr{F}_{t-1}) \rightarrow^p 1. \quad (23)$$

For space, we shall show the central limit theorem for $\hat{W}$ for large $J$ (i.e., $J \rightarrow \infty$). The proof for fixed $J$ is similar and simpler because $d_J(h)$ is finite and summable.

We shall verify the first condition by showing $2^{-2J}n^{-2}\sum_{t=2}^{n} E(W_t^4) \rightarrow 0$. Put $\mu_4 \equiv E(u_t^4)$. By Assumption 1, we have

$$E(W_t^4) = \mu_4 R_\varepsilon^{-4}(0) E\left[\sum_{h=1}^{t-1} d_J(h)u_{t-h}\right]^4,$$
$$= \mu_4^2 R_\varepsilon^{-4}(0)\sum_{h=1}^{t-1} d_J^4(h) + 6\mu_4 R_\varepsilon^{-2}(0)\sum_{h_1=2}^{t-1}\sum_{h_2=1}^{h_1-1} d_J^2(h_1)d_J^2(h_2) \leq 3\mu_4^2 R_\varepsilon^{-4}(0)\left[\sum_{h=1}^{n-1} d_J^2(h)\right]^2.$$

It follows from Lemma 2 that $2^{-2J}n^{-2}\sum_{t=1}^{n} E(W_t^4) = O(n^{-1})$. This show (22).

Next, given Lemma 2, it suffices for expression (23) to establish the sufficient condition

$$2^{-2J}\text{var}[n^{-1}\sum_{t=2}^{n} E(W_t^2|\mathscr{F}_{t-1})] \rightarrow 0.$$

By the definition of $W_t$, we have

$$\begin{aligned}
E(W_t^2|\mathscr{F}_{t-1}) &= R_\varepsilon^{-1}(0)\left[\sum_{h=1}^{t-1} d_J(h)u_{t-h}\right]^2, \\
&= E(W_t^2) + R_\varepsilon^{-1}(0)\sum_{h=1}^{t-1} d_J(h)[u_{t-h}^2 - R_\varepsilon(0)] \\
&\quad + 2R_\varepsilon^{-1}(0)\sum_{h_1=2}^{t-1}\sum_{h_2=1}^{h_1-1} d_J(h_1)d_J(h_2)u_{t-h_1}u_{t-h_2}, \\
&= E(W_t^2) + R_\varepsilon^{-1}(0)A_t + 2R_\varepsilon^{-1}(0)B_t.
\end{aligned}$$

It follows that

$$\begin{aligned}
n^{-1}\sum_{t=2}^{n}[E(W_t^2|\mathscr{F}_{t-1}) - E(W_t^2)] &= R_\varepsilon^{-1}(0)n^{-1}\sum_{t=2}^{n} A_t + 2R_\varepsilon^{-1}(0)n^{-1}\sum_{t=2}^{n} B_t \\
&= R_\varepsilon^{-1}(0)\hat{A} + 2R_\varepsilon^{-1}(0)\hat{B}. \quad (24)
\end{aligned}$$

Whence, it suffices to show $2^{-2J}[\mathrm{var}(\hat{A})+\mathrm{var}(\hat{B})]\to 0$. First, noting that $A_t$ is a weighted sum of independent zero-mean variables $\{u_{t-h}^2-R_\varepsilon(0)\}$, we have $E(A_t^2)=[\mu_4-R_\varepsilon^2(0)]\sum_{h=1}^{t-1}d_J^4(h)$. It follows by Minkowski's inequality and Lemma 1(iv) that

$$E(\hat{A}^2)\le\left\{n^{-1}\sum_{t=2}^{n}[E(A_t^2)]^{1/2}\right\}^2\le[\mu_4-R_\varepsilon^2(0)]\left[\sum_{h=1}^{n-1}d_J^4(h)\right]=O(2^J).\quad(25)$$

Next, we consider $\mathrm{var}(\hat{B})$. For all $t\ge s$, we have

$$\begin{aligned}E(B_tB_s)&=R_\varepsilon^2(0)\sum_{m_2=2}^{t-1}\sum_{h_2=1}^{m_2-1}\sum_{m_1=2}^{s-1}\sum_{h_1=1}^{m_1-1}d_J(m_1)d_2(h_1)d_J(m_2)d_J(h_2)\delta_{t-h_1,s-h_2}\delta_{t-m_1,s-m_2},\\&=R_\varepsilon^2(0)\sum_{m=2}^{t-1}\sum_{h=1}^{l-1}d_J(t-s+m)d_J(t-s+h)d_J(m)d_J(h),\end{aligned}$$

where $\delta_{j,h}=1$ if $h=j$ and $\delta_{j,h}=0$ otherwise. It follows that

$$\begin{aligned}E(\hat{B}^2)&\le 2n^{-2}\sum_{t=3}^{n}\sum_{s=2}^{t}E(B_tB_s)\le 2R_\varepsilon^2(0)n^{-1}\sum_{\tau=0}^{n-1}\sum_{m=2}^{n-1}\sum_{h=1}^{m-1}|d_J(\tau+m)d_J(\tau+h)d_J(m)d_J(h)|,\\&\le 2R_\varepsilon^2(0)n^{-1}\left[\sum_{\tau=0}^{n-1}d_J^2(\tau)\right]\left[\sum_{h=1}^{n-1}|d_J(h)|\right]^2=O(2^{3J}/n),\end{aligned}\quad(26)$$

by Lemma 1(iv). Combining (24)–(26) yields $2^{-2J}[\mathrm{var}(\hat{A})+\mathrm{var}(\hat{B})]=O(2^{-J}+2^J/n)\to 0$ given $J\to\infty$, $2^J/n\to 0$. Thus, condition (23) holds. By [37, pp.10–11], $[V_n(J)]^{-1/2}n^{1/2}\hat{W}\to^d N(0,1)$. This completes the proof of Theorem 8. □

*Proof* (Theorem 2) We shall show for large $J$ only; the proof for fixed $J$ is similar. Here we explicitly denote $\hat{f}_X(0;J)$ as the spectral estimator (20) with the finest scale $J$. Recalling the definition of $\mathscr{E}(J)$, we have

$$\begin{aligned}\mathscr{E}(\hat{J})-\mathscr{E}(J)&=[V_n(\hat{J})]^{-1/2}n^{1/2}\pi\{\hat{f}_X(0;\hat{J})-(2\pi)^{-1}\}-[V_n(J)]^{-1/2}n^{1/2}\pi\{\hat{f}_X(0;J)-(2\pi)^{-1}\},\\&=[V_n(\hat{J})/V_n(J)]^{1/2}[V_n(J)]^{-1/2}n^{1/2}\pi\{\hat{f}_X(0;\hat{J})-\hat{f}_X(0;J)\}+\{[V_n(J)/V_n(\hat{J})]^{-1/2}-1\}\mathscr{E}(J).\end{aligned}$$

Note that we have for any given constants $C_0>0$ and $\varepsilon>0$,

$$\begin{aligned}P\left(|V_n(J)/V_n(\hat{J})-1|>\varepsilon\right)&\le P\left(|V_n(J)/V_n(\hat{J})-1|>\varepsilon,\,C_02^{J/2}|2^{\hat{J}}/2^J-1|\le\varepsilon\right)\\&\quad+P\left(C_02^{J/2}|2^{\hat{J}}/2^J-1|>\varepsilon\right).\end{aligned}\quad(27)$$

We now study $|d_{\hat{J}}(h) - d_J(h)|$. We show that

$$|d_{\hat{J}}(h) - d_J(h)| \leq \sum_{j=\min(J,\hat{J})+1}^{\max(J,\hat{J})} |\lambda(2\pi h/2^j)|.$$

Note that given $C_0 2^{J/2}|2^{\hat{J}}/2^J - 1| \leq \varepsilon$, we have that

$$|d_{\hat{J}}(h) - d_J(h)| \leq \sum_{j=\log_2[2^J(1-\varepsilon/(C_0 2^{J/2}))]+1}^{\log_2[2^J(1+\varepsilon/(C_0 2^{J/2}))]} |\lambda(2\pi h/2^j)|,$$

and the lower and upper bounds in the summation are now non stochastic. Since $|\sum_{m=-\infty}^{\infty} \psi(2\pi h/2^j + 2\pi m)| \leq C$, we have that

$$|d_{\hat{J}}(h) - d_J(h)| \leq C \sum_{j=\log_2[2^J(1-\varepsilon/(C_0 2^{J/2}))]+1}^{\log_2[2^J(1+\varepsilon/(C_0 2^{J/2}))]} |\hat{\psi}(2\pi h/2^j)|.$$

Since $\sum_{h=-\infty}^{\infty} |\hat{\psi}(2\pi h/2^j)| \leq C 2^j$, then

$$\sum_{h=1}^{n-1} |d_{\hat{J}}(h) - d_J(h)| \leq C \sum_{j=\log_2[2^J(1-\varepsilon/(C_0 2^{J/2}))]+1}^{\log_2[2^J(1+\varepsilon/(C_0 2^{J/2}))]} 2^j \leq C 2^{J/2}\varepsilon/C_0.$$

A similar argument allows us to show that

$$\sum_{h=1}^{n-1} |d_{\hat{J}}(h) - d_J(h)|^2 \leq C 2^{J/2}\varepsilon/C_0. \tag{28}$$

We show that $V_n(\hat{J})/V_n(J) \to^p 1$. Note that

$$|V_n(\hat{J}) - V_n(J)| \leq \sum_{h=1}^{n-1} |d_{\hat{J}}(h) - d_J(h)|^2 + 2\left(\sum_{h=1}^{n-1} d_J^2(h)\right)^{1/2} \left(\sum_{h=1}^{n-1} [d_{\hat{J}}(h) - d_J(h)]^2\right)^{1/2}.$$

Since $\sum_{h=1}^{n-1} d_J^2(h) = O(2^J)$, by (27) and (28), we have the announced result that $V_n(\hat{J})/V_n(J) \to^p 1$.

Because $\mathscr{E}(J) = O_P(1)$ by Theorem 1 and since $V_n(\hat{J})/V_n(J) \to^p 1$, we have $\mathscr{E}(\hat{J}) - \mathscr{E}(J) \to^p 0$ provided $[V_n(J)]^{-1/2} n^{1/2} \pi\{\hat{f}_{\hat{J}}(0) - \hat{f}_J(0)\} \to^p 0$, which we shall show below. The asymptotic normality of $\mathscr{E}(\hat{J})$ follows from a standard application of Slutsky's theorem and Theorem 1.

Because $V_n(J) = O(2^J)$, it suffices to show $\hat{f}_X(0; \hat{J}) - \hat{f}_X(0; J) = o_P(2^{J/2}/n^{1/2})$. Write

$$\pi\{\hat{f}_X(0; \hat{J}) - \hat{f}_X(0; J)\} = \hat{R}_X^{-1}(0) \sum_{h=1}^{n-1} [d_{\hat{J}}(h) - d_J(h)] R_X(h).$$

Given $|d_{\hat{J}}(h) - d_J(h)| \leq \sum_{j=\min(\hat{J},J)}^{\max(\hat{J},J)} |\lambda(2\pi h/2^j)|$, we have, under the null hypothesis, the following inequality

$$E \sum_{h=1}^{n-1} \left| d_{\hat{J}}(h) - d_J(h) \right| |R_X(h)| \leq \sup_h E(R_X^2(h))^{1/2} \sum_{h=1}^{n-1} \left( E \left| d_{\hat{J}}(h) - d_J(h) \right|^2 \right)^{1/2} = o(2^{J/2}/n^{1/2}).$$

We obtain $\hat{f}_X(0; \hat{J}) - \hat{f}_X(0; J) = o_P(2^{J/2}/n^{1/2})$. This completes the proof of Theorem 2. □

*Proof* (Theorem 3) Recall $\hat{R}_X(h) = n^{-1} \sum_{t=|h|+1}^{n} (X_t - \bar{X})(X_{t-|h|} - \bar{X})$ and $\tilde{R}_X(h) = n^{-1} \sum_{t=|h|+1}^{n} (X_t - \mu)(X_{t-|h|} - \mu)$. We study $\hat{f}_X(0; J) - f_X(0)$. Write

$$\hat{f}_X(0; J) - f_X(0) = \{\hat{f}_X(0; J) - \tilde{f}_X(0; J)\} + \{\tilde{f}_X(0; J) - E[\tilde{f}_X(0; J)]\} + \{E[\tilde{f}_X(0; J)] - f_X(0)\}, \tag{29}$$

where

$$\tilde{f}_X(0; J) = \frac{\tilde{R}_X(0)}{2\pi} + \frac{1}{\pi} \sum_{h=1}^{n-1} d_J(h) \tilde{R}_X(h). \tag{30}$$

We can show that $E[\hat{f}_X(0; J) - \tilde{f}_X(0; J)]^2 = O(n^{-2} + 2^{2J}/n^2)$, meaning that replacing $\bar{X}$ by $\mu$ has to impact asymptotically. For the second term in (29), we show that

$$\begin{aligned} E[\tilde{f}_X(0; J) - E\tilde{f}_X(0; J)]^2 = & \frac{\text{var}[\tilde{R}_X(0)]}{4\pi^2} + \pi^{-2} \sum_{h=1}^{n-1} \sum_{m=1}^{n-1} d_J(h) d_J(m) \text{cov}(\tilde{R}_X(h), \tilde{R}_X(m)) \\ & + \pi^{-2} \sum_{h=1}^{n-1} d_J(h) \text{cov}(\tilde{R}_X(0), \tilde{R}_X(h)). \end{aligned} \tag{31}$$

From [38, p. 313], we have

$$\begin{aligned} \frac{(n-l)(n-m)}{n^2} \text{cov}\left[\tilde{R}_X(h), \tilde{R}_X(m)\right] = n^{-1} \sum_{u=-\infty}^{\infty} & w_n(u, h, m)[R_X(u) R_X(u+m-h) \\ & + R_X(u+m) R_X(u-h) + \kappa(0, h, u, u+m)], \end{aligned}$$

where the function $w_n(u, h, m)$ is defined in [38]. We write

$$E\left\{\tilde{f}_X(0; J) - E[\tilde{f}_X(0; J)]\right\}^2 = \frac{\text{var}(\tilde{R}_X(0))}{4\pi^2} + \frac{A_n}{\pi^2} + \frac{B_n}{\pi^2}, \tag{32}$$

where

$$A_n = E\left\{[\tilde{R}_X(0) - E(\tilde{R}_X(0))]\sum_{h=1}^{n-1} d_J(h)[\tilde{R}_X(h) - E(\tilde{R}_X(h))]\right\},$$

$$B_n = \sum_{h=1}^{n-1}\sum_{m=1}^{n-1} d_J(h)d_J(m)\text{cov}[\tilde{R}_X(h), \tilde{R}_X(m)].$$

It follows that

$$\begin{aligned}(n/2^{J+1})B_n &= 2^{-(J+1)}\sum_{h=1}^{n-1}\sum_{m=1}^{n-1} d_J(h)d_J(m)\sum_{u=-\infty}^{\infty} w_n(u, h, m)R_X(u)R_X(u+m-h)\\ &\quad +2^{-(J+1)}\sum_{h=1}^{n-1}\sum_{m=1}^{n-1} d_J(h)d_J(m)\sum_{u=-\infty}^{\infty} w_n(u, h, m)R_X(u+m)R_X(u-h)\\ &\quad +2^{-(J+1)}\sum_{h=1}^{n-1}\sum_{m=1}^{n-1} d_J(h)d_J(m)\sum_{u=-\infty}^{\infty} w_n(u, h, m)\kappa(0, h, u, u+m),\\ &= B_{1n} + B_{2n} + B_{3n}.\end{aligned} \tag{33}$$

Following a reasoning similar to [42, proof of Theorem 4.2], we can show that

$$B_{1n} = 2^{-1}D_\psi(2\pi)^2 f_X^2(0)[1+o(1)],\ B_{2n} \to 0,\ B_{3n} \to 0,$$

and $|A_n| = O(2^{J/2}/n) = o(2^J/n)$. It follows that

$$n/2^{J+1}B_n \to 2\pi^2 D_\psi f_X^2(0), \tag{34}$$

and

$$(n/2^{J+1})E[\tilde{f}_X(0; J) - E\tilde{f}_X(0; J)]^2 \to 2D_\psi f_X^2(0). \tag{35}$$

We consider the bias term $E[\tilde{f}_X(J)] - f_X(0)$ in (29). Using the definition of $\tilde{f}_X(0; J)$ in (30), we can decompose

$$\begin{aligned}
E[\tilde{f}_X(J)] - f_X(0) &= \pi^{-1}\sum_{h=1}^{n-1} d_J(h)(1-h/n)R_X(h) - \pi^{-1}\sum_{h=1}^{\infty} R_X(h),\\
&= \pi^{-1}\sum_{h=1}^{n-1}(1-h/n)\,[d_J(h)-1]\,R_X(h) - \pi^{-1}\sum_{h=1}^{n-1}(h/n)R_X(h) - \pi^{-1}\sum_{h=n}^{\infty} R_X(h),\\
&= B_{4n} - B_{5n} - B_{6n},\ \text{say.}
\end{aligned} \tag{36}$$

Following a reasoning similar to [42, proof of Theorem 4.2], we can show that

$$B_{4n} = -2^{-q(J+1)}\lambda_q f_X^{(q)}(0)[1+o(1)],$$

$|B_{5n}| \le n^{-\min(1,q)} \sum_{h=1}^{n-1} l^q |R_X(h)| = O(n^{-\min(1,q)})$ and also that $|B_{6n}| \le 2n^{-q} \sum_{h=n}^{\infty} h^q |R_X(h)| = o(n^{-q})$. The bias term is then

$$E\tilde{f}_X(0;J) - f_X(0) = -2^{-q(J+1)}\lambda_q f_X^{(q)}(0) + o(2^{-qJ}) + O(n^{-\min(1,q)}). \tag{37}$$

Now, combining (29), (35), (37) we obtain

$$E\{[\hat{f}_X(0;J) - f_X(0)]^2\} = \frac{2^{J+1}}{n}2D_\psi f_X^2(0) + 2^{-2q(J+1)}\lambda_q^2[f_X^{(q)}(0)]^2 + o(2^J/n + 2^{-2qJ}).$$

The desired result follows by using $2^{J+1}/n^{\frac{1}{2q+1}} \to c$. This completes the proof of Theorem 3. □

*Proof* (Corollary 1) The result follows immediately from Theorem 2 because the conditions of Corollary 1 imply $2^{\hat{J}}/2^J - 1 = o_P(n^{-1/2(2q+1)}) = o_P(2^{-J/2})$, where the nonstochastic finest scale $J$ is given by $2^{J+1} \equiv \max\{[q\lambda_q^2\alpha(q)n/D_\psi]^{1/(2q+1)}, 0\}$. The latter satisfies the conditions of Theorem 2. □

To prove Theorems 4–6, we first state a useful lemma.

**Lemma 3** *Suppose Assumptions 1 and 2 hold, $J \to \infty$, and $2^J/n \to 0$. Define*

$$b_J(h,m) = a_J(h,m) + a_J(-h,-m) + a_J(h,-m) + a_J(-h,m),$$

*where $a_J(h,m) = 2\pi \sum_{j=0}^{J}\sum_{k=1}^{2^j} \hat{\psi}_{jk}(2\pi h)\hat{\psi}_{jk}^*(2\pi m)$. Then*

*(i) $b_J(h,m)$ is real-valued, $b_J(0,m) = b_J(h,0) = 0$ and $b_J(h,m) = b_J(m,h)$;*
*(ii) $\sum_{h=1}^{n-1}\sum_{m=1}^{n-1} h^\nu |b_J(h,m)| = O(2^{(1+\nu)J})$ for $0 \le \nu \le \frac{1}{2}$;*
*(iii) $\sum_{h=1}^{n-1}\{\sum_{m=1}^{n-1} |b_J(h,m)|\}^2 = O(2^J)$;*
*(iv) $\sum_{h_1=1}^{n-1}\sum_{h_2=1}^{n-1}\{\sum_{m=1}^{n-1} |b_J(h_1,m)b_J(h_2,m)|\}^2 = O\{(J+1)2^J\}$;*
*(v) $\sum_{h=1}^{n-1} b_J(h,h) = (2^{J+1}-1)\{1+O((J+1)/2^J + 2^{J(2\tau-1)}/n^{2\tau-1})\}$, where $\tau$ is in Assumption 3;*
*(vi) $\sum_{h=1}^{n-1}\sum_{m=1}^{n-1} b_J^2(h,m) = 2(2^{J+1}-1)\{1+o(1)\}$;*

*(vii)* $\sup_{1\le h,m\le n-1} |b_J(h,m)| \le C(J+1)$;
*(viii)* $\sup_{1\le h\le n-1} \sum_{m=1}^{n-1} |b_J(h,m)| \le C(J+1)$.

*Proof* (Lemma 3) See [50, Appendix B] for (i)–(vi) and [43, Lemma A.1] for (vii) and (viii). □

*Proof* (Theorem 4) Let $Z_t = \varepsilon_t - 1$ be such that $E(Z_t) = 0$. Under $\mathbb{H}_0^{\mathscr{A}}$, $\varepsilon_t = e_t$. We consider $\bar{R}_Z(h) = n^{-1}\sum_{t=|h|+1}^{n} Z_t Z_{t-|h|}$ and $\bar{\alpha}_{ejk} = \sum_{h=1-n}^{n-1} \bar{R}_Z(h)\hat{\psi}_{jk}^*(2\pi h)$. Let $\bar{\mathscr{A}}(J)$ defined as $\mathscr{A}(J)$ but using $\bar{\alpha}_{ejk}$. Let $\bar{\rho}_Z(h) = \bar{R}_Z(h)/\bar{R}_Z(0)$. Because $\bar{\rho}_Z(-h) = \bar{\rho}_Z(h)$ and $\bar{\alpha}_{ejk}$ is real-valued, we have

$$2\pi n \sum_{j=0}^{J}\sum_{k=1}^{2^j} \bar{\alpha}_{ejk}^2 = n\sum_{h=1}^{n-1}\sum_{m=1}^{n-1} b_J(h,m)\bar{\rho}_Z(h)\bar{\rho}_Z(m),$$

where the equality follows from re-indexing and the definition of $b_J(h,m)$. We have $\bar{R}_Z(0) - \sigma_Z^2 = O_P(n^{-1/2})$, since under $\mathbb{H}_0^{\mathscr{A}}$ we have that $\{e_t\}$ is iid. Also, we have

$$\begin{aligned} n\sum_{h=1}^{n-1}\sum_{m=1}^{n-1} b_J(h,m)\bar{\rho}_Z(h)\bar{\rho}_Z(m) &= \sigma_Z^{-4} n\sum_{h=1}^{n-1}\sum_{m=1}^{n-1} b_J(h,m)\bar{R}_Z(h)\bar{R}_Z(m) \\ &\quad +[\bar{R}_Z^{-2}(0) - \sigma_Z^{-4}]n\sum_{h=1}^{n-1}\sum_{m=1}^{n-1} b_J(h,m)\bar{R}_Z(h)\bar{R}_Z(m), \quad (38) \\ &= \sigma_Z^{-4} n\sum_{h=1}^{n-1}\sum_{m=1}^{n-1} b_J(h,m)\bar{R}_Z(h)\bar{R}_Z(m) + O_P(2^J/n^{1/2}), \end{aligned}$$

where the second term is of the indicated order of magnitude because

$$E\left[\sum_{h=1}^{n-1}\sum_{m=1}^{n-1} |b_J(h,m)\bar{R}_Z(h)\bar{R}_Z(m)|\right] \le Cn^{-1}\sum_{h=1}^{n-1}\sum_{m=1}^{n-1} |b_J(h,m)| = O(2^J/n).$$

given $E[\bar{R}_Z^2(h)] \le Cn^{-1}$ and Lemma 3(ii). We now focus on the first term in (38). We have,

$$\begin{aligned} n\sum_{h=1}^{n-1}\sum_{m=1}^{n-1} b_J(h,m)\bar{R}_Z(h)\bar{R}_Z(m) &= n^{-1}\sum_{h=1}^{n-1}\sum_{m=1}^{n-1} b_J(h,m)\sum_{t=h+1}^{n}\sum_{s=m+1}^{n} Z_t Z_{t-h} Z_s Z_{s-m}, \\ &= n^{-1}\sum_{h=1}^{n-1}\sum_{m=1}^{n-1} b_J(h,m)\left(\sum_{t=1}^{n}\sum_{s=1}^{n} - \sum_{t=1}^{h}\sum_{s=m+1}^{n} - \sum_{t=1}^{n}\sum_{s=1}^{m}\right) Z_t Z_{t-h} Z_s Z_{s-m}, \\ &= \hat{C}_n + \hat{D}_{1n} - \hat{D}_{2n} - \hat{D}_{3n}, \quad (39) \end{aligned}$$

where

$$\hat{C}_n = n^{-1}\sum_{h=1}^{n-1}\sum_{m=1}^{n-1} b_J(h,m)\left(\sum_{t=2}^{n}\sum_{s=1}^{t-1}+\sum_{s=2}^{n}\sum_{t=1}^{s-1}\right) Z_t Z_{t-h} Z_s Z_{s-m},$$

$$= 2n^{-1}\sum_{h=1}^{n-1}\sum_{m=1}^{n-1} b_J(h,m)\sum_{t=2}^{n}\sum_{s=1}^{t-1} Z_t Z_{t-h} Z_s Z_{s-m}, \text{ given } b_J(h,m) = b_J(m,h),$$

$$\hat{D}_{1n} = n^{-1}\sum_{h=1}^{n-1}\sum_{m=1}^{n-1} b_J(h,m)\sum_{t=1}^{n} Z_t^2 Z_{t-h} Z_{t-m},$$

$$\hat{D}_{2n} = n^{-1}\sum_{h=1}^{n-1}\sum_{m=1}^{n-1} b_J(h,m)\sum_{t=1}^{h}\sum_{s=m+1}^{n} Z_t Z_{t-h} Z_s Z_{s-m},$$

$$\hat{D}_{3n} = n^{-1}\sum_{h=1}^{n-1}\sum_{m=1}^{n-1} b_J(h,m)\sum_{t=1}^{n}\sum_{s=1}^{m} Z_t Z_{t-h} Z_s Z_{s-m}.$$

In order to prove Theorem 4, we first state Proposition 1 that shows that $\hat{C}_n$ represents the dominant term.

**Proposition 1** *Suppose Assumptions 1–3 hold, $J \to \infty$, and $2^{2J}/n \to 0$. Then*

$$2^{-J/2}\left\{2\pi n \sum_{j=0}^{J}\sum_{k=1}^{2^j} \bar{\alpha}_{jk}^2 - (2^{J+1}-1)\right\} = 2^{-J/2}\sigma^{-4}\hat{C}_n + o_P(1).$$

We now decompose $\hat{C}_n$ into the terms with $t-s > q$ and $t-s \le q$, for some $q \in \mathbb{Z}^+$:

$$\hat{C}_n = 2n^{-1}\sum_{h=1}^{n-1}\sum_{m=1}^{n-1} b_J(h,m)\left(\sum_{t=q+2}^{n}\sum_{s=1}^{t-q-1}+\sum_{t=2}^{n}\sum_{s=\max(t-q,1)}^{t-1}\right) Z_t Z_{t-h} Z_s Z_{s-m},$$

$$= \hat{D}_n + \hat{D}_{4n}. \tag{40}$$

Furthermore, we decompose

$$\hat{D}_n = 2n^{-1}\left(\sum_{h=1}^{q}\sum_{m=1}^{q}+\sum_{h=1}^{q}\sum_{m=q+1}^{n-1}+\sum_{h=q+1}^{n-1}\sum_{m=1}^{n-1}\right) b_J(h,m)\sum_{t=q+2}^{n}\sum_{s=1}^{t-q-1} Z_t Z_{t-h} Z_s Z_{s-m},$$

$$= \hat{U}_n + \hat{D}_{5n} + \hat{D}_{6n}, \text{ say,} \tag{41}$$

where $\hat{D}_{5n}$ and $\hat{D}_{6n}$ are the contributions from $m > q$ and $h > q$, respectively.

Proposition 2 shows that $\hat{C}_n$ can be approximated arbitrarily well by $\hat{U}_n$ under a proper condition on $q$.

**Proposition 2** *Suppose Assumptions 1–3 hold,* $J \to \infty$, $2^{2J}/n \to 0$, *and* $q \equiv q_n \to \infty$, $q/2^J \to \infty$, $q^2/n \to 0$. *Then* $2^{-J/2}\hat{C}_n = 2^{-J/2}\hat{U}_n + o_P(1)$.

It is much easier to show the asymptotic normality of $\hat{U}_n$ than of $\hat{C}_n$, because for $\hat{U}_n$, $\{Z_t Z_{t-h}\}$ and $\{Z_s Z_{s-m}\}$ are independent given $t - s > q$ and $0 < h, m \le q$.

**Proposition 3** *Suppose Assumptions 1–3 hold, and* $J \to \infty$, $2^{2J}/n \to 0$, $q/2^J \to \infty$, $q^2/n \to 0$. *Let* $\lambda_n^2 = E(\hat{U}_n^2)$. *Then* $4(2^{J+1} - 1)\sigma^8/\lambda_n^2 \to 1$, *and* $\lambda_n^{-1}\hat{U}_n \to^d N(0, 1)$.

Propositions 1–3 and Slutsky's Theorem imply $\bar{\mathscr{A}}(J) \to^d N(0, 1)$. Propositions 4 and 5 show that parameter estimation does not have impact on the asymptotic distribution of the test statistic.

**Proposition 4** $n \sum_{j=0}^{J} \sum_{k=1}^{2^j} (\hat{\alpha}_{jk} - \bar{\alpha}_{jk})^2 = O_P(2^J/n) + O_P(1)$.

**Proposition 5** $n \sum_{j=0}^{J} \sum_{k=1}^{2^j} (\hat{\alpha}_{jk} - \bar{\alpha}_{jk})\bar{\alpha}_{jk} = o_P(2^{J/2})$.

The proof of Theorem 4 will be completed provided Propositions 1–5 are shown. The proofs of Propositions 1–3 are very similar to the proofs of Propositions 1–3 in [50], for proving the asymptotic normality of a wavelet-based test statistic for serial correlation. These proofs are then omitted (but for the interested reader all the detailed proofs are available from the authors).

*Proof* (Proposition 4) A standard Taylor's expansion gives

$$D_t^{-1}(\hat{\boldsymbol{\theta}}) = D_t^{-1}(\boldsymbol{\theta}_0) + \left\{ \frac{\partial}{\partial \boldsymbol{\theta}} D_t^{-1}(\boldsymbol{\theta}_0) \right\}^{\top} (\hat{\boldsymbol{\theta}} - \boldsymbol{\theta}_0) + \frac{1}{2}(\hat{\boldsymbol{\theta}} - \boldsymbol{\theta}_0)^{\top} \frac{\partial^2}{\partial \boldsymbol{\theta} \partial \boldsymbol{\theta}^{\top}} D_t^{-1}(\bar{\boldsymbol{\theta}})(\hat{\boldsymbol{\theta}} - \boldsymbol{\theta}_0),$$

where $\bar{\boldsymbol{\theta}}$ lies between $\hat{\boldsymbol{\theta}}$ and $\boldsymbol{\theta}_0$. We have

$$\begin{aligned} \hat{R}_Z(h) - \ddot{R}_Z(h) &= n^{-1} \sum_{t=|h|+1}^{n} (\hat{Z}_t - Z_t)(\hat{Z}_{t-|h|} - Z_{t-|h|}) + n^{-1} \sum_{t=|h|+1}^{n} Z_t(\hat{Z}_{t-|h|} - Z_{t-|h|}) \\ &\quad + n^{-1} \sum_{t=|h|+1}^{n} (\hat{Z}_t - Z_t) Z_{t-|h|}, \\ &= \hat{A}_1(h) + \hat{A}_2(h) + \hat{A}_3(h). \end{aligned}$$

We write $\hat{\alpha}_{jk} - \tilde{\alpha}_{jk} = \hat{B}_{1jk} + \hat{B}_{2jk} + \hat{B}_{3jk}$ where

$$\hat{B}_{1jk} = \sum_{h=1-n}^{n-1} \hat{A}_1(h)\hat{\psi}_{jk}(2\pi h), \quad \hat{B}_{2jk} = \sum_{h=1-n}^{n-1} \hat{A}_2(h)\hat{\psi}_{jk}(2\pi h), \quad \hat{B}_{3jk} = \sum_{h=1-n}^{n-1} \hat{A}_3(h)\hat{\psi}_{jk}(2\pi h).$$

Then $(\hat{\alpha}_{jk} - \tilde{\alpha}_{jk})^2 \leq 4(\hat{B}_{1jk}^2 + \hat{B}_{2jk}^2 + \hat{B}_{3jk}^3)$ We first study the term involving $\hat{B}_{1jk}$. We write

$$2\pi \sum_{j=0}^{J} \sum_{k=1}^{2^j} \hat{B}_{1jk}^2 = \sum_{h=1-n}^{n-1} \sum_{m=1-n}^{n-1} a_J(h,m)\hat{A}_1(h)\hat{A}_1(m) = \sum_{h=1}^{n-1} \sum_{m=1}^{n-1} b_J(h,m)\hat{A}_1(h)\hat{A}_1(m).$$

We have

$$2\pi \sum_{j=0}^{J} \sum_{k=1}^{2^j} \hat{B}_{1jk}^2 \leq \left(\sup \hat{A}_1(h)\right)^2 |\sum_{h=1}^{n-1} \sum_{m=1}^{n-1} b_J(h,m)| = O_P(2^J/n).$$

We now study the term involving $\hat{B}_{2jk}$. Let

$$\hat{a}_{21}(h) = n^{-1} \sum_{t=|h|+1}^{n} Z_t X_{t-|h|} \left\{ \frac{\partial}{\partial \boldsymbol{\theta}} D_{t-|h|}^{-1}(\boldsymbol{\theta}_0) \right\}^\top,$$

$$\hat{a}_{22}(h) = n^{-1} \sum_{t=|h|+1}^{n} Z_t X_{t-|h|} \frac{\partial^2}{\partial \boldsymbol{\theta} \partial \boldsymbol{\theta}^\top} D_{t-|h|}^{-1}(\bar{\boldsymbol{\theta}}_0).$$

We write $\hat{A}_2(h) = \hat{A}_{21}(h) + \hat{A}_{22}(h)$, where

$$\hat{A}_{21}(h) = \hat{a}_{21}(h)(\hat{\boldsymbol{\theta}} - \boldsymbol{\theta}_0),$$

$$\hat{A}_{22}(h) = \frac{1}{2}(\hat{\boldsymbol{\theta}} - \boldsymbol{\theta}_0)^\top \hat{a}_{22}(h)(\hat{\boldsymbol{\theta}} - \boldsymbol{\theta}_0).$$

Then

$$\hat{B}_{2jk} = \sum_{h=1-n}^{n-1} \hat{A}_2(h)\hat{\psi}_{jk}(2\pi h),$$

$$= \left[ \sum_{h=1-n}^{n-1} \hat{a}_{21}(h)\hat{\psi}_{jk}(2\pi h) \right] (\hat{\boldsymbol{\theta}} - \boldsymbol{\theta}_0) + \frac{1}{2}(\hat{\boldsymbol{\theta}} - \boldsymbol{\theta}_0)^\top \left[ \sum_{h=1-n}^{n-1} \hat{a}_{22}(h)\hat{\psi}_{jk}(2\pi h) \right] (\hat{\boldsymbol{\theta}} - \boldsymbol{\theta}_0).$$

We obtain

$$2\pi n \sum_{j=0}^{J} \sum_{k=1}^{2^j} B_{2jk}^2 \leq 4n||\hat{\boldsymbol{\theta}} - \boldsymbol{\theta}_0||^2 \sum_{j=0}^{J} \sum_{k=1}^{2^j} \left\| \sum_{h=1-n}^{n-1} \hat{a}_{21}(h)\hat{\psi}_{jk}(2\pi h) \right\|^2$$

$$+2n||\hat{\boldsymbol{\theta}} - \boldsymbol{\theta}_0||^4 \sum_{j=0}^{J} \sum_{k=1}^{2^j} \left\| \sum_{h=1-n}^{n-1} \hat{a}_{22}(h)\hat{\psi}_{jk}(2\pi h) \right\|^2 = O_P(2^J/n),$$

since

$$2\pi \sum_{j=0}^{J} \sum_{k=1}^{2^j} \left\| \sum_{h=1-n}^{n-1} \hat{a}_{21}(h)\hat{\psi}_{jk}(2\pi h) \right\|^2 = \left| \sum_{h=1}^{n-1} \sum_{m=1}^{n-1} b_J(h,m)\hat{a}_{21}(h)\hat{a}_{21}^{\top}(m) \right|,$$

$$\leq \left(\sup ||\hat{a}_{21}(h)||\right)^2 \left( \sum_{h=1}^{n-1} \sum_{m=1}^{n-1} |b_J(h,m)| \right) = O_P(2^J/n).$$

We now study the term involving $\hat{B}_{3jk}$.

$$\hat{a}_{31}(h) = n^{-1} \sum_{t=|h|+1}^{n} Z_{t-|h|} X_t \left( \frac{\partial}{\partial \boldsymbol{\theta}} D_t^{-1}(\boldsymbol{\theta}_0) \right)^{\top},$$

$$\hat{a}_{32}(h) = n^{-1} \sum_{t=|h|+1}^{n} Z_{t-|h|} X_t \frac{\partial^2}{\partial \boldsymbol{\theta} \partial \boldsymbol{\theta}^{\top}} D_t^{-1}(\bar{\boldsymbol{\theta}}_0).$$

We write $\hat{A}_3(h) = \hat{A}_{31}(h) + \hat{A}_{32}(h)$, where

$$\hat{A}_{31}(h) = \hat{a}_{31}(h)(\hat{\boldsymbol{\theta}} - \boldsymbol{\theta}_0),$$

$$\hat{A}_{32}(h) = \frac{1}{2}(\hat{\boldsymbol{\theta}} - \boldsymbol{\theta}_0)^{\top} \hat{a}_{32}(h)(\hat{\boldsymbol{\theta}} - \boldsymbol{\theta}_0).$$

We obtain

$$2\pi n \sum_{j=0}^{J} \sum_{k=1}^{2^j} B_{3jk}^2 \leq 4n \sum_{j=0}^{J} \sum_{k=1}^{2^j} \left[ \sum_{h=1-n}^{n-1} \hat{A}_{31}(h)\hat{\psi}_{jk}(2\pi h) \right]^2 + 4n \sum_{j=0}^{J} \sum_{k=1}^{2^j} \left[ \sum_{h=1-n}^{n-1} \hat{A}_{32}(h)\hat{\psi}_{jk}(2\pi h) \right]^2.$$

We write $\hat{a}_{31}(h) = E[\hat{a}_{31}(h)] + \left\{\hat{a}_{31}(h) - E[\hat{a}_{31}(h)]\right\}$. We have

$$n \sum_{j=0}^{J} \sum_{k=1}^{2^j} \left[ \sum_{h=1-n}^{n-1} \hat{A}_{31}(h)\hat{\psi}_{jk}(2\pi h) \right]^2 \leq 2n||\hat{\boldsymbol{\theta}} - \boldsymbol{\theta}_0||^2 \left\{ \sum_{j=0}^{J} \sum_{k=1}^{2^j} \left\| \sum_{h} E[\hat{a}_{31}(h)]\hat{\psi}_{jk}(2\pi h) \right\|^2 \right.$$

$$\left. + \sum_{j=0}^{J} \sum_{k=1}^{2^j} \left\| \sum_{h} [\hat{a}_{31}(h) - E\hat{a}_{31}(h)]\hat{\psi}_{jk}(2\pi h) \right\|^2 \right\}.$$

Since we can interpret $E\hat{a}_{31}(h) = \text{cov}(Z_{t-|h|}, X_t \frac{\partial}{\partial \boldsymbol{\theta}} D_t^{-1}(\boldsymbol{\theta}_0))$ as a cross-correlation function, we have that

$$\sum_{j=0}^{J} \sum_{k=1}^{2^j} \left\| \sum_{h} E[\hat{a}_{31}(h)]\hat{\psi}_{jk}(2\pi h) \right\|^2 = O(1).$$

Also,

$$\sum_{j=0}^{J}\sum_{k=1}^{2^j} E\left\|\sum_h [\hat{a}_{31}(h) - E\hat{a}_{31}(h)]\hat{\psi}_{jk}(2\pi h)\right\|^2$$

$$= |\sum_{h=1}^{n-1}\sum_{m=1}^{n-1} b_J(h,m)E[(\hat{a}_{31}(h) - E\hat{a}_{31}(h))(\hat{a}_{31}(m) - E\hat{a}_{31}(m))^\top]|,$$

$$\leq \sum_{h=1}^{n-1}\sum_{m=1}^{n-1} |b_J(h,m)| \left\{E\|\hat{a}_{31}(h) - E\hat{a}_{31}(h)\|^2\right\}^{1/2} \left\{E\|\hat{a}_{31}(m) - E\hat{a}_{31}(m)\|^2\right\}^{1/2},$$

$$\leq \sup_h E\|\hat{a}_{31}(h) - E\hat{a}_{31}(h)\|^2 \sum_h\sum_m |b_J(h,m)| = O(2^J/n).$$

This shows Proposition 4. □

*Remark 1* Proposition 4 is established under a general stationary process for $\{Z_t\}$, that is, the result is established without assuming the null hypothesis.

*Proof* (Proposition 5) We write $(\hat{\alpha}_{jk} - \tilde{\alpha}_{jk})\tilde{\alpha}_{jk} = \hat{C}_{1jk} + \hat{C}_{2jk} + \hat{C}_{3jk}$, where

$$\hat{C}_{1jk} = \sum_{h=1-n}^{n-1} \hat{A}_1(h)\hat{\psi}_{jk}(2\pi h)\tilde{\alpha}_{jk}, \tag{42}$$

$$\hat{C}_{2jk} = \sum_{h=1-n}^{n-1} \hat{A}_2(h)\hat{\psi}_{jk}(2\pi h)\tilde{\alpha}_{jk}, \tag{43}$$

$$\hat{C}_{3jk} = \sum_{h=1-n}^{n-1} \hat{A}_3(h)\hat{\psi}_{jk}(2\pi h)\tilde{\alpha}_{jk}. \tag{44}$$

By the Cauchy–Schwarz inequality and the fact that $n\sum_{j=0}^{J}\sum_{k=1}^{2^j}\tilde{\alpha}_{jk}^2 = O_P(2^J)$ under the null hypothesis, we have that

$$n\sum_{j=0}^{J}\sum_{k=1}^{2^j}\sum_{h=1-n}^{n-1} \hat{A}_1(h)\hat{\psi}_{jk}(2\pi h)\tilde{\alpha}_{jk} = O_P(2^J/n^{1/2}).$$

Since $n\sum_{j=0}^{J}\sum_{k=1}^{2^j} B_{2jk}^2 = O_P(2^J/n)$, we have that

$$n\sum_{j=0}^{J}\sum_{k=1}^{2^j}\hat{C}_{2jk} = O_P(2^J/n^{1/2}).$$

We write

$$\hat{C}_{3jk} = \left[\sum_{h=1-n}^{n-1} \hat{a}_{31}(h)\hat{\psi}_{jk}(2\pi h)\tilde{\alpha}_{jk}\right](\hat{\boldsymbol{\theta}} - \boldsymbol{\theta}_0) + \frac{1}{2}(\hat{\boldsymbol{\theta}} - \boldsymbol{\theta}_0)^\top \left\{\sum_{h=1-n}^{n-1} \hat{a}_{32}(h)\hat{\psi}_{jk}(2\pi h)\tilde{\alpha}_{jk}\right\}(\hat{\boldsymbol{\theta}} - \boldsymbol{\theta}_0).$$

We have

$$2\pi n \sum_{j=0}^{J}\sum_{k=1}^{2^j}\sum_{h=1-n}^{n-1} E[\hat{a}_{31}(h)]\hat{\psi}_{jk}(2\pi h)\tilde{\alpha}_{jk} = n\sum_{h=1}^{n-1}\sum_{m=1}^{n-1} E[\hat{a}_{31}(h)]\tilde{R}(m)b_J(h,m).$$

Since

$$\begin{aligned}
&n^2 E\left\|\sum_{h=1}^{n-1}\sum_{m=1}^{n-1} E[\hat{a}_{31}(h)]\tilde{R}(m)b_J(h,m)\right\|^2 \\
&\le Cn\sum_{h_1=1}^{n-1}\sum_{h_2=1}^{n-1}\sum_{m=1}^{n-1} E[\hat{a}_{31}(h_1)]E[\hat{a}_{31}(h_2)]^\top |b_J(h_1,m)b_J(h_2,m)|, \\
&\le Cn\left(\sum_{h_1=1}^{n-1}\sum_{h_2=1}^{n-1}[\sum_{m=1}^{n-1}|b_J(h_1,m)b_J(h_2,m)|]^2\right)^{1/2} = O(n(J+1)^{1/2}2^{(J+1)/2}).
\end{aligned}$$

Then

$$n\sum_{h=1}^{n-1}\sum_{m=1}^{n-1} E[\hat{a}_{31}(h)]\tilde{R}(m)b_J(h,m) = O_P(n^{1/2}J^{1/4}2^{J/4}).$$

Wc havc

$$\begin{aligned}
&E\|n\sum_{j=0}^{J}\sum_{k=1}^{2^j}\sum_{h}[\hat{a}_{31}(h) - E\hat{a}_{31}(h)]\hat{\psi}_{jk}(h)\tilde{\alpha}_{jk}\| \\
&\le \frac{n}{2\pi}\sum_{h=1}^{n-1}\sum_{m=1}^{n-1} E\left[\|\hat{a}_{31}(h) - E\hat{a}_{31}(h)\|\|\tilde{R}(m)|\right]|b_J(h,m)|, \\
&\le \frac{n}{2\pi}\sum_{h=1}^{n-1}\sum_{m=1}^{n-1}\left[E\|\hat{a}_{31}(h) - E\hat{a}_{31}(h)\|^2\right]^{1/2}[E\tilde{R}^2(m)]^{1/2}|b_J(h,m)|, \\
&= O(n\, n^{-1/2}\, n^{-1/2}\, 2^{J+1}) = O(2^{J+1}).
\end{aligned}$$

This completes the proof of Proposition 5 and so Theorem 4. □

*Proof* (Theorem 5) We write

$$\mathscr{A}(\hat{J}) - \mathscr{A}(J) = [D_n(\hat{J})]^{-1/2}\left\{2\pi n \sum_{j=J}^{\hat{J}}\sum_{k=1}^{2^j}\hat{\alpha}_{jk}^2 - [C_n(\hat{J}) - C_n(J)]\right\} - \left\{1 - [D_n(J)]^{1/2}/[D_n(\hat{J})]^{1/2}\right\}\mathscr{A}(J),$$

where $C_n(J) = 2^{J+1} - 1$, $D_n(J) = 4(2^{J+1} - 1)$. Given $\mathscr{A}(J) = O_P(1)$ by Theorem 4, it suffices for $\mathscr{A}(\hat{J}) - \mathscr{A}(J) \to^p 0$ and $\mathscr{A}(\hat{J}) \to^d N(0, 1)$ to establish

$$(i)\ \ [D_n(J)]^{-1/2}\left\{2\pi n \sum_{j=J}^{\hat{J}}\sum_{k=1}^{2^j}\hat{\alpha}_{jk}^2 - [C_n(\hat{J}) - C_n(J)]\right\} \to^p 0,$$

$$(ii)\ \ D_n(\hat{J})/D_n(J) \to^p 1.$$

We first show (i). Decompose

$$\begin{aligned} 2\pi n \sum_{j=J}^{\hat{J}}\sum_{k=1}^{2^j}\hat{\alpha}_{jk}^2 &= 2\pi n \sum_{j=J}^{\hat{J}}\sum_{k=1}^{2^j}(\hat{\alpha}_{jk} - \tilde{\alpha}_{jk})^2 + 2\pi n \sum_{j=J}^{\hat{J}}\sum_{k=1}^{2^j}\tilde{\alpha}_{jk}^2 + 4\pi n \sum_{j=J}^{\hat{J}}\sum_{k=1}^{2^j}(\hat{\alpha}_{jk} - \tilde{\alpha}_{jk})\tilde{\alpha}_{jk}, \\ &= \hat{G}_1 + \hat{G}_2 + 2\hat{G}_3. \end{aligned} \tag{45}$$

For the first term in (45), we write

$$\hat{G}_1 = 2\pi n \sum_{j=0}^{\hat{J}}\sum_{k=1}^{2^j}(\hat{\alpha}_{jk} - \tilde{\alpha}_{jk})^2 - 2\pi n \sum_{j=0}^{J}\sum_{k=1}^{2^j}(\hat{\alpha}_{ijk} - \tilde{\alpha}_{jk})^2 = \hat{G}_{11} - \hat{G}_{12}. \tag{46}$$

By Proposition 4, we have $[D_n(J)]^{-1/2}\hat{G}_{12} \to^p 0$. For the first term in (46), we have for any given constants $M > 0$ and $\varepsilon > 0$,

$$P\left(\hat{G}_{11} > \varepsilon\right) \leq P\left(\hat{G}_{11} > \varepsilon, C_0 2^{J/2}|2^{\hat{J}}/2^J - 1| \leq \varepsilon\right) + P\left(C_0 2^{J/2}|2^{\hat{J}}/2^J - 1| > \varepsilon\right). \tag{47}$$

For any given constants $C_0, \varepsilon > 0$, the second term in (47) vanishes to 0 as $n \to \infty$ given $2^{J/2}|2^{\hat{J}}/2^J - 1| \to^p 0$. For the first term, given $C_0 2^{J/2}|2^{\hat{J}}/2^J - 1| \leq \varepsilon$, we have for $n$ sufficiently large,

$$\begin{aligned} [D_n(J)]^{-1/2}\hat{G}_{11} &\leq [D_n(J)]^{-1/2} 2\pi n \sum_{j=0}^{[\log_2 2^J(1+\varepsilon/(C_0 2^{J/2}))]}\ \sum_{k=1}^{2^j}(\hat{\alpha}_{jk} - \tilde{\alpha}_{jk})^2, \\ &\leq [D_n(J)]^{-1/2} 2\pi n \sum_{j=0}^{J+1}\sum_{k=1}^{2^j}(\hat{\alpha}_{jk} - \tilde{\alpha}_{jk})^2 = o_P(1). \end{aligned}$$

by Proposition 4. Therefore, we have

$$[D_n(J)]^{-1/2}\hat{G}_1 = o_P(1). \tag{48}$$

Next, we consider $\hat{G}_2$ in (45). We write

$$\hat{G}_2 = n\sum_{h=1}^{n-1}\sum_{m=1}^{n-1}\tilde{R}_e(h)\tilde{R}_e(m)\left[b_{\hat{J}}(h,m) - b_J(h,m)\right].$$

Since for $n$ sufficiently large,

$$\begin{aligned}
&\sum_{h=1-n}^{n-1}\sum_{m=1-n}^{n-1}\left|a_J(h,m) - a_{\hat{J}}(h,m)\right| \\
&\leq C\sum_{h=1-n}^{n-1}\sum_{m=1-n}^{n-1}\sum_{j=[\log_2 2^J(1-\varepsilon/(C_0 2^{J/2}))]}^{[\log_2 2^J(1+\varepsilon/(C_0 2^{J/2}))]}\left|c_j(h,m)\hat{\psi}(2\pi h/2^j)\hat{\psi}^*(2\pi m/2^j)\right|, \\
&\leq C\sum_{j=[\log_2 2^J(1-\varepsilon/(C_0 2^{J/2}))]}^{[\log_2 2^J(1+\varepsilon/(C_0 2^{J/2}))]} 2^j\left[2^{-j}\sum_{h=1-\bar{T}}^{\bar{T}-1}|\hat{\psi}(2\pi h/2^j)|\right], \\
&\times\left[\sum_{r=-\infty}^{\infty}|\hat{\psi}(2\pi h/2^j + 2\pi r)|\right], \\
&\leq C2^J\varepsilon/(C_0 2^{J/2}),
\end{aligned}$$

given Assumption 3 and

$$c_j(h,m) = \begin{cases} 1 \text{ if } m-h = 2^j r \text{ for some } r \in \mathbb{Z}, \\ 0 \text{ otherwise.} \end{cases} \tag{49}$$

Cf. [61, (6.19), p.392]. Therefore

$$\begin{aligned}
E|\hat{G}_2| &\leq n\sum_{h=1-n}^{n-1}\sum_{m=1-n}^{n-1} E|\tilde{R}_e(h)\tilde{R}_e(m)||a_{\hat{J}}(h,m) - a_J(h,m)|, \\
&\leq n\sup_h \operatorname{var}\tilde{R}_e(h)\sum_{h=1-n}^{n-1}\sum_{m=1-n}^{n-1}|a_{\hat{J}}(h,m) - a_J(h,m)|, \\
&\leq C2^{J/2}\varepsilon/C_0.
\end{aligned}$$

Then $[D_n(J)]^{-1/2}\left\{\hat{G}_2 - [C(\hat{J}) - C(J)]\right\} \to^p 0$. Note that $[D_n(J)]^{-1/2}[C_n(\hat{J}) - C_n(J)] = o_P(1)$. Next, by the Cauchy-Schwarz inequality and (48), we have

$$[D_n(J)]^{-1/2}|\hat{G}_3| \leq \left([D_n(J)]^{-1/2}\hat{G}_1\right)^{1/2}\left([D_n(J)]^{-1/2}|\hat{G}_2|\right)^{1/2} = o_P(1).$$

Summarizing, we obtain result (i), that is

$$[D_n(J)]^{-1/2}\left\{2\pi n\sum_{j=J}^{\hat{J}}\sum_{k=1}^{2^j}\hat{\alpha}_{jk}^2 - [C_n(\hat{J}) - C_n(J)]\right\} = o_P(1).$$

We now show (ii), that is $D_n(\hat{J})/D_n(J) = 1 + o_P(1)$. Using the fact that $2^{\hat{J}}/2^J = 1 + o_P(2^{-J/2})$ and $J \to \infty$ such that $2^J/n \to 0$, one shows easily that

$$\frac{D_n(\hat{J})}{D_n(J)} = \frac{2^{\hat{J}+1}-1}{2^{J+1}-1} \to^p 1.$$

This shows (ii). This completes the proof of Theorem 5. □

*Proof* (Theorem 6) We first show $Q(\hat{f}, f) = Q(\tilde{f}, f) + o_P(2^J/T + 2^{-2qJ})$. Write

$$\begin{aligned} Q(\hat{f}, f) - Q(\tilde{f}, f) &= Q(\hat{f}, \tilde{f}) + 2\int_{-\pi}^{\pi}[\hat{f}(\omega) - \tilde{f}(\omega)][\tilde{f}(\omega) - f(\omega)]d\omega, \\ &= \hat{Q}_1 + 2\hat{Q}_2. \end{aligned} \tag{50}$$

For the first term in (50), by Parseval's identity, Proposition 4 (which can be shown to continues to hold given Assumptions 2–3 and 7–10; See Remark 1), and $D_n(J) \propto O(2^{J+1})$, we have

$$\hat{Q}_1 = \sum_{j=0}^{J}\sum_{k=1}^{2^j}(\hat{\alpha}_{jk} - \tilde{\alpha}_{jk})^2 = O_P[n^{-1} + 2^J n^{-2}] = o_P(2^J/n), \tag{51}$$

as $n \to \infty$. For the second term, we have $\hat{Q}_2 = o_P(2^J/n + 2^{-2qJ})$ by the Cauchy-Schwarz inequality, (51) and the fact that $Q(\tilde{f}, f) = O_P(2^J/n + 2^{-2qJ})$, which follows by Markov's inequality and $EQ(\tilde{f}, f) = O(2^J/n + 2^{-2qJ})$. The latter is to be shown below.

To compute $E[Q(\tilde{f}, f)]$, we write

$$E[Q(\tilde{f}, f)] = E[Q(\tilde{f}, E\tilde{f})] + Q[E(\tilde{f}), f]. \tag{52}$$

We first consider the second term in (52). Put $B(\omega) \equiv \sum_{j=J+1}^{\infty}\sum_{k=1}^{2^j}\alpha_{jk}\Psi_{jk}(\omega)$. Then

$$Q[E(\tilde{f}), f] = \int_{-\pi}^{\pi} B^2(\omega)d\omega + \sum_{j=0}^{J}\sum_{k=1}^{2^j}(E\tilde{\alpha}_{jk} - \alpha_{jk})^2. \tag{53}$$

We evaluate directly $\int_{-\pi}^{\pi} B^2(\omega)d\omega$. Using the orthonormality of the wavelet basis, we have that

$$\int_{-\pi}^{\pi} B^2(\omega)d\omega = \sum_{j=J+1}^{\infty}\sum_{k=1}^{2^j}\alpha_{jk}^2.$$

Replacing $\alpha_{jk} = \sum_{h=-\infty}^{\infty} R_e(h)\hat{\psi}_{jk}(2\pi h)$ and since $\hat{\psi}_{jk}(2\pi h) = e^{-i2\pi hk/2^j} 2^{-j/2}\hat{\psi}(2\pi h/2^j)$,

$$\begin{aligned}\sum_{j=J+1}^{\infty}\sum_{k=1}^{2^j}\alpha_{jk}^2 &= \sum_{j=J+1}^{\infty}\sum_{h=-\infty}^{\infty}\sum_{m=-\infty}^{\infty} R_e(h)R_e(m)\{2^{-j}\sum_{k=1}^{2^j} e^{i2\pi(m-h)k/2^j}\}\hat{\psi}(2\pi h/2^j)\hat{\psi}^*(2\pi m/2^j),\\ &= \sum_{j=J+1}^{\infty}\sum_{h=-\infty}^{\infty}\sum_{m=-\infty}^{\infty} c_j(h,m)R_e(h)R_e(m)\hat{\psi}(2\pi h/2^j)\hat{\psi}^*(2\pi m/2^j),\end{aligned}$$

where $c_j(h,m) = 2^{-j}\sum_{k=1}^{2^j} e^{i2\pi(m-h)k/2^j}$ is as in (49). By a change of variables,

$$\sum_{j=J+1}^{\infty}\sum_{k=1}^{2^j}\alpha_{jk}^2 = \sum_{j=J+1}^{\infty}\sum_{h=-\infty}^{\infty}\sum_{r=-\infty}^{\infty} R_e(h)R_e(h+2^jr)\hat{\psi}(2\pi h/2^j)\hat{\psi}^*(2\pi h/2^j+2\pi r). \quad (54)$$

We evaluate separately the case corresponding to $r = 0$ and $r \neq 0$ in (54).

$$\begin{aligned}&\sum_{j=J+1}^{\infty}\sum_{h=-\infty}^{\infty} R_e^2(h)\hat{\psi}(2\pi h/2^j)\hat{\psi}^*(2\pi h/2^j)\\ &= \sum_{j=J+1}^{\infty}\sum_{h=-\infty}^{\infty} R_e^2(h)|\hat{\psi}(2\pi h/2^j)|^2,\\ &= \sum_{j=J+1}^{\infty}\sum_{h=-\infty}^{\infty} R_e^2(h)|2\pi h/2^j|^{2q}\frac{|\hat{\psi}(2\pi h/2^j)|^2}{|2\pi h/2^j|^{2q}},\\ &= \lim_{z\to 0}\frac{|\hat{\psi}(z)|^2}{|z|^{2q}}[1+o(1)](2\pi)^{2q}\sum_{j=J+1}^{\infty}\sum_{h=-\infty}^{\infty}|h|^{2q}R_e^2(h)(2^{-2q})^j,\\ &= \lim_{z\to 0}\frac{|\hat{\psi}(z)|^2}{|z|^{2q}}[1+o(1)](2\pi)^{2q}\sum_{j=J+1}^{\infty}(2^{-2q})^j\sum_{h=-\infty}^{\infty}|h|^{2q}R_e^2(h),\\ &= (2\pi)^{2q+1}\lim_{z\to 0}\frac{|\hat{\psi}(z)|^2}{|z|^{2q}}[1+o(1)]\frac{2^{-2q(J+1)}}{1-2^{-2q}}\int_{-\pi}^{\pi}[f_e^{(q)}(\omega)]^2d\omega,\end{aligned}$$

where $f_e^{(q)}(\cdot)$ is defined in Sect. 4.3 and $o(1)$ is uniform in $\omega \in [-\pi, \pi]$. It follows that

$$\int_{-\pi}^{\pi} B^2(\omega)d\omega = 2^{-2q(J+1)}\vartheta_q\int_{-\pi}^{\pi}\left[f_e^{(q)}(\omega)\right]^2 d\omega + o(2^{-2qJ}). \quad (55)$$

It may be show that the term corresponding to $r \neq 0$ is $o(2^{-2qJ})$.

For the second term in (53), we have

$$\sum_{j=0}^{J}\sum_{k=1}^{2^j}(E\tilde{\alpha}_{jk}-\alpha_{jk})^2 = \sum_{j=0}^{J}\sum_{k=1}^{2^j}\left[n^{-1}\sum_{h=1-n}^{n-1}|h|R_e(h)\hat{\psi}_{jk}(2\pi h)+\sum_{|h|\geq n}R_e(h)\hat{\psi}_{jk}(2\pi h)\right]^2,$$
$$\leq 4Cn^{-2}\sum_{h=1}^{n-1}\sum_{m=1}^{n-1}|hmR_e(h)R_e(m)b_J(h,m)|,$$
$$= O[(J+1)/n^2), \quad (56)$$

given Lemma 3(vii) and $\sum_{h=-\infty}^{\infty}|hR_e(h)| \leq C$ as implied by Assumption 10.

Finally, we consider the variance factor in (52). We write

$$E[Q(\tilde{f},E\tilde{f})] = \sum_{h=1-n}^{n-1}\sum_{m=1-n}^{n-1}b_J(h,m)\mathrm{cov}[\tilde{R}_e(h),\tilde{R}_e(m)],$$
$$= \sum_{h=1-n}^{n-1}\sum_{m=1-n}^{n-1}b_J(h,m)n^{-1}\sum_{l}\left[1-\frac{\eta(l)+m}{n}\right]$$
$$\times\left[R_e(l)R_e(l+m-h)+R_e(l+m)R_e(l-h)+\kappa(l,h,m-h)\right],$$
$$\equiv V_{1n}+V_{2n}+V_{3n}, \text{ say,}$$

where the function $\eta(l)$ satisfies

$$\eta(l) \equiv \begin{cases} l, & \text{if } l>0, \\ 0, & \text{if } h-m\leq l\leq 0, \\ -l+h-m, & \text{if } -(n-h)+1\leq l\leq h-m. \end{cases}$$

For more details see [61, p. 326]. Given Assumption 9 and Lemma 3(vii), we have $|V_{2n}| \leq C(J+1)n^{-1}$ and $|V_{3n}| \leq C(J+1)n^{-1}$. For the first term $V_{1n}$, we can write

$$V_{1n} = \sum_{h=1-n}^{n-1}b_J(h,h)n^{-1}\sum_{l=-\infty}^{\infty}(1-|l|/n)R_e^2(l)+\sum_{h}\sum_{|r|=1}^{n-1}b_J(h,h+r)n^{-1}\sum_{l=-\infty}^{\infty}R_e(l)R_e(l+r),$$
$$= n^{-1}(2^{J+1}-1)\sum_{h=-\infty}^{\infty}R_e^2(h)+O[(J+1)/n],$$

where we have used Lemma 3(v) for the first term, which corresponds to $h=m$; the second term corresponds to $h\neq m$ and it is $O[(J+1)/T]$ given $\sum_{h=-\infty}^{\infty}|R(h)| \leq C$ and Lemma 3(v). It follows that as $J\to\infty$

$$E[Q(\tilde{f},E\tilde{f})] = \frac{2^{J+1}}{n}\int_{-\pi}^{\pi}f_e^2(\omega)d\omega+o(2^J/n). \quad (57)$$

Collecting (55)–(57) and $J \to \infty$, we obtain

$$E[Q(\hat{f}, f)] = \frac{2^{J+1}}{n} \int_{-\pi}^{\pi} f_e^2(\omega) d\omega + 2^{-2qJ} \vartheta_q \int_{-\pi}^{\pi} [f_e^{(q)}(\omega)]^2 d\omega + o(2^J/n + 2^{-2qJ}).$$

This shows the Theorem. □

*Proof* (Corollary 2) The result follows immediately from Theorem 6 because Assumption 9 implies $2^{\hat{J}}/2^J - 1 = o_P(T^{-1/2(2q+1)}) = o_P(2^{-J/2})$, where the non-stochastic finest scale $J$ is given by $2^{J+1} \equiv \max\{[2\alpha\vartheta_q \zeta_0(q) T]^{1/(2q+1)}, 0\}$. The latter satisfies the conditions of Theorem 6. □

## References


1. Admati, A. R., & Pfleiderer, P. (1988). A theory of intraday patterns: Volume and price variability. *Review of Financial Studies*, *1*, 3–40.
2. Andrews, D. W. K. (1991). Heteroskedasticity and autocorrelation consistent covariance matrix estimation. *Econometrica*, *59*, 817–858.
3. Andrews, D. W. K. (2001). Testing when a parameter is on the boundary of the maintained hypothesis. *Econometrica*, *69*, 683–734.
4. Bauwens, L., & Giot, P. (2000). The logarithmic ACD model: An application to the bid-ask quote process of three NYSE stocks. *Annals of Economics and Statistics*, *60*, 117–149.
5. Bauwens, L., & Giot, P. (2001). *Econometric modelling of stock market intraday activity. Advances studies in theoretical and applied econometrics*. Boston: Kluwer.
6. Bauwens, L., & Giot, P. (2003). Asymmetric ACD Models: Introducing price information in ACD models. *Empir. Econ.*, *28*, 709–731.
7. Beltrao, K., & Bloomfield, P. (1987). Determining the bandwidth of a kernel spectrum estimate. *Journal of Time Series Analysis*, *8*, 21–38.
8. Bera, A. K., & Higgins, M. L. (1992). A test for conditional heteroskedasticity in time series models. *Journal of Time Series Analysis*, *13*, 501–519.
9. Bizer, D. S., & Durlauf, S. N. (1990). Testing the positive theory of government finance. *Journal of Monetary Economics*, *26*, 123–141.
10. Box, G. E. P., & Pierce, D. (1970). Distribution of residual autocorrelations in autoregressive integrated moving average time series models. *Journal of the American Statistical Association*, *65*, 1509–1526.
11. Bühlmann, P. (1996). Locally adaptive lag-window spectral estimation. *Journal of Time Series Analysis*, *17*, 247–270.
12. Daubechies, I. (1992). Ten lectures on wavelets. In *CBS-NSF regional conferences in applied mathematics* (Vol. 61). Philadelphia: Society for Industrial and Applied Mathematics.
13. Drost, F. C., & Werker, B. J. M. (2000). *Efficient estimation in semiparametric time series: The ACD model*. Working paper, Tilburg University.
14. Drost, F. C., & Werker, B. J. M. (2004). Semiparametric duration models. *Journal of Business & Economic Statistics*, *22*, 40–50.
15. Duchesne, P. (2006a). On testing for serial correlation with a wavelet-based spectral density estimator in multivariate time series. *Econometric Theory*, *22*, 633–676.
16. Duchesne, P. (2006b). Testing for multivariate autoregressive conditional heteroskedasticity using wavelets. *Computational Statistics & Data Analysis*, *51*, 2142–2163.
17. Duchesne, P., Li, L., & Vandermeerschen, J. (2010). On testing for serial correlation of unknown form using wavelet thresholding. *Computational Statistics & Data Analysis*, *54*, 2512–2531.

18. Duchesne, P., & Pacurar, M. (2008). Evaluating financial time series models for irregularly spaced data: A spectral density approach. *Computers & Operations Research (Special Issue: Applications of OR in Finance), 35*, 130–155.
19. Easley, D., & O'Hara, M. (1992). Time and the process of security price adjustment. *The Journal of Finance, 19*, 69–90.
20. Engle, R. F. (2000). The econometrics of ultra-high frequency data. *Econometrica, 68*, 1–22.
21. Engle, R. F., & Russell, J. R. (1997). Forecasting the frequency of changes in quoted foreign exchange prices with the autoregressive conditional duration model. *Journal of Empirical Finance, 4*, 187–212.
22. Engle, R. F., & Russell, J. R. (1998). Autoregressive conditional duration: A new model for irregularly spaced transaction data. *Econometrica, 66*, 1127–1162.
23. Escanciano, J. C. (2006). Goodness-of-fit tests for linear and nonlinear time series models. *Journal of the American Statistical Association, 101*, 531–541.
24. Fan, Y., & Gençay, R. (2010). Unit root tests with wavelets. *Econometric Theory, 26*, 1305–1331.
25. Francq, C., Roy, R., & Zakoïan, J.-M. (2005). Diagnostic checking in ARMA models with uncorrelated errors. *Journal of the American Statistical Association, 100*, 532–544.
26. Francq, C., & Zakoïan, J.-M. (2009). Testing the nullity of GARCH coefficients: Correction of the standard tests and relative efficiency comparisons. *Journal of the American Statistical Association, 104*, 313–324.
27. Fernandes, M., & Grammig, J. (2005). Non-parametric specification tests for conditional duration models. *Journal of Econometrics, 127*, 35–68.
28. Gao, H. (1993). *Wavelet estimation of spectral densities in time series analysis*. Ph.D Dissertation, Department of Statistics, University of California, Berkeley.
29. Gençay, R., & Signori, D. (2012). *Multi-scale tests for serial correlation*. Technical report, Department of Economics, Simon Fraser University.
30. Gençay, R., Yazgan, E., & Ozkan, H. (2012). *A test of structural change of unknown location with wavelets*. Technical report, Department of Economics, Simon Fraser University.
31. Ghysels, E., & Jasiak, J. (1994). *Stochastic volatility and time deformation: an application to trading volume and leverage effects*. Working paper, C.R.D.E., Université de Montréal.
32. Ghysels, E., & Jasiak, J. (1998). *Long-term dependence in trading*. Working paper, Dept. of Economics, Penn State University and York University.
33. Ghysels, E., & Jasiak, J. (1998). GARCH for irregularly spaced financial data: The ACD-GARCH model. *Studies in Nonlinear Dynamics and Econometrics, 2*, 133–149.
34. Grammig, J., Hujer, R., Kokot, S., & Maurer, K.-O. (1998). *Modeling the Deutsche Telekom IPO using a new ACD specification, an application of the Burr-ACD model using high frequency IBIS data*. Working paper, Department of Economics, Johann Wolfgang Goethe-University of Frankfurt.
35. Granger, C. (1966). The typical spectral shape of an economic variable. *Econometrica, 34*, 150–161.
36. Granger, C. W. J., & Newbold, P. (1986). *Forecasting economic time series* (2nd ed.). New York: Academic Press.
37. Hall, P., & Heyde, C. C. (1980). *Martingale limit theory and its application*. New York: Academic Press.
38. Hannan, E. (1970). *Multiple time series*. New York: Wiley.
39. Hautsch, N. (2006). *Testing the conditional mean function of autoregressive conditional duration models*. Working Paper, University of Copenhagen.
40. Hernandez, E., & Weiss, G. (1996). *A first course on wavelets*. New York: CRC Press.
41. Higgins, M. L. (2000). Testing for autoregressive conditional duration. Presented at World Congress of Econometric Society, Seattle.
42. Hong, Y. (2001). *Wavelet-based estimation for heteroskedasticity and autocorrelation consistent variance-covariance matrices*. Working paper, Department of Economics and Department of Statistical Science, Cornell University.

43. Hong, Y., & Kao, C. (2004). Wavelet-based testing for serial correlation of unknown form in panel models. *Econometrica*, *72*, 1519–1563.
44. Hong, Y., & Lee, J. (2001). One-sided testing for ARCH effect using wavelets. *Econometric Theory*, *17*, 1051–1081.
45. Hong, Y., & Lee, Y.-J. (2011). Detecting misspecifications in autoregressive conditional duration models and non-negative time-series processes. *Journal of Time Series Analysis*, *32*, 1–32.
46. Jasiak, J. (1999). Persistence in intertrade durations. *Finance*, *19*, 166–195.
47. Jensen, M. J. (2000). An alternative maximum likelihood estimator of long memory processes using compactly supported wavelets. *Journal of Economic Dynamics and Control*, *24*, 361–387.
48. Kyle, A. (1985). Continuous time auctions and insider trading. *Econometrica*, *53*, 1315–1336.
49. Lee, J. H. H., & King, M. L. (1993). A locally most mean powerful based score test for ARCH and GARCH regression disturbances. *Journal of Business & Economic Statistics, 11*, 17–27 (Correction 1994, 12, p. 139).
50. Lee, J., & Hong, Y. (2001). Testing for serial correlation of unknown form using wavelet methods. *Econometric Theory*, *17*, 386–423.
51. Li, L., Yao, S., & Duchesne, P. (2014). On wavelet-based testing for serial correlation of unknown form using Fan's adaptive Neyman method. *Computational Statistics & Data Analysis*, *70*, 308–327.
52. Li, W. K. (2004). *Diagnostic checks in time series*. New York: Chapman & Hall/CRC.
53. Li, W. K., & Mak, T. K. (1994). On the square residual autocorrelations in non-linear time series with conditional heteroskedasticity. *Journal of Time Series Analysis*, *15*, 627–636.
54. Li, W. K., & Yu, L. H. (2003). On the residual autocorrelation of the autoregressive conditional duration model. *Economics Letters*, *79*, 169–175.
55. Mallat, S. G. (1989). A theory for multiresolution signal decomposition: The wavelet representation. *IEEE Transactions on Pattern Analysis and Machine Intelligence*, *11*, 674–693.
56. Meitz, M., & Teräsvirta, T. (2006). Evaluating models of autoregressive conditional duration. *Journal of Business & Economic Statistics*, *24*, 104–124.
57. Neumann, M. H. (1996). Spectral density estimation via nonlinear wavelet methods for stationary non-Gaussian time series. *Journal of Time Series Analysis*, *17*, 601–633.
58. Newey, W. K., & West, K. D. (1994). Automatic lag selection in covariance matrix estimation. *Review of Financial Studies*, *61*, 631–653.
59. Pacurar, M. (2008). Autoregressive conditional duration models in finance: A survey of the theoretical and empirical literature. *Journal of Economic Surveys*, *22*, 711–751.
60. Percival, D. B., & Walden, A. T. (2000). *Wavelet methods for time series analysis, Cambridge Series in Statistical and Probabilistic Mathematics*. Cambridge: Cambridge University Press.
61. Priestley, M. B. (1981). *Spectral analysis and time series*. London: Academic Press.
62. Priestley, M. B. (1996). Wavelets and time-dependent spectral analysis. *Journal of Time Series Analysis*, *17*, 85–103.
63. Ramsey, J. (1999). The contribution of wavelets to the analysis of economic and financial data. *Philosophical Transactions of the Royal Society A: Mathematical, Physical and Engineering*, *537*, 2593–2606.
64. Robinson, P. M. (1991). Automatic frequency domain inference on semiparametric and nonparametric models. *Econometrica*, *59*, 1329–1363.
65. Stock, J. (1988). Estimating continuous time processes subject to time deformation. *Journal of the American Statistical Association*, *83*, 77–85.
66. Tauchen, G., & Pitts, M. (1983). The price variability-volume relationship on speculative markets. *Econometrica*, *51*, 485–505.
67. Tsay, R. S. (2005). *Analysis of financial time series* (2nd ed.). New York: Wiley.
68. Tsay, R. S. (2013). *An introduction to analysis of financial data with R*. New York: Wiley.
69. Vidakovic, B. (1999). *Statistical modeling by wavelets*. New York: Wiley.
70. Wang, Y. (1995). Jump and sharp cusp detection by wavelets. *Biometrika*, *82*, 385–397.
71. Watson, N. W. (1993). Measures of fit for calibrated models. *Journal of Political Economy*, *101*, 1011–1041.

72. Xue, Y., Gençay, R., & Fagan, S. (2010). *Jump detection with wavelets*. Technical report, Department of Economics, Simon Fraser University.
73. Zhang, M. Y., Russell, J. R., & Tsay, R. S. (2001). A nonlinear autoregressive conditional duration model with applications to financial transaction data. *Journal of Econometrics*, *104*, 179–207.
74. Zhu, K., & Li, W. K. (2015). A bootstrapped spectral test for adequacy in weak ARMA models. *Journal of Econometrics*, *187*, 113–130.

# Diagnostic Checking for Weibull Autoregressive Conditional Duration Models

**Yao Zheng, Yang Li, Wai Keung Li and Guodong Li**

**Abstract** We derive the asymptotic distribution of residual autocorrelations for the Weibull autoregressive conditional duration (ACD) model, and this leads to a portmanteau test for the adequacy of the fitted Weibull ACD model. The finite-sample performance of this test is evaluated by simulation experiments and a real data example is also reported.

**Keywords** Autoregressive conditional duration model · Weibull distribution · Model diagnostic checking · Residual autocorrelation

**Mathematics Subject Classification (2010)** Primary 62M10 · 91B84; Secondary 37M10

## 1 Introduction

First proposed by Engle and Russell [3], the autoregressive conditional duration (ACD) model has become very popular in the modeling of high-frequency financial data. ACD models are applied to describe the duration between trades for a frequently traded stock such as IBM and it provides useful information on the intraday market activity. Note that the ACD model for durations is analogous to the commonly used generalized autoregressive conditional heteroscedastic (GARCH) model [1, 2] for stock returns. Driven by the strong similarity between the ACD and GARCH models,

Y. Zheng (✉) · Y. Li · W.K. Li · G. Li
Department of Statistics and Actuarial Science,
University of Hong Kong, Pokfulam Road, Hong Kong
e-mail: zheng.yao@hku.hk

Y. Li
e-mail snliyang@connect.hku.hk

W.K. Li
e-mail: hrntlwk@hku.hk

G. Li
e-mail: gdli@hku.hk

 
W.K. Li et al. (eds.), *Advances in Time Series Methods and Applications*,
Fields Institute Communications 78, DOI 10.1007/978-1-4939-6568-7_4

various extensions to the original ACD model of Engle and Russell [3] have been suggested. However, despite the great variety of ACD specifications, the question of model diagnostic checking has received less attention.

The approach used by Engle and Russell [3] and widely adopted by subsequent authors to assess the adequacy of the estimated ACD model consists of applying the Ljung–Box Q-statistic [7] to the residuals from the fitted time series model and to its squared sequence. The latter case is commonly known as the McLeod–Li test [8]. As pointed out by Li and Mak [5] in the context of GARCH models, this approach is questionable, because this test statistic does not have the usual asymptotic chi-square distribution under the null hypothesis when it is applied to residuals of an estimated GARCH model. Following Li and Mak [5], Li and Yu [6] derived a portmanteau test for the goodness-of-fit of the fitted ACD model when the errors follow the exponential distribution.

In this paper, we consider a portmanteau test for checking the adequacy of the fitted ACD model when the errors have a Weibull distribution. This paper has similarities to [6] since the two papers both follow the approach by Li and Mak [5] to construct the portmanteau test statistic. Besides the difference in the distribution of the error term, the functional form of the ACD model in the present paper is more general than that of [6], because the latter only discusses the ACD model with an ARCH-like form of the conditional mean duration.

The remainder of this paper is organized as follows. Section 2 presents the portmanteau test for the Weibull ACD model estimated by the maximum likelihood method. In Sect. 3, two Monte Carlo simulations are performed to study the finite-sample performance of the diagnostic tool and an illustrative example is reported to demonstrate its usefulness.

## 2 A Portmanteau Test

### 2.1 Basic Definitions and the ML Estimation

Consider the autoregressive conditional duration (ACD) model,

$$x_i = \psi_i \varepsilon_i, \qquad \psi_i = \omega + \sum_{j=1}^{p} \alpha_j x_{i-j} + \sum_{j=1}^{q} \beta_j \psi_{i-j}, \tag{1}$$

where $t_0 < t_1 < \cdots < t_n < \cdots$ are arrival times, $x_i = t_i - t_{i-1}$ is an interval, $\omega > 0$, $\alpha_j \geq 0, \beta_j \geq 0$, and the innovations $\{\varepsilon_t\}$ are identically and independently distributed ($i.i.d.$) nonnegative random variables with mean one [3].

For ACD model at (1), we assume that the innovation $\varepsilon_i$ has the density of a standardized Weibull distribution,

$$f_\gamma(x) = \gamma c_\gamma x^{\gamma-1}\exp\{-c_\gamma x^\gamma\}, \qquad x \geq 0,$$

where $c_\gamma = [\Gamma(1+\gamma^{-1})]^\gamma$, $\Gamma(\cdot)$ is the Gamma function, and $E(\varepsilon_i) = 1$. The Weibull distribution has a decreasing (increasing) hazard function if $\gamma < 1$ ($\gamma > 1$) and reduces to the standard exponential distribution if $\gamma = 1$. We denote this model by WACD$(p, q)$ in this paper.

Let $\boldsymbol{\alpha} = (\alpha_1, \ldots, \alpha_p)'$, $\boldsymbol{\beta} = (\beta_1, \ldots, \beta_q)'$ and $\boldsymbol{\theta} = (\omega, \boldsymbol{\alpha}', \boldsymbol{\beta}')'$. Denote by $\boldsymbol{\lambda} = (\gamma, \boldsymbol{\theta}')'$ the parameter vector of the Weibull ACD model, and its true value $\boldsymbol{\lambda}_0 = (\gamma_0, \boldsymbol{\theta}_0')'$ is an interior point of a compact set $\Lambda \subset \mathbb{R}^{p+q+2}$. The following assumption gives some constraints on the parameter space $\Lambda$.

**Assumption 1** $\omega > 0$, $\alpha_j > 0$ for $1 \leq j \leq p$, $\beta_j > 0$ for $1 \leq j \leq q$, $\sum_{j=1}^p \alpha_j + \sum_{j=1}^q \beta_j < 1$, and Polynomials $\sum_{j=1}^p \alpha_j x^j$ and $1 - \sum_{j=1}^q \beta_j x^j$ have no common root.

Given nonnegative observations $x_1, \ldots, x_n$, the log-likelihood function of the Weibull ACD model is

$$\begin{aligned} L_n(\boldsymbol{\lambda}) &= \sum_{i=1}^n \left\{ \log f_\gamma\left(\frac{x_i}{\psi_i(\boldsymbol{\theta})}\right) - \log \psi_i(\boldsymbol{\theta}) \right\} \\ &= \sum_{i=1}^n \left\{ -\gamma \log[\psi_i(\boldsymbol{\theta})] - c_\gamma \left[\frac{x_i}{\psi_i(\boldsymbol{\theta})}\right]^\gamma \right\} + (\gamma - 1)\sum_{i=1}^n \log(x_i) + n\log(\gamma \cdot c_\gamma). \end{aligned}$$

Note that the above functions all depend on unobservable values of $x_i$ with $i \leq 0$, and some initial values are hence needed for $x_0, x_{-1}, \ldots, x_{1-p}$ and $\psi_0(\boldsymbol{\theta}), \psi_{-1}(\boldsymbol{\theta}), \ldots, \psi_{1-q}(\boldsymbol{\theta})$. We simply set them to be $\bar{x} = n^{-1}\sum_{i=1}^n x_i$, and denote the corresponding functions respectively by $\widetilde{\psi}_i(\boldsymbol{\theta})$ and $\widetilde{L}_n(\boldsymbol{\lambda})$. Thus, the MLE can be defined as

$$\widetilde{\boldsymbol{\lambda}}_n = (\widetilde{\gamma}_n, \widetilde{\boldsymbol{\theta}}_n')' = \underset{\boldsymbol{\lambda} \in \Lambda}{\operatorname{argmax}}\, \widetilde{L}_n(\boldsymbol{\lambda}).$$

Let

$$c_1(x, \gamma) = -\frac{\partial \log f_\gamma(x)}{\partial x}x - 1 = -\gamma(1 - c_\gamma x^\gamma)$$

and

$$c_2(x, \gamma) = \frac{\partial \log f_\gamma(x)}{\partial \gamma} = -c_\gamma x^\gamma \log(x) + \log(x) - c_\gamma' x^\gamma + \gamma^{-1} + c_\gamma'/c_\gamma,$$

where $c_\gamma' = \partial c_\gamma/\partial\gamma$. It can be verified that $E[c_1(\varepsilon_i, \gamma_0)] = 0$ and $E[c_2(\varepsilon_i, \gamma_0)] = 0$. Denote $\kappa_1 = \text{var}[c_1(\varepsilon_i, \gamma_0)]$, $\kappa_2 = \text{var}[c_2(\varepsilon_i, \gamma_0)]$, $\kappa_3 = \text{cov}[c_1(\varepsilon_i, \gamma_0), c_2(\varepsilon_i, \gamma_0)]$ and

$$\Sigma = \begin{pmatrix} \kappa_2 & \kappa_3 E[\psi_i^{-1}(\boldsymbol{\theta}_0)\partial\psi_i(\boldsymbol{\theta}_0)/\partial\boldsymbol{\theta}'] \\ \kappa_3 E[\psi_i^{-1}(\boldsymbol{\theta}_0)\partial\psi_i(\boldsymbol{\theta}_0)/\partial\boldsymbol{\theta}] & \kappa_1 E\{\psi_i^{-2}(\boldsymbol{\theta}_0)[\partial\psi_i(\boldsymbol{\theta}_0)/\partial\boldsymbol{\theta}][\partial\psi_i(\boldsymbol{\theta}_0)/\partial\boldsymbol{\theta}']\} \end{pmatrix}.$$

If Assumption 1 holds, then $\widetilde{\boldsymbol{\lambda}}_n$ converges to $\boldsymbol{\lambda}_0$ in almost surely sense as $n \to \infty$, and $\sqrt{n}(\widetilde{\boldsymbol{\lambda}}_n - \boldsymbol{\lambda}_0) \to_d N(0, \Sigma^{-1})$ as $n \to \infty$; see Engle and Russell [3] and Francq and Zakoian [4].

Denote by $\{\widetilde{\varepsilon}_i\}$ the residual sequence from the fitted Weibull ACD model, where $\widetilde{\varepsilon}_i = x_i/\widetilde{\psi}_i(\widetilde{\boldsymbol{\theta}}_n)$. For the quantities in the information matrix $\Sigma$, $\kappa_1$, $\kappa_2$, $\kappa_3$, $E[\psi_i^{-1}(\boldsymbol{\theta}_0)\partial\psi_i(\boldsymbol{\theta}_0)/\partial\boldsymbol{\theta}]$, and $E[\psi_i^{-2}(\boldsymbol{\theta}_0)(\partial\psi_i(\boldsymbol{\theta}_0)/\partial\boldsymbol{\theta})(\partial\psi_i(\boldsymbol{\theta}_0)/\partial\boldsymbol{\theta}')]$, we can estimate them respectively by

$$\widetilde{\kappa}_1 = \frac{1}{n}\sum_{i=1}^{n}[c_1(\widetilde{\varepsilon}_i, \widetilde{\gamma}_n)]^2, \quad \widetilde{\kappa}_2 = \frac{1}{n}\sum_{i=1}^{n}[c_2(\widetilde{\varepsilon}_i, \widetilde{\gamma}_n)]^2, \quad \widetilde{\kappa}_3 = \frac{1}{n}\sum_{i=1}^{n}c_1(\widetilde{\varepsilon}_i, \widetilde{\gamma}_n)c_2(\widetilde{\varepsilon}_i, \widetilde{\gamma}_n),$$

$$\frac{1}{n}\sum_{i=1}^{n}\frac{1}{\widetilde{\psi}_i(\widetilde{\boldsymbol{\theta}}_n)}\frac{\partial\widetilde{\psi}_i(\widetilde{\boldsymbol{\theta}}_n)}{\partial\boldsymbol{\theta}} \quad \text{and} \quad \frac{1}{n}\sum_{i=1}^{n}\frac{1}{\widetilde{\psi}_i^2(\widetilde{\boldsymbol{\theta}}_n)}\frac{\partial\widetilde{\psi}_i(\widetilde{\boldsymbol{\theta}}_n)}{\partial\boldsymbol{\theta}}\frac{\partial\widetilde{\psi}_i(\widetilde{\boldsymbol{\theta}}_n)}{\partial\boldsymbol{\theta}'}.$$

The above estimators are all consistent, and hence a consistent estimator of the information matrix $\Sigma$. Moreover,

$$\sqrt{n}(\widetilde{\boldsymbol{\theta}}_n - \boldsymbol{\theta}_0) \to_d N(0, \Sigma_1^{-1}) \quad \text{as } n \to \infty, \tag{2}$$

where

$$\Sigma_1 = \kappa_1 \cdot E\left[\frac{1}{\psi_i^2(\boldsymbol{\theta}_0)}\frac{\partial\psi_i(\boldsymbol{\theta}_0)}{\partial\boldsymbol{\theta}}\frac{\partial\psi_i(\boldsymbol{\theta}_0)}{\partial\boldsymbol{\theta}'}\right] - \frac{\kappa_3^2}{\kappa_2}\cdot E\left[\frac{1}{\psi_i(\boldsymbol{\theta}_0)}\frac{\partial\psi_i(\boldsymbol{\theta}_0)}{\partial\boldsymbol{\theta}}\right]E\left[\frac{1}{\psi_i(\boldsymbol{\theta}_0)}\frac{\partial\psi_i(\boldsymbol{\theta}_0)}{\partial\boldsymbol{\theta}'}\right].$$

## 2.2 The Main Result

This subsection derives asymptotic distributions of the residual autocorrelations from the estimated Weibull ACD model, and hence a portmanteau test for checking the adequacy of this model. Note that the residuals are nonnegative, and the residual autocorrelations here are also the absolute residual autocorrelations.

Without confusion, we denote $\widetilde{\psi}_i(\widetilde{\boldsymbol{\theta}}_n)$ and $\psi_i(\boldsymbol{\theta}_0)$ respectively by $\widetilde{\psi}_i$ and $\psi_i$ for simplicity. Consider the residual sequence $\{\widetilde{\varepsilon}_i\}$ with $\widetilde{\varepsilon}_i = x_i/\widetilde{\psi}_i$. Note that $n^{-1}\sum_{i=1}^{n}\widetilde{\varepsilon}_i = 1 + o_p(1)$ and then, for a positive integer $k$, the lag-$k$ residual autocorrelation can be defined as

$$\widetilde{r}_k = \frac{\sum_{i=k+1}^{n}(\widetilde{\varepsilon}_i - 1)(\widetilde{\varepsilon}_{i-k} - 1)}{\sum_{i=1}^{n}(\widetilde{\varepsilon}_i - 1)^2}.$$

We next consider the asymptotic distributions of the first $K$ residual autocorrelations, $\widetilde{R} = (\widetilde{r}_1, \ldots, \widetilde{r}_K)'$, where $K$ is a predetermined positive integer.

Denote $\widetilde{\psi}_i(\widetilde{\boldsymbol{\theta}}_n)$ and $\psi_i(\boldsymbol{\theta}_0)$ respectively by $\widetilde{\psi}_i$ and $\psi_i$, and let $\widetilde{\varepsilon}_i = x_i/\widetilde{\psi}_i$. Let $\widetilde{C} = (\widetilde{C}_1, \ldots, \widetilde{C}_K)'$ and $C = (C_1, \ldots, C_K)'$, where

$$\widetilde{C}_k = \frac{1}{n}\sum_{i=k+1}^{n}(\widetilde{\varepsilon}_i - 1)(\widetilde{\varepsilon}_{i-k} - 1) \quad \text{and} \quad C_k = \frac{1}{n}\sum_{i=k+1}^{n}(\varepsilon_i - 1)(\varepsilon_{i-k} - 1).$$

By the $\sqrt{n}$-consistency of $\widetilde{\boldsymbol{\theta}}_n$ at (2) and the ergodic theorem, it follows that $n^{-1}\sum_{i=1}^{n}(\widetilde{\varepsilon}_i - 1)^2 = \sigma_{\gamma_0}^2 + o_p(1)$, where $\sigma_{\gamma_0}^2 = \text{var}(\varepsilon_i)$, and thus it suffices to derive the asymptotic distribution of $\widetilde{C}$.

By the Taylor expansion, it holds that

$$\widetilde{C} = C + H'(\widetilde{\boldsymbol{\theta}}_n - \boldsymbol{\theta}_0) + o_p(n^{-1/2}), \tag{3}$$

where $H = (H_1, \ldots, H_K)$ with $H_k = -E[\psi_i^{-1}(\varepsilon_{i-k} - 1)\partial\psi_i/\partial\boldsymbol{\theta}]$. Moreover,

$$\sqrt{n}(\widetilde{\boldsymbol{\theta}}_n - \boldsymbol{\theta}_0) = A\Sigma^{-1} \cdot \frac{1}{\sqrt{n}}\sum_{i=1}^{n}\left[c_2(\varepsilon_i, \gamma_0), \frac{c_1(\varepsilon_i, \gamma_0)}{\psi_i}\frac{\partial\psi_i}{\partial\boldsymbol{\theta}'}\right]' + o_p(1), \tag{4}$$

where the $c_j(\varepsilon_i, \gamma_0)$ is as defined in Sect. 2.1, and the matrix $A = (0, \mathbf{I})$ with $\mathbf{I}$ being the $(p+q+1)$-dimensional identity matrix. Note that $E[\varepsilon_i c_2(\varepsilon_i, \gamma_0)] = 0$ and $E[\varepsilon_i c_1(\varepsilon_i, \gamma_0)] = 1$. By (3), (4), the central limit theorem and the Cramér-Wold device, it follows that

$$\sqrt{n}\widetilde{R} \rightarrow_d N(0, \Omega) \quad \text{as } n \rightarrow \infty,$$

where $\Omega = \mathbf{I} - \sigma_{\gamma_0}^{-4}H'\Sigma_1^{-1}H$, $\sigma_{\gamma_0}^2 = \text{var}(\varepsilon_i)$, $H = (H_1, \ldots, H_K)$ with $H_k = -E[\psi_i^{-1}(\varepsilon_{i-k} - 1)\partial\psi_i/\partial\boldsymbol{\theta}]$, and $\Sigma_1$ is as defined in Sect. 2.1.

Let $\widetilde{\sigma}_{\gamma_0}^2 = n^{-1}\sum_{i=1}^{n}(\widetilde{\varepsilon}_i - 1)^2$, $\widetilde{H}_k = -n^{-1}\sum_{i=1}^{n}\widetilde{\psi}_i^{-1}(\widetilde{\varepsilon}_{i-k} - 1)\partial\widetilde{\psi}_i/\partial\boldsymbol{\theta}$ and $\widetilde{H} = (\widetilde{H}_1, \ldots, \widetilde{H}_K)$. Then we have $\widetilde{H} = H + o_p(1)$ and hence a consistent estimator of $\Omega$ can be constructed, denoted by $\widetilde{\Omega}$. Let $\widetilde{\Omega}_{kk}$ be the diagonal elements of $\widetilde{\Omega}$, for $1 \leq k \leq K$. We therefore can check the significance of $\widetilde{r}_k$ by comparing its absolute value with $1.96\sqrt{\widetilde{\Omega}_{kk}/n}$, where the significance level is 5 %.

To check the significance of $\widetilde{R} = (\widetilde{r}_1, \ldots, \widetilde{r}_K)'$ jointly, we can construct a portmanteau test statistic,

$$Q(K) = n\widetilde{R}'\widetilde{\Omega}^{-1}\widetilde{R},$$

and it will be asymptotically distributed as $\chi_K^2$, the chi-square distribution with $K$ degrees of freedom.

## 3 Numerical Studies

### 3.1 Simulation Experiments

This subsection conducts two Monte Carlo simulation experiments to check the finite-sample performance of the proposed portmanteau test in the previous section.

The first experiment evaluates the sample approximation for the asymptotic variance of residual autocorrelations $\Omega$, and the data generating process is

$$x_i = \psi_i \varepsilon_i, \qquad \psi_i = 0.1 + \alpha x_{i-1} + \beta \psi_{i-1},$$

where $\varepsilon_i$ follows the standardized Weibull distribution with the parameter of $\gamma$. We consider $\gamma = 0.8$ and 1.2, corresponding to a heavy-tailed distribution and a light-tailed one, and $(\alpha, \beta)' = (0.2, 0.6)'$ and $(0.4, 0.5)'$. The sample size is set to $n = 200$, 500 or 1000, and there are 1000 replications for each sample size. As shown in Table 1, the asymptotic standard deviations (ASDs) of the residual autocorrelations at lags 2, 4 and 6 are close to their corresponding empirical standard deviations (ESDs) when the sample size is as small as $n = 500$.

In the second experiment, we check the size and power of the proposed portmanteau test $Q(K)$ using the data generating process,

$$x_i = \psi_i \varepsilon_i, \qquad \psi_i = 0.1 + 0.3x_{i-1} + \alpha_2 x_{i-2} + 0.3\psi_{i-1},$$

where $\alpha_2 = 0$, 0.15 or 0.3, and $\varepsilon_i$ follows the standardized Weibull distribution with $\gamma = 0.8$ or 1.2. All the other settings are preserved from the previous experiment. We fit the model of orders (1, 1) to the generated data; hence, the case with $\alpha_2 = 0$ corresponds to the size and those with $\alpha_2 > 0$ to the power. The rejection rates of test statistic $Q(K)$ with $K = 6$ are given in Table 2. For comparison, the corresponding rejection rates of the Ljung–Box statistics for the residual series and its squared process are also reported, denoted by $Q_1^*(K)$ and $Q_2^*(K)$. The critical value is the upper 5th percentile of the $\chi_6^2$ distribution for all these tests. As shown in the table,

**Table 1** Empirical standard deviations (ESD) and asymptotic standard deviations (ASD) of residual autocorrelations at lags 2, 4 and 6

| | $n$ | | $\boldsymbol{\theta} = (0.1, 0.2, 0.6)'$ | | | $\boldsymbol{\theta} = (0.1, 0.4, 0.5)'$ | | |
|---|---|---|---|---|---|---|---|---|
| | | | 2 | 4 | 6 | 2 | 4 | 6 |
| $\gamma = 0.8$ | 200 | ESD | 0.1025 | 0.1061 | 0.1065 | 0.0610 | 0.0660 | 0.0635 |
| | | ASD | 0.0605 | 0.0655 | 0.0673 | 0.0625 | 0.0658 | 0.0675 |
| | 500 | ESD | 0.0402 | 0.0415 | 0.0431 | 0.0389 | 0.0419 | 0.0416 |
| | | ASD | 0.0387 | 0.0411 | 0.0424 | 0.0402 | 0.0418 | 0.0427 |
| | 1000 | ESD | 0.0284 | 0.0289 | 0.0301 | 0.0280 | 0.0297 | 0.0305 |
| | | ASD | 0.0277 | 0.0291 | 0.0298 | 0.0285 | 0.0297 | 0.0301 |
| $\gamma = 1.2$ | 200 | ESD | 0.0847 | 0.0862 | 0.0889 | 0.0632 | 0.0656 | 0.0658 |
| | | ASD | 0.0604 | 0.0652 | 0.0673 | 0.0629 | 0.0659 | 0.0674 |
| | 500 | ESD | 0.0386 | 0.0414 | 0.0421 | 0.0395 | 0.0433 | 0.0410 |
| | | ASD | 0.0387 | 0.0409 | 0.0422 | 0.0401 | 0.0418 | 0.0426 |
| | 1000 | ESD | 0.0277 | 0.0290 | 0.0296 | 0.0276 | 0.0301 | 0.0292 |
| | | ASD | 0.0276 | 0.0289 | 0.0297 | 0.0284 | 0.0296 | 0.0301 |

**Table 2** Rejection rates of the test statistics $Q(K)$, $Q_1^*(K)$ and $Q_2^*(K)$ with $K = 6$ and $\gamma = 0.8$ or 1.2

| | $n$ | $\alpha_2 = 0$ | | $\alpha_2 = 0.15$ | | $\alpha_2 = 0.3$ | |
|---|---|---|---|---|---|---|---|
| | | 0.8 | 1.2 | 0.8 | 1.2 | 0.8 | 1.2 |
| $Q(K)$ | 200 | 0.101 | 0.107 | 0.110 | 0.131 | 0.196 | 0.305 |
| | 500 | 0.085 | 0.089 | 0.147 | 0.172 | 0.414 | 0.633 |
| | 1000 | 0.080 | 0.092 | 0.205 | 0.314 | 0.709 | 0.934 |
| $Q_1^*(K)$ | 200 | 0.021 | 0.022 | 0.041 | 0.052 | 0.133 | 0.207 |
| | 500 | 0.013 | 0.018 | 0.076 | 0.082 | 0.329 | 0.558 |
| | 1000 | 0.016 | 0.008 | 0.115 | 0.203 | 0.639 | 0.899 |
| $Q_2^*(K)$ | 200 | 0.046 | 0.022 | 0.059 | 0.048 | 0.084 | 0.139 |
| | 500 | 0.051 | 0.024 | 0.080 | 0.072 | 0.149 | 0.314 |
| | 1000 | 0.052 | 0.022 | 0.088 | 0.135 | 0.209 | 0.617 |

the test $Q(K)$ is oversized when $n = 1000$, while the other two tests are largely undersized for some $\gamma$. Furthermore, we found that increasing the sample size to 9000 could result in $Q(K)$ having sizes of 0.058 and 0.053 for $\gamma = 0.8$ and 1.2, while the sizes of the other two tests do not become closer to the nominal value even for very large $n$. For the power simulations, it can be seen clearly that $Q(K)$ is the most powerful test among the three and $Q_2^*(K)$ is the least powerful one. Moreover, the powers are interestingly observed to have smaller values when the generated data are heavy-tailed ($\gamma = 0.8$).

## 3.2 An Empirical Example

As an illustrative example, this subsection considers the trade durations of the US IBM stock on fifteen consecutive trading days starting from November 1, 1990. The data are truncated from a larger data set which consists of the diurnally adjusted IBM trade durations data from November 1, 1990, to January 31, 1991, adjusted

**Table 3** Model diagnostic checking results for the adjusted durations for IBM stock traded in first fifteen trading days of November 1990: $p$ values for $Q(K)$, $Q_1^*(K)$ and $Q_2^*(K)$ with $K = 6$, 12 and 18, at the 5 % significance level

| $K$ | $q = 1$ | | | $q = 2$ | | | $q = 3$ | | |
|---|---|---|---|---|---|---|---|---|---|
| | $Q(K)$ | $Q_1^*(K)$ | $Q_2^*(K)$ | $Q(K)$ | $Q_1^*(K)$ | $Q_2^*(K)$ | $Q(K)$ | $Q_1^*(K)$ | $Q_2^*(K)$ |
| 6 | 0.0081 | 0.0123 | 0.4827 | 0.0560 | 0.0938 | 0.3778 | 0.3915 | 0.5010 | 0.5172 |
| 12 | 0.0225 | 0.0233 | 0.4313 | 0.1157 | 0.1372 | 0.3890 | 0.4933 | 0.5427 | 0.5315 |
| 18 | 0.0012 | 0.0022 | 0.0723 | 0.0116 | 0.0190 | 0.0727 | 0.0815 | 0.1200 | 0.1211 |

and analyzed by Tsay [9, Chap. 5]. Focusing on positive durations, we have 12,532 diurnally adjusted observations.

We consider the WACD$(p, q)$ models with $p = 1$ and $q = 1, 2$ or 3. The major interest is on whether the models fit the data adequately. To this end, the $p$ values for $Q(K)$, $Q_1^*(K)$ and $Q_2^*(K)$ with $K = 6$, 12 and 18 at the 5 % significance level are reported in Table 3. It can be seen that the WACD$(1, 3)$ model fits the data adequately according to all the test statistics. The fitted WACD$(1, 1)$ model is clearly rejected by both $Q(K)$ and $Q_1^*(K)$ with $K = 6$, 12 and 18. For the fitted WACD$(1, 2)$ model, both $Q(K)$ and $Q_1^*(K)$ suggest an adequate fit of the data with $K = 6$ or 12, but not with $K = 18$. While for the data, $Q(K)$ and $Q_1^*(K)$ always lead to the same conclusions, the fact that the $p$ value for $Q(K)$ is always smaller than that for $Q_1^*(K)$ confirms that $Q(K)$ is more powerful than $Q_1^*(K)$. In contrast, $Q_2^*(K)$ fails to detect any inadequacy of the fitted WACD models.

**Acknowledgements** We are grateful to the co-editor and two anonymous referees for their valuable comments and constructive suggestions that led to the substantial improvement of this paper.

## References

1. Bollerslev, T. (1986). Generalized autoregressive conditional heteroskedasticity. *Journal of Econometrics, 31*, 307–327.
2. Engle, R. F. (1982). Autoregression conditional heteroscedasticity with estimates of the variance of U.K. inflation. *Econometrica, 50*, 987–1008.
3. Engle, R. F., & Russell, J. R. (1998). Autoregressive conditional duration: A new model for irregularly spaced transaction data. *Econometrica, 66*, 1127–1162.
4. Francq, C., & Zakoian, J. M. (2004). Maximum likelihood estimation of pure GARCH and ARMA-GARCH processes. *Bernoulli, 10*, 605–637.
5. Li, W. K., & Mak, T. K. (1994). On the squared residual autocorrelations in non-linear time series with conditoinal heteroskedasticity. *Journal of Time Series Analysis, 15*, 627–636.
6. Li, W. K., & Yu, P. L. H. (2003). On the residual autocorrelation of the autoregressive conditional duration model. *Economic Letters, 79*, 169–175.
7. Ljung, G. M., & Box, G. E. P. (1978). On a measure of lack of fit in time series models. *Biometrika, 65*, 297–303.
8. McLeod, A. I., & Li, W. K. (1983). Diagnostic checking ARMA time series models using squared residual autocorrelations. *Journal of Time Series Analysis, 4*, 269–273.
9. Tsay, R. S. (2010). *Analysis of financial time series* (3rd ed.). New York: Wiley.

# Diagnostic Checking for Partially Nonstationary Multivariate ARMA Models

**M.T. Tai, Y.X. Yang and S.Q. Ling**

**Abstract** This paper studies the residual autocorrelation functions (ACFs) of partially nonstationary multivariate autoregressive moving-average (ARMA) models. The limiting distributions of the full rank estimators and the Gaussian reduced rank estimators are derived. Using these results, we derive the limiting distributions of the residual ACFs under full rank and reduce rank estimations. Based on these limiting distributions, we construct the portmanteau statistics for model checking. It is shown that these statistics asymptotically follow $\chi^2$-distributions. Simulations are carried out to assess their performances in finite samples and two real examples are given.

**Keywords** Limiting distributions · Autoregressive model · Autoregressive moving-average model · Partially nonstationary · Portmanteau statistics

**Mathematics Subject Classification (2010)** Primary 91B84 · 37M10 · Secondary 62M10

## 1 Introduction

It is well known that model diagnostic checking is an essential and important step in time series modeling. Box and Pierce [2] used the asymptotic distribution of residual autoregressive functions (ACFs) to devise a portmanteau statistic for model checking. More general cases were studied by McLeod [8]. McLeod and Li [9] proposed a new statistic based on the squared residual ACFs for model checking. Based on the $m$th root of the determinant of the $m$th autocorrelation matrix, Peña and Rodríguez [10, 11] proposed a powerful portmanteau test for ARMA model. Gallagher and Fisher [3] introduced a data-adaptive weighted portmanteau test for ARMA model.

M.T. Tai · Y.X. Yang (✉) · S.Q. Ling
Department of Mathematics, Hong Kong University of Science and Technology, Clear Water Bay, Hong Kong
e-mail: yyangaj@ust.hk

S.Q. Ling
e-mail: maling@ust.hk



W.K. Li et al. (eds.), *Advances in Time Series Methods and Applications*, Fields Institute Communications 78, DOI 10.1007/978-1-4939-6568-7_5

All these results are for the univariate time series models. Li and McLeod [6] studied the residual ACFs of the multivariate stationary ARMA model and proposed a portmanteau test for model checking. A general diagnostic checking approach for stationary multivariate time series models, including linear and nonlinear models, was proposed by Li and Ling [5]. Mahdi and McLeod [7] studied an improved multivariate portmanteau test for stationary ARMA model. However, until now, there has been little research on the diagnostic checking of nonstationary multivariate time series models. The main difficulties are too many unknown parameters in the model and its complicated structures. We refer to Li [4] for more references in this area.

This paper studies the residual ACFs of partially nonstationary multivariate autoregressive (AR) and autoregressive moving-average (ARMA) models. Ahn and Reinsel [1] and Yap and Reinsel [13] derived the limiting distributions of the full rank estimators and the Gaussian reduced rank estimators for the two models. Using these results, we derive the limiting distributions of the residual ACFs under full rank and reduce rank estimations. Based on these limiting distributions, we construct the portmanteau statistics for model checking. It is shown that these statistics asymptotically follow $\chi^2$-distributions. Simulations are carried out to assess their performances in finite samples and two real examples are given.

The paper is organised as follows. Sections 2 and 3 presents our models. Section 4 states our main results. Simulation results are reported in Sect. 5. Section 6 gives two real examples. Throughout this paper, we use the following notations: $I_k$ denotes the identity matrix of order $k$; $\|\cdot\|$ denotes the Euclid norm; $\stackrel{D}{\rightarrow}$ denotes convergence in distribution; $o_p(1)$ denotes a series of random numbers converging to zero in probability and $\otimes$ denotes the Kronecker product.

## 2 Partially Nonstationary Multivariate AR Models

An $m$-dimensional AR process $\{Y_t\}$ with order $p$ is defined as

$$\Phi(B)Y_t = \varepsilon_t, \tag{1}$$

where $\Phi(B) = I_m - \sum_{j=1}^{p} \Phi_j B^j$ is a matrix polynomial in $B$ of degree $p$, det $\{\Phi(B)\} = 0$ has $d < m$ roots equal to unity, all the other roots lie outside the unit circle and $\varepsilon_t = (\varepsilon_{1t}, \ldots, \varepsilon_{mt})'$ are independent and identically distributed (i.i.d.) white noises with $E(\varepsilon_t) = 0$, $\text{var}(\varepsilon_{it}) = \sigma_{ii}^2$ for $i = 1, \ldots, m$, $\text{cov}(\varepsilon_t) = \Omega_\varepsilon$, and $E\|\varepsilon_t\|^{2+\iota} < \infty$ for some $\iota > 0$. Model (1) is called the partially nonstationary multivariate AR model.

Denote $\Phi(1) = I_m - \sum_{j=1}^{p} \Phi_j$, $C = -\Phi(1)$, $\Phi_j^* = -\sum_{k=j+1}^{p} \Phi_k$ and $r = m - d$. We assume the rank of $\Phi(1)$ is $r$, so that each component of the first difference $W_t := Y_t - Y_{t-1}$ is stationary. We can rewrite (1) as

$$W_t = CY_{t-1} + \sum_{j=1}^{p-1} \Phi_j^* W_{t-j} + \varepsilon_t. \tag{2}$$

We call model (2) multivariate full rank AR model. Denote $F = (C, \Phi_1^*, \Phi_2^*, \ldots, \Phi_{p-1}^*)$ and $X_{t-1} = (Y_{t-1}', W_{t-1}', \ldots, W_{t-p+1}')'$. We can further write (2) as

$$W_t = FX_{t-1} + \varepsilon_t. \tag{3}$$

Given $\{Y_{-p+1}, \ldots, Y_n\}$, the least square estimator (LSE) of $F$ is

$$\hat{F} := \left(\sum_{t=1}^{n} W_t X_{t-1}'\right)\left(\sum_{t=1}^{n} X_{t-1} X_{t-1}'\right)^{-1}.$$

The residual of model (3) is defined as $\hat{\varepsilon}_t = W_t - \hat{F}X_{t-1}$.

To see the asymptotic properties of $\hat{F}$, we first introduce some notations. Note that $\sum_{i=1}^{p} \Phi_i$ has Jordan canonical form $J = \text{diag}(I_d, \Lambda_r)$ due to the assumption $\text{rank}\{\Phi(1)\} = r$. Let $P$ and $Q = P^{-1}$ be $m \times m$ matrices $Q$ such that $Q(\sum_{j=1}^{p} \Phi_j)$ $P = J$. Partition $Q = [Q_1, Q_2]'$ and $P = [P_1, P_2]'$ such that $Q_1$ and $P_1$ are $m \times d$ matrices and $Q_2$ and $P_2$ are $m \times r$ matrices. We define $X_t^* = Q^* X_t$ with $Q^* = \text{diag}(Q, I_{m(p-1)})$, $Z_t = [Z_{1,t}, Z_{2,t}]$ with $Z_{1,t} = Q_1' Y_t$ and $Z_{2,t} = Q_2' Y_t$. We partition $X_t^*$ into nonstationary and stationary part, i.e. $X_t^* = [Z_{1,t}', U_t']'$ such that $U_t = [Z_{2,t}', W_t', \ldots, W_{t-p+2}']'$ is $[r + m(p-1)] \times 1$ matrix. Denote $D^* = \text{diag}(D, \sqrt{n} I_{m(p-1)})$ with $D = \text{diag}(nI_d, \sqrt{n} I_r)$, $P^* = \text{diag}(P, I_{m(p-1)})$ and $a_t = Q\varepsilon_t$. Then

$$\begin{aligned}
&Q(\hat{F} - F)P^* D^* \\
&= \left(\sum_{t=1}^{n} a_t X_{t-1}^{*'}\right) D^{*-1} \left(D^{*-1} \sum_{t=1}^{n} X_{t-1}^* X_{t-1}^{*'} D^{*-1}\right)^{-1} \\
&= \left[\left(\frac{1}{n}\sum_{t=1}^{n} a_t Z_{1,t-1}'\right)\left(\frac{1}{n^2} Z_{1,t-1} Z_{1,t-1}'\right)^{-1}, \left(\frac{1}{\sqrt{n}}\sum_{t=1}^{n} a_t U_{t-1}'\right)\left(\frac{1}{n}\sum_{t=1}^{n} U_{t-1} U_{t-1}'\right)^{-1}\right] + o_p(1).
\end{aligned}$$

Ahn and Reinsel [1] gave the following result:

$$(\hat{F} - F)P^* D^* \xrightarrow{D} P[M, N],$$

as $n \to \infty$, where

$$M = \Omega_a^{1/2} \left(\int_0^1 B_d(u) dB_m(u)'\right)' \left(\int_0^1 B_d(u) B_d'(u) du\right)^{-1} \Omega_{a_1}^{-1/2} \Psi_{11}^{-1}$$

with $\Omega_a = cov(a_t) = Q\Omega_\varepsilon Q'$, $\Omega_{a_1} = [I_d, 0]\Omega_a [I_d, 0]'$, $B_m(u)$ being an $m$-dimensional standard Brownian motion, $B_d(u) = \Omega_{a_1}^{-1/2} [I_d, 0] \Omega_a^{1/2} B_m(u)$ being a $d$-dimensional standard Brownian motion and $\Psi_{11} = [I_d, 0](\sum_{k=0}^{\infty} \Psi_k)[I_d, 0]'$.

The partially nonstationary multivariate AR model (2) has rank deficient coefficient matrix $C$, and it is more suitable to estimate $C$ with the reduced rank structure. Since the rank of $C$ is $r$, it may be expressed as $C = AB$, where $A$ and $B$ are $m \times r$ and $r \times m$ matrices, respectively. To obtain a unique parameterization, we normalise $B$ so that $B = [I_r, B_0]$, where $B_0$ is an $r \times (m - r)$ unknown matrix. Hence,

$$C = AB = A[I_r, B_0].$$

Model (2) can then be written as

$$W_t = A[I_r, B_0]Y_{t-1} + \sum_{j=1}^{p-1} \Phi_j^* W_{t-j} + \varepsilon_t. \tag{4}$$

Model (4) is called multivariate reduced rank AR model. Let $\beta = (\beta_0', \alpha')'$, where $\beta_0 = \text{vec}(B_0')$ and $\alpha = \text{vec}[(A, \Phi_1^*, \ldots, \Phi_{p-1}^*)']$. Since model (4) is no longer linear under reduced rank structure, we cannot use the same method as in the full rank case. Define

$$\varepsilon_t(\beta) = W_t - A[I_r, B_0]Y_{t-1} - \sum_{j=1}^{p-1} \Phi_j^* W_{t-j}.$$

Given $\{Y_{-p+1}, \ldots, Y_n\}$, the Gaussian estimator of $\beta$, denoted by $\hat{\beta}$, is the estimator that maximize the log-likelihood function:

$$L_n(\beta, \Omega_\varepsilon) = -\frac{n}{2} \log |\Omega_\varepsilon| - \frac{1}{2} \sum_{t=1}^{n} \varepsilon_t'(\beta) \Omega_\varepsilon^{-1} \varepsilon_t(\beta).$$

We denote the residual $\varepsilon_t(\hat{\beta})$ by $\hat{\varepsilon}_t$ for simplicity. Ahn and Reinsel [1] gave the following results:

$$\begin{aligned} n(\hat{B}_0 - B_0) &= \left(A'\Omega_\varepsilon^{-1}A\right)^{-1} A'\Omega_\varepsilon^{-1} \left(\frac{1}{n}\sum_{t=1}^{n} \varepsilon_t Z_{1,t-1}'\right)\left(\frac{1}{n^2}\sum_{t=1}^{n} Z_{1,t-1}Z_{1,t-1}'\right)^{-1} P_{21}^{-1} + o_P(1) \\ &\xrightarrow{D} \left(A'\Omega_\varepsilon^{-1}A\right)^{-1} A'\Omega_\varepsilon^{-1} PMP_{21}^{-1}, \end{aligned}$$

$$\begin{aligned} \sqrt{n}(\hat{\alpha} - \alpha) &= \frac{1}{\sqrt{n}} \sum_{t=1}^{n} \left[ I_m \otimes \left( \frac{1}{n} \sum_{t=1}^{n} \tilde{U}_{t-1}\tilde{U}_{t-1}' \right)^{-1} \tilde{U}_{t-1} \right] \varepsilon_t + o_P(1) \\ &\xrightarrow{D} N\left(0, \Omega_\varepsilon \otimes \Gamma_{\tilde{U}}^{-1}\right), \end{aligned}$$

as $n \to \infty$, where $\Gamma_{\tilde{U}}^{-1} = \text{cov}(\tilde{U}_t)$ and $\tilde{U}_{t-1} = [(BY_{t-1}', W_{t-1}', \ldots, W_{t-p+1}')]'$.

## 3 Partially Nonstationary Multivariate ARMA Models

An $m$-dimensional nonstationary multivariate ARMA process $Y_t$ is defined as

$$\Phi(B)Y_t = \Theta(B)\varepsilon_t, \tag{5}$$

where $\Phi(B)$ is defined in the same way as in (1), $\Theta(B) = I_m - \sum_{i=1}^{q} \Theta_i B^i$ is a matrix polynomial in $B$ of $q$ and $\det\{\Theta(B)\} = 0$ has all its roots lying outside the unit circle. The assumptions on the noises $\varepsilon_t$ are the same as in (1). Model (5) is called the partially nonstationary multivariate ARMA model.

Using a similar argument as in (2), we can rewrite (5) as

$$W_t = CY_{t-1} + \sum_{j=1}^{p-1} \Phi_j^* W_{t-j} - \sum_{j=1}^{q} \Theta_j \varepsilon_{t-j} + \varepsilon_t. \tag{6}$$

Let $P = [P_1, P_2]$, $Q = [Q_1, Q_2]$ and $Z_t = [Z_{1,t}, Z_{2,t}]$ being defined as in Sect. 2. Note that $Z_t = QY_{t-1}$. We rewrite $CY_{t-1}$ as

$$CY_{t-1} = C\{PZ_{t-1}\} = C\{[P_1, P_2][Z_{1,t-1}, Z_{2,t-1}]'\} = CP_1Z_{1,t-1} + CP_2Z_{2,t-1}.$$

Then model (6) has the following form:

$$W_t = CP_1Z_{1,t-1} + CP_2Z_{2,t-1} + \sum_{j=1}^{p-1} \Phi_j^* W_{t-j} - \sum_{j=1}^{q} \Theta_j \varepsilon_{t-j} + \varepsilon_t. \tag{7}$$

Let $\tau = (\tau_0', \tau_1')$, where $\tau_0 = \text{vec}(CP_1)$ and $\tau_1 = \text{vec}(CP_2, \Phi_1^*, \ldots, \Phi_{p-1}^*, \Theta_1, \ldots, \Theta_q)$. Define

$$\varepsilon_t(\tau) = W_t - CP_1Z_{1,t-1} - CP_2Z_{2,t-1} - \sum_{j=1}^{p-1} \Phi_j^* W_{t-j} + \sum_{j=1}^{q} \Theta_j \varepsilon_{t-j}(\tau).$$

Assume that $\{Y_1, \ldots, Y_n\}$ is a sample from model (3.1) with sample size $n$. Given the initial value $\{Y_{1-p}, \ldots, Y_0\}$ and $\varepsilon_t(\tau) \equiv 0$ for $t \leq 0$, the Gaussian estimator of $\tau$, denoted by $\hat{\tau} = (\hat{\tau}_0', \hat{\tau}_1')$, is the estimator that maximize the log-likelihood function:

$$L_n(\tau, \Omega_\varepsilon) = -\frac{n}{2} \log |\Omega_\varepsilon| - \frac{1}{2} \sum_{t=1}^{n} \varepsilon_t'(\tau) \Omega_\varepsilon^{-1} \varepsilon_t(\tau).$$

Denote the residual $\varepsilon_t(\hat{\tau})$ by $\hat{\varepsilon}_t$. Let $\hat{C}$ be the estimator of $C$. Yap and Reinsel [13] gave the following results:

$$n(\hat{C}-C)P_1 = P\left(\frac{1}{n}\sum_{t=1}^{n} b_t Z'_{1,t-1}\right)\left(\frac{1}{n^2}\sum_{t=1}^{n} Z_{1,t-1}Z'_{1,t-1}\right)^{-1} + o_p(1)$$
$$\xrightarrow{D} PM_0,$$

$$\sqrt{n}\,(\hat{\tau}_1-\tau_1) = \left(\frac{1}{n}\sum_{t=1}^{n} U^*_{t-1}\Omega_\varepsilon^{-1}U^{*'}_{t-1}\right)^{-1}\left(\frac{1}{\sqrt{n}}\sum_{t=1}^{n} U^*_{t-1}\Omega_\varepsilon^{-1}\varepsilon_t\right) + o_p(1)$$
$$\xrightarrow{D} N(0, V^{-1}),$$

as $n \to \infty$, where $b_t = Q\Theta(1)\varepsilon_t$,

$$M_0 = \Omega_b^{1/2}\left(\int_0^1 B_d(u)dB_m(u)'\right)'\left(\int_0^1 B_d(u)B'_d(u)du\right)^{-1}\Omega_{b_1}^{-1/2}\Psi_{22}^{-1}$$

with $\Omega_b = cov(b_t)$, $\Omega_{b1} = [I_d, 0]\Omega_b[I_d, 0]'$, $B_m(u)$ denotes an $m$-dimensional standard Brownian motion, $B_d(u) = \Omega_{b_1}^{-1/2}[I_d, 0]\Omega_b^{1/2}B_m(u)$ is a $d$-dimensional standard Brownian motion and $\Psi_{22} = [Q'_t\Phi^*(1)P_1]^{-1}$, $V = E(U^*_{t-1}\Omega_\varepsilon^{-1}U^*_{t-1})$ and

$$U^*_{t-1} = \begin{bmatrix} -\left(Q_2^{-1'} \otimes I_m\right)\frac{\partial\varepsilon'_t}{\partial \text{vec}C} \\ -\frac{\partial\varepsilon'_t}{\partial \text{vec}[\Phi^*_1,\ldots,\Phi^*_{p-1},\Theta_1,\ldots,\Theta_q]} \end{bmatrix}.$$

The partially nonstationary multivariate ARMA model (6) has rank deficient coefficient matrix $C$. Similar to the reduced rank model (4) of the partially nonstationary multivariate AR model (2), the reduced rank model (6) can be written as

$$W_t = A[I_r, B_0]Y_{t-1} + \sum_{j=1}^{p-1}\Phi^*_j W_{t-j} - \sum_{j=1}^{q}\Theta_j\varepsilon_{t-j} + \varepsilon_t. \tag{8}$$

Similar to the AR model case, model (6) and (8) are called multivariate full rank and reduced rank ARMA model, respectively.

Let $\delta = (\delta'_0, \delta'_1)'$, where $\delta_0 = \text{vec}(B_0)$ and $\delta_1 = \text{vec}[A, \Phi^*_1, \ldots, \Phi^*_{p-1}, \Theta_1, \ldots, \Theta_q]$. Define

$$\varepsilon_t(\delta) = W_t - A[I_r, B_0]Y_{t-1} - \sum_{j=1}^{p-1}\Phi^*_j W_{t-j} + \sum_{j=1}^{q}\Theta_j\varepsilon_{t-j}(\delta).$$

Given the observations $\{Y_1, \ldots, Y_n\}$ and initial value $\{Y_{1-p}, \ldots, Y_0\}$ and $\varepsilon_t(\delta) = 0$ for $t \le 0$, the Gaussian estimator of $\delta$, denoted by $\hat{\delta}$, is the estimator that maximize the log-likelihood function:

$$L_n(\delta, \Omega_\varepsilon) = -\frac{n}{2}\log|\Omega_\varepsilon| - \frac{1}{2}\sum_{t=1}^{n}\varepsilon_t'(\delta)\Omega_\varepsilon^{-1}\varepsilon_t(\delta).$$

Yap and Reinsel [13] obtained the following results:

$$n(\hat{B}_0 - B_0) = \Sigma\left(\frac{1}{n}\sum_{t=1}^{n}\varepsilon_t Z_{1,t-1}\right)\left(\frac{1}{n^2}\sum_{t=1}^{n}Z_{1,t-1}Z_{1,t-1}'\right)^{-1}P_{21}^{-1} + o_p(1)$$
$$\xrightarrow{D} \Sigma\Theta^{-1}(1)PM_0P_{21}^{-1},$$

$$\sqrt{n}(\hat{\delta}_1 - \delta_1) = \left[\frac{1}{n}\sum_{t=1}^{n}\tilde{U}_{t-1}^{*}\Omega_\varepsilon^{-1}\tilde{U}_{t-1}^{*\prime}\right]^{-1}\left[\frac{1}{\sqrt{n}}\sum_{t=1}^{n}\tilde{U}_{t-1}^{*}\Omega_\varepsilon^{-1}\varepsilon_t\right] + o_p(1)$$
$$\xrightarrow{D} N(0, V^{*-1}),$$

as $n \to \infty$, where $\Sigma = (A'\Theta^{-1}(1)'\Omega_\varepsilon^{-1}\Theta^{-1}(1)A)^{-1}A'\Theta^{-1}(1)'\Omega_\varepsilon^{-1}$, $\tilde{U}_{t-1}^{*} = -\partial\varepsilon_t'/\partial\delta_1$ and $V^* = E(\tilde{U}_{t-1}^{*}\Omega_\varepsilon^{-1}\tilde{U}_{t-1}^{*\prime})$.

## 4 Main Results

Let $\hat{\varepsilon}_t$ be the residual in model (2). The corresponding residual autocovariance matrix is defined by

$$\hat{R}_l = \frac{1}{n}\sum_{t=1}^{n-l}\hat{\varepsilon}_t\hat{\varepsilon}_{t+l}', \tag{9}$$

where $l$ is an integer. The residual autocorrelation matrix is then defined as

$$\tilde{R}_l = \hat{V}_0^{-1/2}\hat{R}_l\hat{V}_0^{-1/2}, \tag{10}$$

where

$$\hat{V}_0 = diag\left(\frac{1}{n}\sum_{t=1}^{n}\hat{\varepsilon}_{1,t}^2, \ldots, \frac{1}{n}\sum_{t=1}^{n}\hat{\varepsilon}_{m,t}^2\right).$$

Let $\hat{r}_M = \text{vec}[\tilde{R}_1, \tilde{R}_2, \ldots, \tilde{R}_M]$ with $M$ being the number of lags. We are interested in the asymptotic distribution of $\sqrt{n}\hat{r}_M$ under full rank estimation and reduced rank estimation.

The following result gives the limiting distribution of the residual ACF for model (1) under full rank estimation.

**Theorem 4.1** *Under Assumptions of model* (1), *it follows that*

$$\sqrt{n}\hat{r}_M \xrightarrow{D} N(0, \Sigma_1),$$

*where* $\Sigma_1 = \text{diag}(\Sigma_{1,1}, \ldots, \Sigma_{1,M})$ *and* $\Sigma_{1,l} = \Delta^{-1}\Omega_\varepsilon\Delta^{-1} \otimes \Delta^{-1}(\Omega_\varepsilon - D_{1,l}A_1^{-1} D'_{1,l})\Delta^{-1}$ *for* $l = 1, \ldots, M$ *with* $\Delta = diag(\sigma_{11}, \ldots, \sigma_{mm})$, $D_{1,l} = E(\varepsilon_t U'_{t+l-1})$ *and* $A_1 = E(U_{t-1}U'_{t-1})$.

The following results give the limiting distribution of the residual ACFs of model (4) under reduced rank estimation.

**Theorem 4.2** *For model* (4), *under reduced rank estimation, we have*

$$\sqrt{n}\hat{r}_M \xrightarrow{D} N(0, \Sigma_2),$$

*where* $\Sigma_2 = \text{diag}(\Sigma_{2,1}, \ldots, \Sigma_{2,M})$ *and* $\Sigma_{2,l} = (\Delta^{-1} \otimes \Delta^{-1})E[D_{2,l}\Omega_\varepsilon D'_{2,l}](\Delta^{-1} \otimes \Delta^{-1})$ *with* $D_{2,l} = I_M \otimes \varepsilon_t - S_l K_1 A_2^{-1}\tilde{U}'_{t+l-1}$, $A_2 = E[\tilde{U}_{t-1}\tilde{U}'_{t-1}]$, $S_l = E[(I_m \otimes \varepsilon_t)(\tilde{U}'_{t+l-1} \otimes I_m)]$ *and* $K_1$ *is the* $(rm + m^2(p-1)) \times (rm + m^2(p-1))$ *commutation matrix that converts* $vec[(A, \Phi_1^*, \ldots, \Phi_{p-1}^*)']$ *into* $vec[A, \Phi_1^*, \ldots, \Phi_{p-1}^*]$.

The following result gives the limiting distribution of the residual ACF for model (6) under full rank estimation.

**Theorem 4.3** *For model* (6), *under full rank estimation, we have*

$$\sqrt{n}\hat{r}_M \xrightarrow{D} N(0, \Sigma_3),$$

*where*

$$\begin{aligned}
\Sigma_3 &= \text{diag}(\Sigma_{1,1}, \ldots, \Sigma_{1,M}) \\
\Sigma_{3,l} &= (\Delta^{-1} \otimes \Delta^{-1})E[D_{3,l}\Omega_\varepsilon D'_{3,l}](\Delta^{-1} \otimes \Delta^{-1}) \quad \textit{for } l = 1, \ldots, M \\
D_{3,l} &= I_M \otimes \varepsilon_t + \mathfrak{F}_l A_3^{-1} U^*_{t+l-1}\Omega_\varepsilon^{-1}, \\
A_3 &= E[U^*_{t-1}\Omega_\varepsilon^{-1}U^{*'}_{t-1}], \\
\mathfrak{F}_l &= E[(I_m \otimes \varepsilon_t)\Theta^{-1}(B)M_{t+l-1}], \\
M_{t-1} &= \left[ -\left(\tilde{G}'_{t-1} \otimes I_m\right), \sum_{j=1}^{q} \left(\varepsilon'_{t-j} \otimes I_m\right) d_j \right], \\
\tilde{G}_{t-1} &= \left[Z'_{2,t-1}, W'_{t-1}, \ldots, W'_{t-p+1}\right]'
\end{aligned}$$

*and* $d_j$ *is a* $m^2 \times m^2 q$ *matrix that can be blocked into* $q$ $m^2 \times m^2$ *matrices and the* $j$*-th matrix is* $I_{m^2}$ *while other are zero matrices.*

By Theorem 4.3, we can construct the statistic

$$Q_M := n\hat{r}_M' \hat{\Sigma}_3^{-1} \hat{r}_M,$$

where $\hat{\Sigma}_3$ is a consistent estimator of $\Sigma_3$. By Theorem 4.3, we can show that $Q_M$ asymptotically follows the $\chi^2$-distribution with $(M - p - q)m^2$ degree of freedom, i.e. $Q_M \sim \chi^2\big((M - p - q)m^2\big)$. Note that as $l$ large enough, $\mathfrak{F}_l \approx 0$. Thus,

$$\Sigma_{3,l} \approx \Delta^{-1}\Omega_\varepsilon\Delta^{-1} \otimes \Delta^{-1}\Omega_\varepsilon\Delta^{-1} \equiv \Sigma^*.$$

Then the test statistic $Q_M$ can be simple approximated by the test statistic

$$Q_M^* := n\sum_{i=1}^{M}(\text{vec}\tilde{R}_i)'\hat{\Sigma}^{*-1}(\text{vec}\tilde{R}_i) \sim \chi^2\big((M - p - q)m^2\big),$$

where $\hat{\Sigma}^* = \hat{\Delta}^{-1}\hat{\Omega}_\varepsilon\hat{\Delta}^{-1} \otimes \hat{\Delta}^{-1}\hat{\Omega}_\varepsilon\hat{\Delta}^{-1}$ with

$$\hat{\Delta}^2 = \text{diag}\left(\frac{1}{n}\sum_{t=1}^{n}\hat{\varepsilon}_{1t}^2, \ldots, \frac{1}{n}\sum_{t=1}^{n}\hat{\varepsilon}_{mt}^2\right) \text{ and } \hat{\Omega}_\varepsilon = \frac{1}{n}\sum_{t=1}^{n}\hat{\varepsilon}_t\hat{\varepsilon}_t'$$

The $Q_M$ and $Q_M^*$ are called portmanteau test statistics. We compare them with the upper-tailed critical value of $\chi^2\big((M - p - q)m^2\big)$ at an appropriate level. If the statistics are less than the critical value, then the fitted ARMA model is adequate. Note that when $q = 0$, model (6) reduce to the partially nonstationary multivariate AR model and in this case $Q_M^*$ can be used for diagnostic checking for fitted AR models.

**Theorem 4.4** *For model* (8), *under reduced rank estimation, we have*

$$\sqrt{n}\hat{r}_M \xrightarrow{D} N(0, \Sigma_4),$$

*where*

$$\begin{aligned}
\Sigma_4 &= \text{diag}(\Sigma_{2,1}, \ldots, \Sigma_{2,M}),\\
\Sigma_{4,l} &= (\Delta^{-1} \otimes \Delta^{-1})E[D_{4,l}\Omega_\varepsilon D_{4,l}'](\Delta^{-1} \otimes \Delta^{-1}),\\
D_{4,l} &= I_m \otimes \varepsilon_t + \Upsilon_l A_4^{-1}\tilde{U}_{t+l-1}^*\Omega_\varepsilon^{-1},\\
A_4 &= E[\tilde{U}_{t-1}^*\Omega_\varepsilon^{-1}\tilde{U}_{t-1}^{*'}],\\
\Upsilon_l &= E[(I_m \otimes \varepsilon_t)\Theta^{-1}(B)N_{t+l-1}],
\end{aligned}$$

$$N_{t-1} = \left[ -\left(\tilde{H}'_{t-1} \otimes I_m\right), \sum_{j=1}^{q} \left(\varepsilon'_{t-j} \otimes I_m\right) d_j \right],$$

$$\tilde{H}'_{t-1} = \left[(BY_{t-1})', W'_{t-1}, \ldots, W'_{t-p+1}\right]'.$$

Since $\Upsilon_l \approx 0$ as $l$ is large enough, we have $\Sigma_{4,l} \approx \Sigma^*$. Therefore, in this case we can still use $Q^*_M$ with $\hat{\varepsilon}$ being the residual of model (8) to check whether the fitted model is adequate or not.

The proofs of Theorems 4.1–4.4 were given in Tai [12] and the details are omitted.

## 5 Simulation Studies

To study the size and power of test statistics $Q^*_M$ in Sect. 4, we use three models to perform the simulation. The first model is the bivariate AR(1) model

$$Y_t = \Phi_1 Y_{t-1} + \varepsilon_t,$$

where

$$\Omega_\varepsilon = \begin{pmatrix} 1 & \alpha \\ \alpha & 1 \end{pmatrix}$$

with six different values of $\alpha$, $\Phi_1 = A_1, B_1, C_1$ and $D_1$, where

$$A_1 = \begin{pmatrix} 0.60 & 1.00 \\ 0.12 & 0.70 \end{pmatrix}, \quad B_1 = \begin{pmatrix} 0.30 & -0.20 \\ -0.70 & 0.80 \end{pmatrix},$$
$$C_1 = \begin{pmatrix} 0.50 & -0.15 \\ -1.00 & 0.70 \end{pmatrix} \quad \text{and} \quad D_1 = \begin{pmatrix} 0.37 & 0.63 \\ 0.17 & 0.83 \end{pmatrix}.$$

All the matrices $A_1$, $B_1$, $C_1$ and $D_1$ have only one unit root. We choose $M = 15$ and the significance level 0.05 are used. The corresponding critical value is $\chi^2_{2^2(15-1)} = \chi^2_{56} \approx 74.45$. To study the empirical powers of $Q^*_{15}$, the following alternative model is used:

$$Y_t = \Phi_1 Y_{t-1} + \Phi_2 Y_{t-2} + \varepsilon_t,$$

where

$$\Phi_2 = \begin{pmatrix} 0.5 & 0 \\ 0 & -0.4 \end{pmatrix}.$$

**Table 1** Sizes and powers of test statistics $Q^*_{15}$ for bivariate AR(1) model among 1000 simulations

| $\alpha$ | Size | | | | Powers | | | |
|---|---|---|---|---|---|---|---|---|
| | $A_1$ | $B_1$ | $C_1$ | $D_1$ | $A_1$ | $B_1$ | $C_1$ | $D_1$ |
| $n = 200$ | | | | | | | | |
| 0.05 | 66 | 52 | 73 | 53 | 610 | 710 | 829 | 892 |
| −0.1 | 57 | 57 | 94 | 72 | 702 | 562 | 623 | 678 |
| −0.25 | 44 | 63 | 69 | 42 | 797 | 633 | 808 | 698 |
| 0.4 | 48 | 71 | 45 | 59 | 748 | 657 | 723 | 702 |
| −0.6 | 73 | 62 | 63 | 82 | 641 | 682 | 703 | 645 |
| 0.75 | 49 | 64 | 74 | 44 | 576 | 802 | 666 | 687 |
| $n = 400$ | | | | | | | | |
| 0.05 | 62 | 55 | 68 | 57 | 792 | 842 | 893 | 963 |
| −0.1 | 51 | 53 | 72 | 61 | 892 | 748 | 882 | 888 |
| −0.25 | 48 | 61 | 64 | 45 | 821 | 729 | 852 | 877 |
| 0.4 | 48 | 67 | 46 | 55 | 856 | 819 | 821 | 881 |
| −0.6 | 68 | 58 | 61 | 80 | 823 | 781 | 822 | 901 |
| 0.75 | 42 | 57 | 66 | 42 | 756 | 902 | 811 | 748 |
| $n = 500$ | | | | | | | | |
| 0.05 | 53 | 49 | 50 | 48 | 867 | 921 | 951 | 956 |
| −0.1 | 53 | 47 | 59 | 54 | 902 | 873 | 894 | 964 |
| −0.25 | 50 | 54 | 54 | 55 | 934 | 899 | 970 | 905 |
| 0.4 | 52 | 56 | 58 | 44 | 945 | 822 | 934 | 953 |
| −0.6 | 54 | 57 | 56 | 65 | 867 | 832 | 911 | 965 |
| 0.75 | 51 | 51 | 58 | 51 | 878 | 945 | 901 | 845 |

Table 1 reports the sizes and powers of $Q^*_{15}$. From Table 1, we can see that, even when the sample size $n$ is small, the empirical rejection probabilities of the test statistic are close to 5 %. There is little bit inflation or deflation. As the sample size increases to 400 and 500, the performance of the test statistic is improved as evidenced by nearly 5 % of the empirical rejection probabilities for all cases. The test statistic has higher power as the sample size increases, in particular, when the sample size is 500. Furthermore, the nature of AR parameters and the correlation do not have much difference on the sizes and powers of the test statistic. Overall, the test statistic has good performance in all cases.

The second model we consider is the bivariate ARMA(1,1) model

$$Y_t = \Phi_1 Y_{t-1} - \Theta_1 \varepsilon_{t-1} + \varepsilon_t,$$

where $\Phi_1 = A_1, B_1, C_1$ and $D_1$, $\Theta_1 = A_2, B_2, C_2$ and $D_2$ with

$$A_2 = \begin{pmatrix} -0.10 & -0.40 \\ -0.22 & -0.70 \end{pmatrix}, \quad B_2 = \begin{pmatrix} 0.60 & 0.30 \\ 0.12 & 0.70 \end{pmatrix},$$

$$C_2 = \begin{pmatrix} -0.20 & 0.70 \\ 0.30 & -0.30 \end{pmatrix} \quad \text{and} \quad D_2 = \begin{pmatrix} -0.40 & -0.10 \\ 0.10 & -0.90 \end{pmatrix}.$$

All the eigenvalues of the matrices $A_2$, $B_2$, $C_2$ and $D_2$ lie inside the unit circle. We choose $M = 15$ and the significance level 0.05. The corresponding critical value is $\chi^2_{2^2(15-1-1)} = \chi^2_{52)} \approx 69.8$. To study the empirical powers of $Q^*_{15}$, we use the alternative model

$$Y_t = \Phi_1 Y_{t-1} + \Phi_2 Y_{t-2} - \Theta_1 \varepsilon_{t-1} + \varepsilon_t.$$

Table 2 reports the sizes and powers of $Q^*_{15}$ under the null model. The test statistic performs quite well in the empirical rejection probabilities even when the sample size $n$ is small. The performance of the test statistic improves as the sample size increases from 400 to 500. The overspecified ARMA(2,1) model gives a deflation in

**Table 2** Sizes and powers of test statistics $Q^*_{15}$ for bivariate ARMA(1,1) model among 1000 simulations

| $\alpha$ | Size | | | | Powers | | | |
|---|---|---|---|---|---|---|---|---|
| | $A_1, A_2$ | $B_1, B_2$ | $C_1, C_2$ | $D_1, D_2$ | $A_1, A_2$ | $B_1, B_2$ | $C_1, C_2$ | $D_1, D_2$ |
| $n = 200$ | | | | | | | | |
| 0.05 | 67 | 72 | 34 | 42 | 612 | 672 | 712 | 643 |
| −0.1 | 92 | 53 | 68 | 50 | 778 | 743 | 783 | 781 |
| −0.25 | 35 | 46 | 68 | 40 | 592 | 603 | 652 | 576 |
| 0.4 | 82 | 37 | 56 | 71 | 723 | 792 | 792 | 534 |
| −0.6 | 73 | 84 | 87 | 92 | 678 | 782 | 667 | 680 |
| 0.75 | 57 | 56 | 51 | 79 | 630 | 514 | 588 | 583 |
| $n = 400$ | | | | | | | | |
| 0.05 | 61 | 53 | 37 | 57 | 821 | 787 | 843 | 788 |
| −0.1 | 82 | 43 | 62 | 52 | 866 | 834 | 856 | 851 |
| −0.25 | 46 | 51 | 69 | 44 | 785 | 731 | 822 | 781 |
| 0.4 | 77 | 50 | 52 | 61 | 872 | 821 | 810 | 687 |
| −0.6 | 62 | 65 | 68 | 87 | 852 | 840 | 712 | 744 |
| 0.75 | 47 | 54 | 59 | 77 | 785 | 702 | 687 | 674 |
| $n = 500$ | | | | | | | | |
| 0.05 | 53 | 54 | 47 | 59 | 904 | 934 | 956 | 923 |
| −0.1 | 67 | 56 | 53 | 63 | 963 | 904 | 965 | 902 |
| −0.25 | 57 | 51 | 54 | 54 | 932 | 898 | 923 | 882 |
| 0.4 | 56 | 54 | 59 | 49 | 936 | 973 | 890 | 771 |
| −0.6 | 53 | 54 | 62 | 60 | 943 | 887 | 887 | 952 |
| 0.75 | 47 | 52 | 58 | 61 | 909 | 798 | 792 | 890 |

the empirical rejection probabilities. The test statistic has higher power as the sample size increases. Based on the empirical simulation evidences, we find that the finite sample performance of the test statistic seems not to be affected by the MA parts.

Besides from the bivariate cases, we also considered the trivariate AR(1) model

$$Y_t = \Phi_1 Y_{t-1} + \varepsilon_t,$$

where

$$\Omega_\varepsilon = \begin{pmatrix} 1 & \alpha_1 & \alpha_2 \\ \alpha_1 & 1 & \alpha_3 \\ \alpha_2 & \alpha_3 & 1 \end{pmatrix},$$

with $\alpha_1, \alpha_2$ and $\alpha_3$ are chosen among the values $\pm 0.25, \pm 0.5, \pm 0.75$, and $\Phi_1 = A_3, B_3$ and $C_3$ with

$$A_3 = \begin{pmatrix} 0.602 & 0.433 & 0.110 \\ 0.121 & 0.660 & 0.066 \\ 0.103 & 0.166 & 0.838 \end{pmatrix},$$

$$B_3 = \begin{pmatrix} 0.35 & 0.25 & 0 \\ 0 & 0.42 & 0.65 \\ -0.71 & 0.96 & 0.23 \end{pmatrix} \quad \text{and} \quad C_3 = \begin{pmatrix} 1 & 0.48 & 0.55 \\ 0.33 & 0.57 & 0 \\ 0 & -0.32 & 0.63 \end{pmatrix}.$$

For this model, we choose $M = 10$. There is only one unit root for $A_1$ and $B_3$ and there are two unit roots for $C_3$. The upper 5 percent point of $\chi^2_{3^2(10-1)} = \chi^2_{81} \approx 101.88$. To consider the powers of the statistic, the following alternative model is used:

$$Y_t = \Phi_1 Y_{t-1} + \Phi_2 Y_{t-1} + \varepsilon_t,$$

where

$$\Phi_2 = \begin{pmatrix} 0.5 & 0 & 0 \\ 0 & -0.4 & 0 \\ 0 & 0 & 0.1 \end{pmatrix}.$$

The empirical sizes and powers of $Q^*_{10}$ are reported in Table 3. From Table 3, we can see that the size gives an inflation in the empirical rejection probabilities when there is only one unit root in the model, and gives a deflation when there are two unit roots in the model. As sample size increases, the performance of the test statistic improves. Moreover, the test statistic gives a deflation in the empirical rejection probabilities. As we expected, the power of the test statistic increases when $n$ grows from 400 to 500. Again, the finite sample performance of the test statistic seems not to be affected by the nature of AR parameters.

**Table 3** Sizes and powers of test statistics $Q_{10}^*$ for trivariate AR(1) model among 1000 simulations

| $\alpha_1$ | Size | | | | Powers | | | |
|---|---|---|---|---|---|---|---|---|
| | $\alpha_2$ | $\alpha_3$ | $A_3$ | $B_3$ | $C_3$ | $A_3$ | $B_3$ | $C_3$ |
| $n = 200$ | | | | | | | | |
| 0.25 | 0.5 | 0.75 | 62 | 83 | 39 | 623 | 725 | 733 |
| −0.25 | 0.75 | −0.5 | 64 | 43 | 36 | 572 | 596 | 722 |
| 0.5 | −0.25 | 0.75 | 64 | 62 | 43 | 762 | 663 | 719 |
| −0.5 | −0.25 | −0.75 | 57 | 52 | 82 | 802 | 725 | 554 |
| −0.75 | 0.5 | 0.25 | 61 | 72 | 59 | 563 | 706 | 678 |
| −0.75 | −0.5 | −0.25 | 78 | 92 | 48 | 643 | 834 | 726 |
| $n = 400$ | | | | | | | | |
| 0.25 | 0.5 | 0.75 | 62 | 61 | 48 | 748 | 815 | 802 |
| −0.25 | 0.75 | −0.5 | 52 | 49 | 42 | 647 | 652 | 769 |
| 0.5 | −0.25 | 0.75 | 49 | 51 | 50 | 824 | 821 | 803 |
| −0.5 | −0.25 | −0.75 | 57 | 59 | 50 | 923 | 872 | 672 |
| −0.75 | 0.5 | 0.25 | 64 | 62 | 50 | 712 | 911 | 823 |
| −0.75 | −0.5 | −0.25 | 65 | 82 | 45 | 770 | 937 | 871 |
| $n = 500$ | | | | | | | | |
| 0.25 | 0.5 | 0.75 | 57 | 54 | 55 | 892 | 910 | 872 |
| −0.25 | 0.75 | −0.5 | 51 | 52 | 56 | 824 | 848 | 899 |
| 0.5 | −0.25 | 0.75 | 56 | 52 | 56 | 953 | 931 | 918 |
| −0.5 | −0.25 | −0.75 | 57 | 59 | 48 | 972 | 901 | 924 |
| −0.75 | 0.5 | 0.25 | 51 | 55 | 45 | 899 | 953 | 864 |
| −0.75 | −0.5 | −0.25 | 53 | 54 | 43 | 942 | 940 | 918 |

## 6 Two Numerical Examples

In this section, we give two numerical examples to illustrate our methods. The first example considers U.S. monthly data $Y_t$ (in thousands) consisting of housing-starts ($Y_{1t}$) and housing-sold ($Y_{2t}$) from January period 1965 to December 1974. This data set was investigated by Ahn and Reinsel [1] and they fitted the data by a partially stationary bivariate AR(1) model with $d = 1$ unit root, i.e. $Y_t = \Phi_1 Y_{t-1} + \varepsilon_t$, which can be further written as

$$W_t = Y_t - Y_{t-1} = CY_{t-1} + \varepsilon_t.$$

The full rank least squares estimator of $C$ is given by

$$\hat{C} = \begin{pmatrix} -0.537 & 0.951 \\ 0.129 & -0.289 \end{pmatrix}.$$

The eigenvalues of $\hat{C}$ are $-0.785$ and $-0.041$, and $-0.041$ may be considered to be close to 0, which indicates one unit root in the AR operator. To use our test $Q_M^*$, we set $M = 15$. The value of $Q_1^*5$ is 63.4, which is smaller than the critical value $\chi^2_{2^2(15-1)} = \chi^2_{56} \approx 74.45$ at significance level 0.05. Thus, the null hypothesis is accepted and the fitted model is adequate. Based on the reduced rank approach, the estimators of $A$ and $B_0$ are $\tilde{A} = [-0.537, 0.129]'$ and $\tilde{B}_0 = -1.752$. The final reduced rank Gaussian estimator of $C$ is

$$\tilde{C} = \begin{pmatrix} -0.523 & 0.979 \\ 0.141 & -0.265 \end{pmatrix}.$$

Using the reduced rank estimators, we calculate the residuals and the value of $Q_{15}^*$ is 62.3. Therefore fitted reduced rank model is adequate as well.

The second example considers U.S. monthly logarithms of interest rate series $Y_t$ consisting of the Federal Fund rate ($Y_{1t}$), 90-day Treasury Bill rate ($Y_{2t}$) and 1-year Treasury Bill rate ($Y_{3t}$) series from January 1960 to December 1979. Yap and Reinsel [13] fitted this data by a trivariate ARMA(1,1) model, $Y_t = \Phi_1 Y_{t-1} - \Theta_1 \varepsilon_{t-1} + \varepsilon_t$, which can be written as

$$W_t = Y_t - Y_{t-1} = CY_{t-1} - \Theta_1 \varepsilon_{t-1} + \varepsilon_t.$$

The full rank Gaussian estimators are given by

$$\tilde{C} = \begin{pmatrix} -0.203 & 0.243 & -0.003 \\ 0.019 & -0.090 & 0.068 \\ 0.036 & 0.019 & -0.081 \end{pmatrix} \quad \text{and} \quad \tilde{\Theta}_1 = \begin{pmatrix} -0.143 & 0.237 & -0.463 \\ -0.224 & 0.118 & -0.317 \\ -0.125 & 0.037 & -0.330 \end{pmatrix}.$$

The eigenvalues of $\hat{C}$ are $-0.008$, $-0.174$ and $-0.192$. The first root $-0.008$ is close to 0, which indicates that there is one unit root. The value of the test statistic $Q_{10}^*$ is 71.05, which is smaller than the critical value $\chi^2_{3^2(10-1-1)} = \chi^2_{72} \approx 90.66$. Hence the null hypothesis is accepted at significance level 0.05. Based on the reduced rank approach, the estimator of $A$, $B_0$ and $\Theta_1$ are, respectively,

$$\tilde{A} = \begin{bmatrix} -0.199 & 0.250 \\ 0.023 & -0.082 \\ 0.041 & 0.027 \end{bmatrix}, \quad \tilde{B}_0 = \begin{bmatrix} -1.396 \\ -1.147 \end{bmatrix} \quad \text{and} \quad \tilde{\Theta}_1 = \begin{pmatrix} -0.147 & 0.241 & -0.468 \\ -0.231 & 0.127 & -0.334 \\ -0.129 & 0.042 & -0.339 \end{pmatrix}.$$

Then the reduced rank Gaussian estimators of $C$ is

$$\tilde{C} = \begin{pmatrix} -0.199 & 0.250 & -0.009 \\ 0.023 & -0.082 & 0.062 \\ 0.041 & 0.027 & -0.088 \end{pmatrix}.$$

The value of $Q_{10}^*$ is 68.2 and hence the fitted model is also adequate.

## References


1. Ahn, S. K., & Reinsel, G. C. (1990). Estimation for partially nonstationary multivariate autoregressive models. *Journal of the American Statistic Association*, *85*, 813–823.
2. Box, G. E. P., & Pierce, D. A. (1970). Distribution of residual autocorrelations in autoregressive-integrated moving average time series models. *Journal of American Statistical Association*, *65*, 1509–1526.
3. Gallagher, C., & Fisher, T. (2015). On weighted portmanteau tests for time series goodness-of-fit. *Journal of Time Series Analysis*, *36*, 67–83.
4. Li, W. K. (2003). *Diagnostic checks in time series*. London: Chapman & Hall/CRC.
5. Li, W. K., & Ling, S. Q. (1997). Diagnostic checking of nonliear multivariate time series with multivariate arch errors. *Journal of Time Series Analysis*, *18*, 447–464.
6. Li, W. K., & McLeod A. I. (1980). Distribution of the residual autocorrelation in multivariate ARMA time series models. *Joural of the Royal Statistical Society, Series B*, *43*, 231–239.
7. Mahdi, E., & McLeod, A. I. (2012). Improved multivariate portmanteau test. *Journal of Time Series Analysis*, *33*, 211–222.
8. McLeod, A. I. (1978). On the distribution of the residual autocorrelation in Box–Jenkins model. *Joural of the Royal Statistical Society, Series B*, *40*, 296–302.
9. McLeod, A. I., & Li, W. K. (1983). Diagnostic checking ARMA time series models using squared residual autocorrelations. *Time Series Analysis*, *4*, 269–273.
10. Peña, D., & Rodríguez, J. (2002). A powerful portmanteau test of lack of test for time series. *Journal of American Statistical Association*, *97*, 601–610.
11. Peña, D., & Rodríguez, J. (2006). The log of the determinant of the autocorrelation matrix for testing goodness of fit in time series. *Journal of Statistical Planning and Inference*, *8*, 2706–2718.
12. Tai, M. T. (2003). *Portmanteau statistics for partially nonstationary multivariate AR and ARMA models*. M.Phil. thesis. Department of Mathematics in HKUST.
13. Yap, S. F., & Reinsel, G. C. (1995). Estimation and testing for unit roots in a partially nonstationary vector autoregressive moving average model. *Journal of the American Association*, *90*, 253–267.

# The Portmanteau Tests and the LM Test for ARMA Models with Uncorrelated Errors

**Naoya Katayama**

**Abstract** In this article, we investigate the portmanteau tests and the Lagrange multiplier (LM) test for goodness of fit in autoregressive and moving average models with uncorrelated errors. Under the assumption that the error is not independent, the classical portmanteau tests and LM test are asymptotically distributed as a weighted sum of chi-squared random variables that can be far from the chi-squared distribution. To conduct the tests, we must estimate these weights using nonparametric methods. Therefore, by employing the method of Kiefer et al. (Econometrica, 68:695–714, 2000, [11]), we propose new test statistics for the portmanteau tests and the LM test. The asymptotic null distribution of these test statistics is not standard, but can be tabulated by means of simulations. In finite-sample simulations, we demonstrate that our proposed test has a good ability to control the type I error, and that the loss of power is not substantial.

## 1 Introduction

We consider goodness-of-fit tests for univariate autoregressive moving average (ARMA) models with uncorrelated errors. Portmanteau tests and the Lagrange multiplier (LM) test are popular tools in ARMA modeling. Portmanteau test statistics, defined by the sum of squares of the first $m$ residual autocorrelations, are commonly used in time series analysis to describe the goodness of fit. This approach was first presented by Box and Pierce [1] and Ljung and Box [15] for univariate autoregressive (AR) models. McLeod [18] derived the large sample distribution of the residual autocorrelations and the portmanteau statistic for ARMA models. LM tests for ARMA time series models have been investigated by many authors,

N. Katayama (✉)
Faculty of Economics, Kansai University, 3-3-35 Yamate-cho, Suita, Osaka 564-8680, Japan
e-mail: katayama@kansai-u.ac.jp

N. Katayama
Institute of International Business, National Cheng Kung University, No. 1, University Road, Tainan City 70101, Taiwan, ROC



W.K. Li et al. (eds.), *Advances in Time Series Methods and Applications*, Fields Institute Communications 78, DOI 10.1007/978-1-4939-6568-7_6

e.g., Godfrey [6], Newbold [19], and Hosking [7]. The test statistics compare the null hypothesis model ARMA$(p, q)$ against either ARMA$(p + m, q)$ or ARMA$(p, q + m)$. From the viewpoint of finite sample size and power, these two test statistics are often used in combination. Li [14, Chap. 2] reviews several such tests. However, most of these tests impose the restriction that the errors must be independent. This precludes the application of a nonlinear model.

In recent years, the time series literature has been characterized by a growing interest in nonlinear models. Francq et al. [3] reported that many important classes of nonlinear processes admit ARMA models with uncorrelated errors. Some examples include bilinear processes, autoregressive-conditional duration processes, the Markov-switching ARMA model, generalized autoregressive conditionally heteroscedastic (GARCH) model, and hidden Markov models. Francq et al. [3] also reported that, under the Wold decomposition theorem, any purely nondeterministic second-order stationary process admits an infinite-order moving average (MA) representation, where the noise is considered to be white noise. The ARMA model with uncorrelated errors also has this representation, and is regarded as an approximation of the MA model. Therefore, this model covers a very wide class of second-order stationary processes. Fitting nonlinear models is often difficult, whereas fitting ARMA models is easy and computable using statistical software (e.g., SAS, R, SPSS). Additionally, the estimators are easy to interpret. Therefore, ARMA models can be useful tools, even if the true process appears to be nonlinear.

There are now three portmanteau tests for ARMA models with uncorrelated errors: (i) Francq et al. [3] presented an asymptotic distribution of Ljung and Box's [15] portmanteau statistics under the condition of uncorrelated errors. The distribution is given by the weighted sum of a chi-squared random variable that contains the unknown ARMA parameters. Therefore, we have to compute the critical values in each test. (ii) Katayama [9] modified the portmanteau statistic with a correction term that is asymptotically chi-squared. However, these two test statistics require an estimate of the covariance structure of a high-dimensional multivariate process and a large sample size. (iii) Kuan and Lee's [12] portmanteau test is based on the approach developed by Kiefer, Vogelsang, and Bunzel [11] (referred to as KVB). Instead of estimating the asymptotic covariance matrix, Kuan and Lee's [12] portmanteau test statistic employs a random normalizing matrix to eliminate the nuisance parameters of the asymptotic covariance matrix. The asymptotic critical values are tabulated by a series of simulations. We review these test statistics in Sect. 2.

To overcome these weaknesses, we propose a new portmanteau test and an LM test in Sect. 3. Our proposed tests are based on the KVB approach. The test statistics have no use for recursive estimators, and do not require an estimate of the covariance structure of a high-dimensional multivariate process. Therefore, our test statistics have a significantly lower computational cost. We compare the finite sample performance of these test statistics via simulations in Sect. 4. We demonstrate that our proposed test exhibits sufficiently efficient empirical size and power properties with existing portmanteau tests.

In the remainder of this paper, $\Rightarrow$ denotes weak convergence (of associated probability measures) and $\xrightarrow{d}$ denotes the convergence in distribution. Throughout the paper, convergence is described for a sample size $n$ going to infinity. Therefore, we omit the phrase "as $n \to \infty$," except in a few cases. $W_m$ denotes a vector of $m$ independent standard Wiener processes, and $B_m$ is the Brownian bridge with $B_m(\tau) = W_m(\tau) - \tau W_m(1)$ for $\tau \in (0, 1]$. A matrix $A^+$ denotes the MP-inverse of $A$. Let $\partial f(y)/\partial x$ denote $\partial f(x)/\partial x|_{x=y}$, $\nabla_x f(y)$ denote $\partial f(y)/\partial x'$, and $\nabla'_x f(y)$ denote $\partial f(y)/\partial x$. Additionally, $[c]$ denotes the integer part of $c$.

Finally, this paper is based on Katayama [8], which extended new KVB-based tests to the $M$ test and considered not only portmanteau tests and LM tests, but also GMM over-identification tests and the Hausman tests. This paper is available on request.

## 2 Review of the Portmanteau Tests

Suppose that a univariate time series $\{Y_t\}$ is generated by an autoregressive-moving average model ARMA$(p, q)$:

$$Y_t = \sum_{i=1}^{p} a_i^0 Y_{t-i} + \varepsilon_t + \sum_{j=1}^{q} b_j^0 \varepsilon_{t-j}, \quad t = 0, \pm 1, \pm 2, \ldots, \tag{1}$$

where $\{\varepsilon_t\}$ provides white-noise sequences with variance $\sigma_0^2$. It is assumed that the above model is stationary, invertible, and not redundant, so that the polynomials $1 - a_1^0 z - \cdots - a_p^0 z^p = 0$ and $1 + b_1^0 z + \cdots + b_q^0 z^q = 0$ have no common roots, and that all roots are outside the unit circle. We denote the true parameter vector as $\theta_0 = (a_1^0, \ldots, a_p^0, b_1^0, \ldots, b_q^0)'$; this belongs to the parameter space $\Theta \subset \mathbb{R}^{p+q}$. We suppose that $a_p^0 \neq 0$ or $b_q^0 \neq 0$ and any $\theta \in \Theta$ satisfies the conditions of the polynomials. Given a process $\{Y_t\}_{t=1}^n$, as defined in Eq. (1), the nonlinear least-squares estimator of $\theta_0$, $\widehat{\theta}_n = (\widehat{a}_1, \ldots, \widehat{a}_p, \widehat{b}_1, \ldots, \widehat{b}_q)'$, is obtained by minimizing the sum of the squared residuals. The residuals $\widehat{\varepsilon}_t = \varepsilon_t(\widehat{\theta}_n)$ $(t = 1, \ldots, n)$ from the fitted models are given by $\widehat{\varepsilon}_t = Y_t - \widehat{a}_1 Y_{t-1} - \cdots - \widehat{a}_p Y_{t-p} - \widehat{b}_1 \widehat{\varepsilon}_{t-1} - \cdots - \widehat{b}_q \widehat{\varepsilon}_{t-q}$, where the unknown starting values are set to 0: $\widehat{\varepsilon}_0 = \cdots = \widehat{\varepsilon}_{1-q} = Y_0 = \cdots = Y_{1-p} = 0$. Throughout this paper, we assume that:

**Assumption 1** $\{Y_t\}$ is strictly stationary, satisfies the ARMA$(p, q)$ model (1), $E|\varepsilon_t|^{4+2\nu} < \infty$, and $\{\varepsilon_t\}$ is an $\alpha$-mixing of size $-(2+\nu)/\nu$ for some $\nu > 0$.

This assumption is somewhat stronger than Assumption 1′ in Francq et al. [3], because it implies the summability of $\alpha$-mixing coefficients raised to the $\nu/(2+\nu)$th power. Francq and Zakoïan [4] showed that, under this assumption, $\widehat{\theta}_n$ is $\sqrt{n}$-consistent and asymptotically normal. Francq et al. [3] note that Assumption 1 does not require the noise to be independent or a martingale difference sequence (MDS). In Sect. 3, we apply this assumption to establish a functional central limit theorem (FCLT) of near-epoch dependence (NED) in the mixing process $\{\varepsilon_t\}$.

To check the adequacy of the model fit, we examine the residual autocorrelations as follows:

$$\widehat{r}(j) = \frac{\widehat{\gamma}(j)}{\widehat{\gamma}(0)}, \quad \widehat{\gamma}(j) = \frac{1}{n}\sum_{i=j+1}^{n} \widehat{\varepsilon}_i\widehat{\varepsilon}_{i-j}, \qquad j = 0, 1, \ldots, n-1.$$

The vector of residual autocorrelations, $\widehat{r} = [\widehat{r}(1), \ldots, \widehat{r}(q)]'$, is used to test for $H_0$ : $\mathrm{E}[z_t] = 0$ for any $t$, where $z_t = (\varepsilon_{t-1}, \ldots, \varepsilon_{t-m})'\varepsilon_t$. The asymptotic joint distribution of $\widehat{r}$ has been analyzed by Box and Pierce [1] and McLeod [18]. When $\{\varepsilon_t\}$ is independent and identically distributed (i.i.d.), the asymptotic distribution of $\sqrt{n}\widehat{r}$ is a multivariate normal distribution with mean zero and an asymptotic covariance matrix that is approximately idempotent for large $m$. Therefore, both of the abobe papers proposed a portmanteau statistic, $Q_m = n\sum_{i=1}^{m}\widehat{r}(i)^2$ for $p + q < m < n$, which is approximately distributed as $\chi^2_{m-p-q}$. Ljung and Box [15] showed that a better approximation of $Q_m$ can be achieved using the following modified portmanteau statistic:

$$Q_m^* = n(n+2)\sum_{i=1}^{m}\frac{\widehat{r}(i)^2}{n-i}.$$

These statistics have been adopted by many practitioners, and have been modified or extended in various ways (see Li [14] and references therein).

## 2.1 The Portmanteau Test of Francq et al. [3]

The portmanteau tests using $Q_m^*$ are originally chi-squared tests, assuming the error is i.i.d. Francq et al. [3] established the asymptotic distribution of $Q_m^*$ under Assumption 1. The statistic $Q_m^*$ is no longer a chi-squared random variable, but is given by the weighted sums of the chi-squared random variables. Therefore, the present portmanteau test cannot control type I error. Francq et al. [3] established a portmanteau test using the asymptotic distribution of $Q_m^*$. From McLeod [18] and Francq et al. [3], we have:

$$\begin{aligned}
\widehat{r} &= \widehat{\gamma}/\sigma_0^2 + O_p(1/n), \\
\widehat{\gamma} &= \gamma + \sigma_0^2\Lambda_0'\big(\widehat{\theta}_n - \theta_0\big) + O_p(1/n),
\end{aligned} \tag{2}$$

where $\widehat{\gamma} = [\widehat{\gamma}(1), \ldots, \widehat{\gamma}(m)]'$, $\gamma = [\gamma(1), \ldots, \gamma(m)]'$,

$$\gamma(i) = \frac{1}{n} \sum_{j=i+1}^{n} \varepsilon_i \varepsilon_{j-i}, \quad i = 0, 1, \ldots, n-1,$$

$\Lambda_0 = \Lambda(\theta_0) = (\lambda_1, \ldots, \lambda_m)$ is an $m \times (p+q)$ matrix, and $\{\lambda_j\}$ is a $(p+q)$-vector of sequences defined by

$$\frac{\partial \varepsilon_t(\theta_0)}{\partial \theta} = \sum_{j=1}^{\infty} \lambda_j \varepsilon_{t-j}.$$

Note that $\text{rank}\{\Lambda(\theta)\} = p + q$ for any $\theta \in \Theta$. The distribution of $\sqrt{n}\{\gamma', (\widehat{\theta}_n - \theta_0)'\}'$ is asymptotically normal with mean zero and covariance matrix $\Sigma_{\gamma,\theta}$. Estimating this covariance matrix is not easy, as it is the long-run variance of a stationary process. For example, the asymptotic variance of $\sqrt{n}\gamma$ is given by:

$$\Gamma = \sum_{j=-\infty}^{\infty} \mathrm{E}\left(z_t z_{t-j}'\right).$$

When $\{\varepsilon_t\}$ is i.i.d., $\Gamma = \sigma_0^4 \mathbb{I}_m$. However, when $\{\varepsilon_t\}$ is uncorrelated but non-independent, $\Gamma$ is not always simple.

Francq et al. [3] also showed that, when $\{\varepsilon_t\}$ is uncorrelated but non-independent, the asymptotic variance of $\sqrt{n}\widehat{r}$ is no longer idempotent and the asymptotic distribution of $Q_m^*$ is the weighted sum of the chi-squared random variables. Therefore, their proposed portmanteau test with $Q_m^*$ uses critical regions of the non-pivotal distribution with the nonparametric estimator of $\Sigma_{\gamma,\theta}$. Francq et al. [3] referred to their portmanteau test as a modified Ljung–Box (MLB) test. Therefore, we call this the MLB test throughout this paper.

## 2.2 *The Portmanteau Test of Katayama [9]*

Francq et al. [3]'s MLB test must estimate critical values, because the asymptotic distribution is non-pivotal. Recently, Katayama [9] proposed another approach that provides a chi-squared distribution. First, let $D = \Lambda_0'(\Lambda_0 \Gamma^{-1} \Lambda_0')^{-1} \Lambda_0 \Gamma^{-1}$ and $S$ be the square root of $\Gamma$. Katayama [9] assumed that:

**Assumption 2** The matrix $S$ is nonsingular.

This assumption is satisfied for stationary, ergodic, and square-integrable MDSs; see, e.g., Francq and Zakoïan [5, Theorem 5.1]. From (2) and $(\mathbb{I}_m - D)\Lambda_0' = 0$, we have:

$$(\mathbb{I}_m - D)\widehat{\gamma} = (\mathbb{I}_m - D)\gamma + O_p(1/n),$$
$$S^{-1}(\mathbb{I}_m - D)\sqrt{n}\widehat{\gamma} \xrightarrow{d} N(0, \mathbb{I}_m - F(F'F)^{-1}F'),$$

where $F = S^{-1}\Lambda_0'$. Therefore, Katayama [9] proposed that

$$Q_m^{\mathrm{K}} = \widehat{\gamma}' T_n (\mathbb{I}_m - \widehat{D}')\widehat{\Gamma}^{-1}(\mathbb{I}_m - \widehat{D}) T_n \widehat{\gamma},$$

where $\widehat{D}$ is a $\sqrt{n}$-consistent estimator of $D$ and $T_n = \{n(n+2)\}^{1/2}\,\mathrm{diag}\{(n-1)^{-1/2}, \ldots, (n-m)^{-1/2}\}$. The matrix $T_n$ is the small-sample approximation of $\sqrt{n}\mathbb{I}_m$, similar to the weights of $Q_m^*$. The matrix $\widehat{\Gamma}$ is a consistent estimator of $\Gamma$ computed from nonparametric methods. Katayama [9] showed that $Q_m^{\mathrm{K}}$ is approximately $\chi^2_{m-p-q}$. However, simulations indicated that, similarly to the MLB test, the finite-sample properties of $Q_m^{\mathrm{K}}$ result in some size distortions as $m$ increases [9]. This may be due to the difficulty in establishing a non-parametric estimation of $\Gamma$.

## 2.3 The Portmanteau Test of Kuan and Lee [12] and Lee [13]

The main difficulty of conducting the Francq et al. [3] and Katayama [9] portmanteau tests is obtaining nonparametric estimates of $\Sigma_{\gamma,\theta}$ and $\Gamma$. These estimates require an approximation of the covariance matrix of a high-dimensional multivariate process and a large sample size. Following KVB, Kuan and Lee [12] and Lee [13] proposed an alternative approach. Their approach uses random normalized matrices to eliminate the nuisance covariance matrix. Let $\widetilde{\theta}_t$ denote the nonlinear least-squares estimator from subsample $\{y_i\}_{i=1}^t$, and let $\{\widetilde{\varepsilon}_i\}_{i=1}^t$ be the residual sequences given by $\widetilde{\theta}_t$. Define the matrices

$$\widehat{C}_n = \frac{1}{n}\sum_{i,j=1}^{n-1}\sum_{t=1}^{i}\sum_{s=1}^{j}\big\{(\kappa_{ij} - \kappa_{i,j+1}) - (\kappa_{i+1,j} - \kappa_{i+1,j+1})\big\}(\widehat{z}_t - \widehat{\gamma})(\widehat{z}_s - \widehat{\gamma}) \quad (3)$$

$$\widetilde{C}_n = \frac{1}{n}\sum_{i,j=1}^{n-1}\sum_{t=1}^{i}\sum_{s=1}^{j}\big\{(\kappa_{ij} - \kappa_{i,j+1}) - (\kappa_{i+1,j} - \kappa_{i+1,j+1})\big\}(\widetilde{z}_t - \widehat{\gamma})(\widetilde{z}_s - \widehat{\gamma}),$$

with $\widehat{z}_t = \widehat{\varepsilon}_t(\widehat{\varepsilon}_{t-1}, \ldots, \widehat{\varepsilon}_{t-m})'$ and $\widetilde{z}_t = \widetilde{\varepsilon}_t(\widetilde{\varepsilon}_{t-1}, \ldots, \widetilde{\varepsilon}_{t-m})'$. Additionally, $\kappa_{ij} = \kappa(|i-j|/n)$, where $\kappa$ denotes a kernel function. The main idea underlying the KVB approach is to employ a normalizing random matrix instead of estimating the asymptotic variance of $T_n\widehat{\gamma}$. Kuan and Lee [12] and Lee [13] considered two generalized test statistics. These are given by:

$$\widehat{Q}_m^{\mathrm{KL}} = \widehat{\gamma}' T_n \widehat{C}_n^{-1} T_n \widehat{\gamma},$$
$$\widetilde{Q}_m^{\mathrm{KL}} = \widehat{\gamma}' T_n \widetilde{C}_n^{-1} T_n \widehat{\gamma}.$$

Under conditions of the FCLT, Kuan and Lee [12] and Lee [13] showed that

$$T_n\widehat{\gamma} \xrightarrow{d} VW_m(1),$$
$$\widehat{C}_n \Rightarrow S'U_mS,$$
$$\widetilde{C}_n \Rightarrow V'U_mV,$$

where $V$ is the matrix square root of the asymptotic covariance matrix of $\sqrt{n}\widehat{\gamma}$, and $U_m = \int_0^1\int_0^1 \kappa(t-s)dB_m(t)dB_m(s)'$. It follows that

$$\widehat{Q}_m^{\mathrm{KL}} \xrightarrow{d} W_m(1)'V'(S'U_mS)^{-1}VW_m(1),$$
$$\widetilde{Q}_m^{\mathrm{KL}} \xrightarrow{d} W_m(1)'U_m^{-1}W_m(1).$$

Therefore, $\widetilde{Q}_m^{\mathrm{KL}}$ is an asymptotically pivotal distribution, critical values for which can be obtained via simulations. The critical values of $W_m(1)'U_m^{-1}W_m(1)$ are given by KVB (Table II), Lobato [16, Table 1], Kiefer and Vogelsang [10, Tables I and II], and Su [23, Table 1]. Note that Kuan and Lee [12] and Lee [13] assume $V$ is nonsingular. However, this assumption is restrictive, as Francq et al. [3] noted in their Remark 2 that $V$ may be singular. Additionally, because elements of $V$ are nonlinear functions of $\theta_0$, it is difficult to confirm this assumption.

## 3 New Portmanteau Tests and LM Tests Using the KVB Approach

The KVB-based portmanteau statistics proposed by Kuan and Lee [12] and Lee [13] do not estimate asymptotic covariance matrices of $\sqrt{n}\widehat{\gamma}$. However, these statistics contain a recursive estimator, and the assumption on the covariance matrix is restrictive. To solve these problems, in this section, we propose new KVB-based test statistics.

### 3.1 New Portmanteau Tests Using the KVB Approach

We now re-examine (2). Kuan and Lee's [12] approach was based on the asymptotic joint distribution of $\sqrt{n}(\gamma', (\widehat{\theta}_n - \theta_0)')$. However, the the asymptotic distribution of $\sqrt{n}(\widehat{\theta}_n - \theta_0)$ is cumbersome. Therefore, our approach eliminates this estimation effect in a similar manner to Katayama [9].

Let $\mathscr{P}_n^P = \mathbb{I}_m - \widehat{\Lambda}'(\widehat{\Lambda}\widehat{\Lambda}')^{-1}\widehat{\Lambda}$, where $\widehat{\Lambda} = \Lambda(\widehat{\theta}_n)$ and $\mathscr{P}_0^P = \mathbb{I}_m - \Lambda_0'(\Lambda_0\Lambda_0')^{-1}\Lambda_0$. Then, $\mathscr{P}_n^P \xrightarrow{p} \mathscr{P}_0^P$ and $\mathscr{P}_0^P\Lambda_0' = 0$. It follows from (2) that:

$$\mathscr{P}_n^P \widehat{\gamma} = \mathscr{P}_0^P \gamma + o_p(n^{-1/2}). \tag{4}$$

We now construct a KVB-based portmanteau statistic based on this equation. Under Assumption 1, the FCLT for NED functions of some mixing process $\{\varepsilon_t\}$ (Davidson [2], Corollary 29.19) gives:

$$\frac{1}{\sqrt{n}} \sum_{t=1}^{[n\tau]} z_t \Rightarrow S W_m(\tau) \tag{5}$$

for any $\tau \in (0, 1]$. It follows from (4), (5), and the continuous mapping theorem that:

$$T_n \mathscr{P}_n^P \widehat{\gamma} \xrightarrow{d} \Psi_0^P W_m(1) \tag{6}$$

and

$$\mathscr{P}_n^P \widehat{C}_n \mathscr{P}_n^{P'} \Rightarrow \Psi_0^P U_m \Psi_0^{P'}, \tag{7}$$

where $\Psi_0^P = \mathscr{P}_0^P S$ and $\widehat{C}_n$ is given by (3). We define the following portmanteau test statistic:

$$\mathcal{Q}_m^{\text{NEW}} = \widehat{\gamma} T_n \mathscr{P}_n^P \left( \mathscr{P}_n^P \widehat{C}_n \mathscr{P}_n^{P'} \right)^+ \mathscr{P}_n^P T_n \widehat{\gamma}.$$

Since $\mathscr{P}_n^P \widehat{C}_n \mathscr{P}_n^{P'}$ is singular with rank $m - p - q$, we use the MP inverse as a normalizing matrix. Thus, we obtain a new portmanteau test that extends those of Lobato [16] and Su [23] to the estimated parameter case.

**Theorem 1** *Given Assumptions* 1 *and* 2, $\mathcal{Q}_m^{\text{NEW}} \xrightarrow{d} W_{m-p-q}(1)' U_{m-p-q}^{-1} W_{m-p-q}(1)$.

*Proof* The necessary and sufficient condition for the continuity of the MP-inverse matrix is that the rank of the matrices is constant: $\text{rank}(\mathscr{P}_n^P \widehat{C}_n \mathscr{P}_n^{P'}) = \text{rank}(\Psi_0^P U_m \Psi_0^{P'})$; see, e.g., Schott [22, Theorem 5.21]. Because $\text{rank}\, \Lambda(\theta) = p + q$ for any $\theta \in \Theta$, we have $\text{rank}(\mathscr{P}_n^P \widehat{C}_n \mathscr{P}_n^{P'}) = \text{rank}(\mathscr{P}_n^P) = m - p - q$ and $\text{rank}(\Psi_0^P U_m \Psi_0^{P'}) = \text{rank}(\Psi_0^P) = \text{rank}(\mathscr{P}_0^P) = m - p - q$. Therefore, this matrix satisfies the continuity condition of the MP-inverse. It follows from (6) and (7) that

$$\mathcal{Q}_m^{\text{NEW}} \Rightarrow W_m(1)' \Psi_0^{P'} \left( \Psi_0^P U_m \Psi_0^{P'} \right)^+ \Psi_0^P W_m(1).$$

The rest of the proof is similar to that of Equation (9) in Kuan and Lee [12].

As noted by Kuan and Lee [12, Remark 2], we can modify $\widetilde{Q}_m^{\text{KL}}$ using the MP-inverse. However, it is difficult to estimate $\text{rank}(V)$, as $V$ is generally a complicated matrix. Our proposed portmanteau test overcomes this problem without using a recursive estimator.

## 3.2 New LM Test Using the KVB Approach

The LM test as a goodness-of-fit test of ARMA models is a special case of a test for a parameter constraint of a nonlinear regression model. Therefore, we briefly discuss LM tests for nonlinear regression models. Similar to our approach in the previous subsection, the new KVB-based LM test statistic uses a projection matrix. We now consider the following nonlinear regression model:

$$Y_t = f_t(Y^{t-1}; \beta) + \varepsilon_t, \tag{8}$$

where $Y_t$ is the $t$th observation of a dependent variable, $\beta$ is an $r$-dimensional vector of parameters to be estimated, and $f_t$ is a function of $Y^{t-1} = \{Y_j, j < t\}$ and $\beta$ and third-order differentiable with respect to $\beta$. We consider the null hypothesis $\beta_0 = c(\delta_0)$, where $\beta_0$ is a true parameter of $\beta$, $\delta_0$ is an $s$-dimensional constrained vector, and $c$ is a differentiable function from $\mathbb{R}^s$ to $\mathbb{R}^r$ with values in $\mathbb{R}^r$ and $r > s$. We set $e_t(\beta) = Y_t - f_t(Y^{t-1}; \beta)$, and define

$$\mathscr{L}_n(\beta) = -\frac{1}{2n}\sum_{t=1}^{n} e_t(\beta)^2 \tag{9}$$

as a quasi-maximum log-likelihood function. Let $\widehat{\delta}_n$ be a root-$n$ consistent estimator of $\delta_0$ and $\widehat{\beta}_n = c(\widehat{\delta}_n)$ so as to satisfy the first-order condition:

$$\left.\frac{\partial \mathscr{L}_n(c(\delta))}{\partial \delta}\right|_{\delta=\widehat{\delta}_n} = \left.\frac{\partial c(\delta)'}{\partial \delta}\frac{\partial \mathscr{L}_n(\beta)}{\partial \beta}\right|_{\delta=\widehat{\delta}_n,\ \beta=\widehat{\beta}_n} = 0. \tag{10}$$

The classical LM test is:

$$\mathrm{LM} = n\left.\frac{\partial \mathscr{L}_n(\beta)}{\partial \beta'}\mathrm{E}\left[\frac{\partial^2 \mathscr{L}_n(\beta)}{\partial \beta \partial \beta'}\right]^{-1}\frac{\partial \mathscr{L}_n(\beta)}{\partial \beta}\right|_{\beta=\widehat{\beta}_n}.$$

Under standard regularity conditions, and when $\{\varepsilon_t\}$ is i.i.d., this test statistic is asymptotically $\chi^2_{r-s}$ when $\beta_0 = c(\delta_0)$ is true; see, e.g., White [24, Section 10.1]. However, when $\{\varepsilon_t\}$ is not independent but uncorrelated, LM is not always approximately chi-squared, because the asymptotic variance of $\sqrt{n}$ times the score vector does not always coincide with the Fisher information matrix. One modification is to employ a nonparametric estimator of the asymptotic variance. Another is to use the KVB-based LM test statistic given by Kuan and Lee [12] with a recursive estimator.

We now propose another KVB-based LM test statistic with a full sample estimator. To proceed, we further suppose that

$$\left|-\nabla'_\beta \nabla_\beta \mathscr{L}_n(\beta) - \mathrm{E}\left[\nabla'_\beta e_t(\beta)\nabla_\beta e_t(\beta)\right]\right| \xrightarrow{p} 0 \tag{11}$$

uniformly in $\beta$. From (11) and the first-order Taylor series approximation around $\widehat{\delta}_n = \delta_0$, we have that:

$$\frac{\partial \mathscr{L}_n(\widehat{\beta}_n)}{\partial \beta} = \frac{\partial \mathscr{L}_n(\beta_0)}{\partial \beta} + \mathscr{I}_0 C_0'(\widehat{\delta}_n - \delta_0) + o_p(n^{-1/2}), \tag{12}$$

where $\mathscr{I}_0 = -\mathrm{E}[\nabla_\beta' e_t(\beta_0)\nabla_\beta e_t(\beta_0)]$ and $C_0 = \nabla_\delta c(\delta_0)'$. We define the matrices $\mathscr{P}_0^{\mathrm{LM}} = \mathbb{I}_r - \mathscr{I}_0 C_0'(C_0 \mathscr{I}_0 C_0')^{-1} C_0$ and $\mathscr{P}_n^{\mathrm{LM}} = \mathbb{I}_r - \mathscr{I}_n C_n'(C_n \mathscr{I}_n C_n')^{-1} C_n$, where $C_n = \nabla_\delta' c(\widehat{\delta}_n)$ and $\mathscr{I}_n$ denotes a consistent estimator of $\mathscr{I}_0$. These projection matrices are used in a similar way to $Q_m^{\mathrm{NEW}}$. From (12), we have that:

$$\frac{\partial \mathscr{L}_n(\widehat{\beta}_n)}{\partial \beta} = \mathscr{P}_n^{\mathrm{LM}} \frac{\partial \mathscr{L}_n(\widehat{\beta}_n)}{\partial \beta} = \mathscr{P}_0^{\mathrm{LM}} \frac{\partial \mathscr{L}_n(\beta_0)}{\partial \beta} + o_p(n^{-1/2}). \tag{13}$$

The first equality comes from (10), as $C_n \nabla_\beta' \mathscr{L}_n(\widehat{\beta}_n) = 0$. The second equality follows from (12), as $\mathscr{P}_n^{\mathrm{LM}}$ is a consistent estimator of $\mathscr{P}_0^{\mathrm{LM}}$ and $\mathscr{P}_0^{\mathrm{LM}} \mathscr{I}_0 C_0' = 0$. Therefore, if we suppose that $n^{1/2}\nabla_\beta' \mathscr{L}_n(\beta_0) \xrightarrow{d} GW_r(1)$, then (13) implies that $n^{1/2}\nabla_\beta' \mathscr{L}_n(\widehat{\beta}_n) \xrightarrow{d} \mathscr{P}_0^{\mathrm{LM}} GW_r(1)$. We note that the asymptotic variance of $n^{1/2}\nabla_\beta' \mathscr{L}_n(\widehat{\beta}_n)$, $\mathscr{K}_0 = GG'$, is not always equal to $\mathscr{I}_0$.

We define the following new LM test statistic:

$$\mathrm{LM}^{\mathrm{NEW}} = n \frac{\partial \mathscr{L}_n(\widehat{\beta}_n)}{\partial \beta'} \left( \sum_{i=1}^{n-1} \sum_{j=1}^{n-1} k_{ij} \widehat{\varphi}_i \widehat{\varphi}_j' \right)^{+} \frac{\partial \mathscr{L}_n(\widehat{\beta}_n)}{\partial \beta},$$

$$\widehat{\varphi}_j = \frac{1}{\sqrt{n}} \mathscr{P}_n^{\mathrm{LM}} \sum_{i=1}^{j} \left\{ -\frac{\partial e_i(\widehat{\beta}_n)}{\partial \beta} e_i(\widehat{\beta}_n) - \frac{\partial \mathscr{L}_n(\widehat{\beta}_n)}{\partial \beta} \right\}.$$

Theorem 2 gives the limiting distribution of the LM test statistic:

**Theorem 2** *Assume that*

*(i)* $\mathrm{Rank}(C_n \mathscr{I}_n C_n') = \mathrm{rank}(C_0 \mathscr{I}_0 C_0') = s$.

*(ii)* $\frac{1}{\sqrt{n}} \sum_{t=1}^{[n\tau]} \frac{\partial e_t(\beta_0)}{\partial \beta} e_t(\beta_0) \Rightarrow GW_r(\tau)$ *for any* $\tau \in (0, 1]$ *as* $n \to \infty$, *where* $G$ *is a* $r \times r$ *positive definite matrix.*

*(iii) Equation* (11) *or* (12) *holds.*

*Then,* $LM^{\mathrm{NEW}} \Rightarrow W_{r-s}(1)' U_{r-s}^{-1} W_{r-s}(1)$ *as* $n \to \infty$.

*Proof* The proof is similar to that for Theorem 1. Hence, it is omitted here.

This result can be applied to the goodness-of-fit test for ARMA models with uncorrelated errors, e.g., $H_0 : ARMA(p, q)$ against $H_1 : ARMA(p + m, q)$ and $H_0 : ARMA(p, q)$ against $H_1 : ARMA(p, q + m)$, where $p + q = s$ and $m =$

$r - s$. The constrained estimator $\widehat{\theta}_n$ is a quasi-maximum-likelihood estimator of ARMA$(p, q)$ and $\widehat{\beta}_n' = (\widehat{\theta}_n', 0')$. The residuals $\{e_t(\widehat{\beta}_n)\}$ are given by the residuals of ARMA$(p, q)$. The residuals $\{\nabla_\beta e_t(\widehat{\beta}_n)\}$ are derived from the residuals of the alternative model. These statistics can be computed using standard statistical software, such as R and SAS, as they are the same as for ARMA models with i.i.d. errors. The first-order Taylor series approximatin of (12) is obtaiend from the proof of Lemma 5 and Theorem 2 in Francq and Zakoïan [4]. For example, when the null model is AR(1) and the alternative model is AR$(1 + m)$, $\theta_0 = a_1^0$ and $\beta_0 = (1, 0, \ldots, 0)'\theta_0$. $f_t(Y^{t-1}; \beta_0) = a_1^0 Y_{t-1} + a_2^0 Y_{t-2} + \cdots + a_{m+1}^0 Y_{t-m-1}$, $e_t(\widehat{\beta}) = Y_t - \widehat{\theta}_n Y_{t-1}$, $\nabla_\beta e_t$ $(\widehat{\beta}_n) = -(Y_{t-1}, \ldots, Y_{t-m-1})$, and $\mathscr{I}_n$ are given by the sample mean of $\{\nabla_\beta' e_t(\widehat{\beta}_n)$ $\nabla_\beta e_t(\widehat{\beta}_n)\}$.

## 4 Some Simulation Studies

In this section, we examine the empirical size and power of the various portmanteau tests and the LM test to diagnose the goodness of fit of $AR(1)$ models.

### *4.1 Empirical Significance Level*

We first examine the empirical significance level of the following univariate AR(1) models $Y_t = a_1^0 Y_{t-1} + \epsilon_t$, where $\{\epsilon_t\}$ is defined by:

DGP 1 (Gaussian GARCH(1, 1) model): $\epsilon_t = \sigma_t z_t, \sigma_t^2 = 10^{-6} + 0.1\epsilon_{t-1}^2 + 0.8\sigma_{t-1}^2$, where $\{z_t\} \sim i.i.d.N(0, 1)$;

DGP 2 (Non-Gaussian ARCH(1) model): $\epsilon_t = \sigma_t v_t, \sigma_t^2 = 10^{-6} + 0.1\epsilon_{t-1}^2$, where $\{v_t\} \sim i.i.d.$ Skew Normal distribution with location, scale, and shape parameters (0.8, 1.0, 0);

DGP 3 (All-Pass ARMA(1, 1) model): $\epsilon_t = 0.8\epsilon_{t-1} + w_t - 0.8^{-1}w_{t-1}$, where $\{w_t\}$ is i.i.d. Student's $t$ distribution with 10 degrees of freedom;

DGP 4 (Bilinear model): $\epsilon_t = z_{t-1} + 0.5 z_{t-1}\epsilon_{t-2}$.

We selected these data generating processes (DGPs) from Francq et al. [3] and Lobato et al. [17]. DGPs 1 and 2 are MDS examples, and use the R function `garchSim` from the `fGarch` R package with default parameter values. DGPs 3 and 4 are non-MDS examples, where the parameters are given by Lobato et al. [17]. We set $a_1^0 = 0.9$ and considered sample sizes of $n = 200, 400$, and 3000 in each experiment.

Five different test statistics were examined. The first two have to estimate the long-run variance matrices:

(i) $Q_m^{\text{MLB}}$: Francq et al. [3]'s MLB portmanteau test (discussed in Sect. 2.1), where $M = 30$ in step 2 of Francq et al. [3].
(ii) $Q_m^{\text{K}}$: Katayama's [9] modified portmanteau test statistic (discussed in Sect. 2.2).

The remaining three test statistics are based on KVB, where we use sharp original kernels with the $\rho$ value proposed by Phillips et al. [20]:

(iii) $\widetilde{Q}^{\text{KL}}_{m,\rho}$: Kuan and Lee's KVB-based portmanteau statistics with the AR(1) recursive estimator (discussed in Sect. 2.3).
(iv) $Q^{\text{NEW}}_{m,\rho}$: Our proposed portmanteau test, described in Sect. 3.1.
(v) $\text{LM}^{\text{NEW}}_{m,\rho}$: Our proposed LM test, discussed in Sect. 3.2, where the null model is AR(1) and the alternative model is AR$(1+m)$.

The sharp original kernel $\kappa_\rho(x)$ is given by:

$$\kappa_\rho(x) = \begin{cases} (1-|x|)^\rho & |x| \leq 1 \\ 0 & \text{otherwise,} \end{cases}$$

where $\rho$ is a positive integer. When $\rho = 1$, the sharp original kernel is the usual Bartlett kernel. As $\rho$ increases, $\kappa_\rho(x)$ becomes concentrated at the origin with a sharper, more pronounced peak. We investigated the cases $\rho = 1, 8, 16, 32, 48, 64$; for reasons of space, we present the cases $\rho = 1, 16, 64$ here.

The asymptotic distributions of (iii)–(v) are $W_\tau(1)'U_\tau^{-1}W_\tau(1)$, where $\kappa = \kappa_\rho$ and $\tau = m$ or $m-1$. The critical values of the distribution are obtained by simulations. The Brownian motion and Brownian bridge process are approximated using the normalized partial sum of $n = 2000$ i.i.d. $N(0,1)$ random variables, and the simulation involves 30,000 replications. These critical values have also been computed by Su [23].

Tables 1 and 2 present the relative rejection frequencies (in %) for $m = 2, 6, 10, 14$ and $n = 200$, $n = 400$, respectively. The tests using $Q^{\text{MLB}}_m$, $Q^{\text{K}}_m$, and $\widetilde{Q}^{\text{KL}}_{m,\rho}$ seem to have noticeable under-rejection probabilities for larger $m$. Our proposed tests using $Q^{\text{NEW}}_{m,\rho}$ and $\text{LM}^{\text{NEW}}_{m,\rho}$ exhibit relatively stable sizes, which for a finite number of samples is one of the superior features of our proposed tests. The tests using $Q^{\text{NEW}}_{m,\rho}$ and $\text{LM}^{\text{NEW}}_{m,\rho}$ seem to have a slight under-rejection probability for DGP 1 as $m$ increases. The tests using $\text{LM}^{\text{NEW}}_{m,\rho}$ seem to have an over-rejection tendency for some cases when $n = 200$, though this is not observed when $n = 400$.

## 4.2 Empirical Power

We next conducted 3000 replications with $n = 200$ for the univariate AR(2) models defined by: $Y_t = a_1^0 Y_{t-1} + a_2^0 Y_{t-2} + \varepsilon_t$, where $a_1^0 = 0.9$, $a_2^0 = -0.15, -0.3$ and $\{\varepsilon_t\}$ is defined by DGPs $1, 2, \ldots, 4$. We fitted an AR(1) model and conducted the tests to a 5% significance level. Tables 3 and 4 present the empirical powers corresponding to the empirical size in Table 1; Table 3 corresponds to $a_2^0 = -0.15$ and Table 4 to $a_2^0 = -0.30$.

The tests using $\text{LM}^{\text{NEW}}_{m,64}$ were confirmed to be the most powerful in almost all cases. All three KVB-based tests produce an increase in power as $\rho$ increases, which

**Table 1** Empirical significance level of DGPs 1–4 ($n = 200$, $a_0 = 0.9$)

| | $m$ | $Q_m^{\mathrm{MLB}}$ | $Q_m^{\mathrm{K}}$ | $\widetilde{Q}_{m,1}^{\mathrm{KL}}$ | $\widetilde{Q}_{m,16}^{\mathrm{KL}}$ | $\widetilde{Q}_{m,64}^{\mathrm{KL}}$ | $Q_{m,1}^{\mathrm{NEW}}$ | $Q_{m,16}^{\mathrm{NEW}}$ | $Q_{m,64}^{\mathrm{NEW}}$ | $\mathrm{LM}_{m,1}^{\mathrm{NEW}}$ | $\mathrm{LM}_{m,16}^{\mathrm{NEW}}$ | $\mathrm{LM}_{m,64}^{\mathrm{NEW}}$ |
|---|---|---|---|---|---|---|---|---|---|---|---|---|
| DGP1 | 2 | 3.5 | 4.9 | 2.5 | 1.9 | 1.6 | 4.2 | 4.4 | 4.9 | 4.3 | 4.3 | 4.7 |
| | 6 | 2.5 | 4.1 | 1.4 | 1.4 | 1.5 | 3.5 | 3.2 | 3.7 | 3.2 | 3.1 | 3.9 |
| | 10 | 1.6 | 2.8 | 1.1 | 1.0 | 1.4 | 2.4 | 2.9 | 3.7 | 3.3 | 3.5 | 4.4 |
| | 14 | 1.4 | 2.0 | 0.8 | 0.8 | 1.1 | 2.5 | 2.7 | 4.1 | 3.4 | 3.6 | 5.2 |
| DGP2 | 2 | 3.2 | 4.9 | 3.1 | 2.9 | 2.0 | 4.6 | 5.1 | 4.4 | 4.5 | 4.4 | 4.3 |
| | 6 | 2.3 | 4.0 | 2.3 | 2.0 | 1.9 | 5.1 | 5.4 | 5.1 | 5.5 | 5.4 | 4.9 |
| | 10 | 1.7 | 2.7 | 2.0 | 1.9 | 2.0 | 4.6 | 4.9 | 5.8 | 5.6 | 5.9 | 5.9 |
| | 14 | 1.3 | 1.7 | 1.9 | 1.9 | 1.8 | 4.6 | 4.8 | 5.8 | 6.0 | 6.2 | 7.5 |
| DGP3 | 2 | 4.0 | 5.8 | 3.3 | 3.0 | 2.2 | 5.2 | 5.1 | 5.4 | 5.0 | 4.7 | 4.3 |
| | 6 | 2.1 | 3.9 | 2.2 | 2.1 | 1.8 | 4.7 | 4.9 | 4.7 | 5.2 | 5.3 | 4.8 |
| | 10 | 1.6 | 3.3 | 2.0 | 1.9 | 1.8 | 4.5 | 4.4 | 5.3 | 5.2 | 5.5 | 6.1 |
| | 14 | 1.4 | 2.1 | 1.6 | 1.7 | 1.9 | 4.3 | 4.5 | 5.2 | 5.3 | 5.8 | 7.9 |
| DGP4 | 2 | 3.9 | 6.7 | 3.7 | 2.7 | 2.1 | 5.2 | 4.7 | 4.9 | 5.2 | 5.7 | 7.1 |
| | 6 | 2.5 | 4.4 | 2.2 | 2.0 | 1.6 | 4.7 | 4.3 | 4.8 | 4.8 | 5.1 | 5.1 |
| | 10 | 2.0 | 3.3 | 1.1 | 0.9 | 1.4 | 3.8 | 4.0 | 4.8 | 4.4 | 4.0 | 5.1 |
| | 14 | 1.7 | 2.4 | 1.0 | 1.1 | 1.4 | 3.2 | 3.7 | 4.9 | 4.3 | 4.8 | 7.0 |

**Table 2** Empirical significance level of DGPs 1–4 ($n = 400$, $a_0 = 0.9$)

| | $m$ | $Q_m^{\mathrm{MLB}}$ | $Q_m^{\mathrm{K}}$ | $\widetilde{Q}_{m,1}^{\mathrm{KL}}$ | $\widetilde{Q}_{m,16}^{\mathrm{KL}}$ | $\widetilde{Q}_{m,64}^{\mathrm{KL}}$ | $Q_{m,1}^{\mathrm{NEW}}$ | $Q_{m,16}^{\mathrm{NEW}}$ | $Q_{m,64}^{\mathrm{NEW}}$ | $\mathrm{LM}_{m,1}^{\mathrm{NEW}}$ | $\mathrm{LM}_{m,16}^{\mathrm{NEW}}$ | $\mathrm{LM}_{m,64}^{\mathrm{NEW}}$ |
|---|---|---|---|---|---|---|---|---|---|---|---|---|
| DGP1 | 2 | 4.6 | 5.5 | 3.7 | 3.1 | 2.6 | 5.1 | 4.8 | 4.8 | 4.7 | 5.0 | 5.1 |
| | 6 | 3.4 | 4.5 | 2.1 | 1.8 | 1.9 | 3.8 | 3.7 | 4.0 | 3.9 | 4.2 | 4.5 |
| | 10 | 2.6 | 3.5 | 1.6 | 1.3 | 1.4 | 3.6 | 3.4 | 3.5 | 3.4 | 3.2 | 3.7 |
| | 14 | 2.3 | 3.3 | 1.0 | 1.1 | 1.4 | 2.4 | 2.3 | 3.3 | 2.5 | 2.8 | 3.8 |
| DGP2 | 2 | 4.1 | 5.5 | 4.5 | 4.0 | 2.8 | 5.5 | 5.4 | 5.1 | 5.0 | 5.0 | 5.2 |
| | 6 | 3.4 | 4.2 | 2.3 | 2.6 | 2.6 | 5.0 | 5.0 | 4.7 | 5.1 | 4.9 | 4.7 |
| | 10 | 3.3 | 3.5 | 2.4 | 2.5 | 2.3 | 5.1 | 5.0 | 4.7 | 5.5 | 5.5 | 5.5 |
| | 14 | 2.3 | 2.7 | 2.1 | 2.1 | 2.1 | 4.8 | 4.5 | 4.2 | 5.3 | 5.3 | 5.6 |
| DGP3 | 2 | 4.3 | 4.8 | 3.5 | 3.7 | 2.6 | 4.8 | 5.1 | 4.9 | 4.5 | 4.6 | 4.2 |
| | 6 | 3.8 | 4.4 | 2.9 | 2.7 | 2.5 | 5.1 | 5.3 | 4.9 | 5.4 | 4.9 | 5.2 |
| | 10 | 3.4 | 3.8 | 2.4 | 2.5 | 2.7 | 4.9 | 5.4 | 5.2 | 5.6 | 5.4 | 5.1 |
| | 14 | 2.9 | 3.2 | 2.3 | 2.3 | 2.5 | 4.7 | 4.7 | 5.1 | 5.4 | 5.5 | 6.0 |
| DGP4 | 2 | 4.3 | 5.8 | 3.8 | 3.6 | 2.5 | 4.8 | 4.6 | 4.0 | 5.1 | 5.3 | 5.9 |
| | 6 | 3.0 | 4.3 | 2.4 | 2.5 | 2.1 | 5.0 | 5.1 | 4.6 | 5.7 | 5.5 | 5.1 |
| | 10 | 2.8 | 3.9 | 1.6 | 1.4 | 1.6 | 4.5 | 4.6 | 4.0 | 5.0 | 4.8 | 4.6 |
| | 14 | 2.5 | 3.6 | 1.4 | 1.6 | 2.1 | 4.1 | 4.1 | 4.4 | 4.0 | 4.2 | 4.8 |

**Table 3** Empirical power of DGPs 1–4 ($n = 200$, $a_0 = 0.9$, $\alpha_2^0 = -0.15$)

| | $m$ | $Q_m^{\mathrm{MLB}}$ | $Q_m^{\mathrm{K}}$ | $\widetilde{Q}_{m,1}^{\mathrm{KL}}$ | $\widetilde{Q}_{m,16}^{\mathrm{KL}}$ | $\widetilde{Q}_{m,64}^{\mathrm{KL}}$ | $Q_{m,1}^{\mathrm{NEW}}$ | $Q_{m,16}^{\mathrm{NEW}}$ | $Q_{m,64}^{\mathrm{NEW}}$ | $\mathrm{LM}_{m,1}^{\mathrm{NEW}}$ | $\mathrm{LM}_{m,16}^{\mathrm{NEW}}$ | $\mathrm{LM}_{m,64}^{\mathrm{NEW}}$ |
|---|---|---|---|---|---|---|---|---|---|---|---|---|
| DGP1 | 2 | 35.0 | 33.4 | 15.9 | 20.6 | 19.9 | 24.3 | 31.6 | 34.1 | 32.3 | 45.2 | 48.9 |
| | 6 | 18.3 | 18.8 | 5.9 | 7.4 | 11.1 | 12.2 | 16.4 | 22.1 | 13.6 | 17.5 | 23.7 |
| | 10 | 10.9 | 11.6 | 3.2 | 4.4 | 6.6 | 8.7 | 10.2 | 15.2 | 9.1 | 10.6 | 16.3 |
| | 14 | 7.2 | 7.0 | 2.4 | 2.9 | 4.5 | 7.5 | 8.7 | 13.4 | 6.3 | 8.2 | 14.5 |
| DGP2 | 2 | 40.1 | 38.9 | 19.6 | 23.4 | 21.6 | 28.0 | 36.5 | 37.7 | 36.1 | 47.4 | 50.5 |
| | 6 | 21.3 | 20.5 | 10.1 | 12.7 | 14.0 | 17.6 | 20.8 | 24.4 | 17.8 | 21.9 | 26.7 |
| | 10 | 12.9 | 12.5 | 5.6 | 8.1 | 10.1 | 13.4 | 15.4 | 19.9 | 14.0 | 15.8 | 20.9 |
| | 14 | 8.7 | 7.0 | 5.3 | 6.0 | 7.3 | 10.8 | 12.1 | 17.0 | 11.9 | 13.9 | 19.6 |
| DGP3 | 2 | 43.0 | 37.1 | 20.0 | 25.1 | 21.7 | 28.9 | 37.4 | 38.0 | 39.0 | 52.8 | 55.5 |
| | 6 | 22.2 | 22.4 | 8.9 | 12.0 | 14.1 | 17.6 | 21.9 | 26.1 | 18.4 | 22.7 | 26.9 |
| | 10 | 12.5 | 12.5 | 5.7 | 6.4 | 8.4 | 12.6 | 15.5 | 19.3 | 14.1 | 15.6 | 20.8 |
| | 14 | 7.9 | 7.6 | 4.5 | 5.4 | 7.3 | 11.1 | 12.9 | 17.9 | 11.3 | 13.2 | 19.5 |
| DGP4 | 2 | 33.3 | 34.4 | 16.4 | 20.9 | 20.3 | 24.5 | 32.0 | 31.7 | 30.8 | 43.3 | 48.8 |
| | 6 | 19.1 | 17.7 | 7.1 | 9.1 | 11.5 | 15.4 | 17.5 | 20.2 | 16.1 | 19.2 | 25.6 |
| | 10 | 11.7 | 9.2 | 4.8 | 5.9 | 8.0 | 10.2 | 11.2 | 15.3 | 10.5 | 12.7 | 18.7 |
| | 14 | 7.9 | 5.9 | 3.6 | 4.3 | 5.3 | 8.1 | 9.3 | 14.0 | 8.8 | 10.6 | 17.1 |

**Table 4** Empirical power of DGPs 1–4 ($n = 200$, $a_0 = 0.9$, $\alpha_2^0 = -0.30$)

| | $m$ | $Q_m^{\text{MLB}}$ | $Q_m^{\text{K}}$ | $\widetilde{Q}_{m,1}^{\text{KL}}$ | $\widetilde{Q}_{m,16}^{\text{KL}}$ | $\widetilde{Q}_{m,64}^{\text{KL}}$ | $Q_{m,1}^{\text{NEW}}$ | $Q_{m,16}^{\text{NEW}}$ | $Q_{m,64}^{\text{NEW}}$ | $\text{LM}_{m,1}^{\text{NEW}}$ | $\text{LM}_{m,16}^{\text{NEW}}$ | $\text{LM}_{m,64}^{\text{NEW}}$ |
|---|---|---|---|---|---|---|---|---|---|---|---|---|
| DGP1 | 2 | 88.8 | 86.1 | 47.9 | 69.0 | 70.9 | 63.6 | 84.7 | 88.5 | 72.1 | 94.2 | 96.8 |
| | 6 | 75.3 | 58.9 | 23.4 | 36.7 | 55.3 | 38.2 | 55.5 | 74.4 | 38.5 | 57.4 | 76.6 |
| | 10 | 58.4 | 43.7 | 14.2 | 20.3 | 37.5 | 27.1 | 36.5 | 59.3 | 25.5 | 35.5 | 59.8 |
| | 14 | 43.8 | 27.5 | 9.2 | 12.8 | 27.3 | 21.8 | 27.1 | 49.0 | 21.0 | 26.9 | 50.0 |
| DGP2 | 2 | 93.7 | 89.9 | 54.2 | 74.9 | 74.9 | 71.7 | 88.5 | 90.7 | 79.0 | 95.9 | 97.8 |
| | 6 | 84.3 | 71.3 | 33.3 | 48.6 | 64.1 | 51.2 | 68.1 | 81.3 | 51.1 | 68.8 | 82.8 |
| | 10 | 68.6 | 51.9 | 23.9 | 31.6 | 47.0 | 40.5 | 49.5 | 67.2 | 41.3 | 51.3 | 69.2 |
| | 14 | 55.3 | 32.1 | 16.2 | 21.2 | 35.1 | 32.2 | 38.8 | 57.0 | 33.2 | 41.3 | 59.7 |
| DGP3 | 2 | 96.1 | 91.6 | 57.6 | 78.2 | 77.2 | 72.1 | 90.6 | 93.0 | 80.3 | 97.4 | 98.5 |
| | 6 | 87.4 | 74.3 | 35.0 | 50.9 | 65.4 | 52.4 | 70.9 | 85.6 | 52.8 | 71.3 | 85.5 |
| | 10 | 72.0 | 55.5 | 24.5 | 32.9 | 48.6 | 41.0 | 51.8 | 70.4 | 40.8 | 51.9 | 71.5 |
| | 14 | 57.7 | 35.2 | 18.1 | 23.7 | 37.5 | 32.3 | 39.1 | 59.6 | 34.1 | 41.2 | 62.0 |
| DGP4 | 2 | 86.8 | 83.3 | 50.4 | 68.6 | 70.7 | 64.6 | 82.9 | 85.5 | 70.2 | 91.7 | 95.2 |
| | 6 | 74.7 | 57.7 | 28.6 | 40.6 | 56.2 | 43.4 | 58.4 | 74.3 | 44.2 | 59.3 | 76.0 |
| | 10 | 61.4 | 40.0 | 16.8 | 24.1 | 39.7 | 31.8 | 40.3 | 58.0 | 31.8 | 41.3 | 60.2 |
| | 14 | 49.8 | 25.8 | 12.1 | 16.0 | 28.4 | 25.5 | 31.7 | 49.1 | 25.2 | 32.4 | 52.3 |

**Table 5** Empirical size-adjusted power of DGPs 1–4 ($n = 200$, $a_0 = 0.9$, $\alpha_2^0 = -0.15$)

| | $m$ | $Q_m^{\text{MLB}}$ | $Q_m^{\text{K}}$ | $\widetilde{Q}_{m,1}^{\text{KL}}$ | $\widetilde{Q}_{m,16}^{\text{KL}}$ | $\widetilde{Q}_{m,64}^{\text{KL}}$ | $Q_{m,1}^{\text{NEW}}$ | $Q_{m,16}^{\text{NEW}}$ | $Q_{m,64}^{\text{NEW}}$ | $\text{LM}_{m,1}^{\text{NEW}}$ | $\text{LM}_{m,16}^{\text{NEW}}$ | $\text{LM}_{m,64}^{\text{NEW}}$ |
|---|---|---|---|---|---|---|---|---|---|---|---|---|
| DGP1 | 2 | 29.9 | 33.9 | 24.6 | 32.2 | 32.8 | 26.3 | 33.7 | 34.2 | 36.0 | 49.2 | 50.0 |
| | 6 | 27.3 | 21.4 | 14.4 | 20.3 | 25.2 | 16.2 | 21.7 | 27.1 | 18.8 | 23.0 | 27.2 |
| | 10 | 23.1 | 18.2 | 12.6 | 15.0 | 19.1 | 14.4 | 15.7 | 18.3 | 13.2 | 14.3 | 17.8 |
| | 14 | 21.0 | 14.6 | 10.2 | 11.7 | 15.0 | 13.0 | 13.8 | 15.4 | 10.9 | 11.5 | 13.8 |
| DGP2 | 2 | 35.2 | 39.5 | 26.6 | 32.4 | 34.4 | 29.3 | 35.8 | 39.2 | 38.0 | 49.2 | 53.0 |
| | 6 | 31.8 | 23.1 | 15.9 | 19.4 | 23.1 | 17.5 | 20.0 | 24.2 | 16.8 | 21.3 | 26.9 |
| | 10 | 26.0 | 18.5 | 14.5 | 16.1 | 17.1 | 14.5 | 15.6 | 17.7 | 12.9 | 13.5 | 18.3 |
| | 14 | 24.1 | 16.3 | 10.1 | 11.4 | 14.5 | 11.4 | 12.8 | 15.4 | 10.5 | 12.3 | 14.5 |
| DGP3 | 2 | 34.2 | 33.2 | 26.1 | 33.7 | 33.2 | 28.2 | 36.9 | 37.4 | 39.0 | 53.6 | 57.5 |
| | 6 | 32.7 | 25.1 | 17.2 | 21.0 | 26.3 | 18.4 | 22.3 | 27.0 | 18.2 | 21.4 | 27.2 |
| | 10 | 26.2 | 16.9 | 11.9 | 15.5 | 18.7 | 14.1 | 17.0 | 18.3 | 13.5 | 14.9 | 18.1 |
| | 14 | 23.2 | 14.4 | 12.4 | 13.0 | 14.7 | 12.7 | 14.2 | 17.5 | 11.2 | 12.0 | 13.2 |
| DGP4 | 2 | 24.4 | 28.8 | 20.7 | 29.6 | 29.7 | 23.5 | 32.4 | 32.4 | 29.3 | 41.5 | 44.8 |
| | 6 | 24.0 | 19.3 | 14.3 | 17.9 | 21.4 | 16.5 | 18.9 | 21.2 | 16.5 | 19.0 | 25.3 |
| | 10 | 22.3 | 14.4 | 12.6 | 14.5 | 16.6 | 13.1 | 14.3 | 15.7 | 12.2 | 14.7 | 18.7 |
| | 14 | 21.4 | 11.6 | 11.3 | 12.0 | 15.2 | 11.4 | 12.4 | 14.3 | 10.0 | 10.8 | 13.1 |

**Table 6** Empirical size-adjusted power of DGPs 1–4 ($n = 200$, $a_0 = 0.9$, $\alpha_2^0 = -0.30$)

| | $m$ | $Q_m^{\text{MLB}}$ | $Q_m^{\text{K}}$ | $\widetilde{Q}_{m,1}^{\text{KL}}$ | $\widetilde{Q}_{m,16}^{\text{KL}}$ | $\widetilde{Q}_{m,64}^{\text{KL}}$ | $Q_{m,1}^{\text{NEW}}$ | $Q_{m,16}^{\text{NEW}}$ | $Q_{m,64}^{\text{NEW}}$ | $\text{LM}_{m,1}^{\text{NEW}}$ | $\text{LM}_{m,16}^{\text{NEW}}$ | $\text{LM}_{m,64}^{\text{NEW}}$ |
|---|---|---|---|---|---|---|---|---|---|---|---|---|
| DGP1 | 2 | 87.3 | 86.5 | 61.5 | 81.5 | 83.0 | 66.9 | 86.0 | 88.5 | 75.4 | 92.9 | 97.1 |
| | 6 | 86.4 | 62.3 | 41.5 | 60.7 | 75.5 | 44.3 | 63.8 | 78.9 | 46.8 | 57.9 | 80.0 |
| | 10 | 79.1 | 54.4 | 34.1 | 44.9 | 62.5 | 38.8 | 47.2 | 64.7 | 32.7 | 37.2 | 62.2 |
| | 14 | 74.9 | 45.8 | 29.1 | 36.1 | 50.8 | 30.6 | 37.1 | 52.6 | 27.9 | 30.0 | 48.9 |
| DGP2 | 2 | 91.1 | 90.1 | 64.9 | 82.3 | 83.9 | 73.7 | 88.2 | 91.5 | 80.5 | 96.3 | 98.1 |
| | 6 | 91.2 | 74.1 | 47.1 | 62.7 | 76.8 | 50.9 | 66.0 | 81.0 | 48.9 | 66.9 | 82.9 |
| | 10 | 85.4 | 63.1 | 39.6 | 47.4 | 61.6 | 41.7 | 49.8 | 63.8 | 39.0 | 47.2 | 66.1 |
| | 14 | 82.0 | 52.2 | 29.4 | 36.3 | 52.9 | 33.9 | 39.7 | 54.1 | 30.2 | 37.2 | 51.5 |
| DGP3 | 2 | 93.8 | 89.8 | 65.5 | 84.7 | 84.9 | 71.2 | 90.4 | 92.6 | 80.3 | 97.6 | 98.8 |
| | 6 | 93.4 | 77.8 | 49.5 | 66.8 | 81.9 | 53.8 | 71.3 | 86.3 | 52.4 | 69.8 | 85.6 |
| | 10 | 87.6 | 63.4 | 38.6 | 50.6 | 66.4 | 43.5 | 55.1 | 69.2 | 39.6 | 50.5 | 67.8 |
| | 14 | 82.9 | 50.0 | 36.8 | 43.0 | 56.2 | 34.6 | 41.3 | 59.1 | 33.7 | 38.7 | 52.6 |
| DGP4 | 2 | 80.5 | 78.4 | 56.2 | 77.9 | 79.3 | 63.4 | 83.3 | 85.6 | 68.9 | 91.2 | 93.8 |
| | 6 | 82.2 | 60.5 | 42.7 | 58.4 | 70.9 | 44.4 | 60.5 | 75.4 | 44.7 | 58.8 | 75.9 |
| | 10 | 77.9 | 50.1 | 35.0 | 45.9 | 57.8 | 37.7 | 45.8 | 58.9 | 34.6 | 45.1 | 60.2 |
| | 14 | 74.5 | 41.1 | 29.6 | 35.5 | 49.6 | 31.8 | 37.3 | 49.6 | 28.1 | 32.9 | 45.0 |

is consistent with the asymptotic power envelope under the local alternatives given by Phillips et al. [20, 21]. Tests using $Q_m^{\text{MLB}}$ and $Q_{m,64}^{\text{NEW}}$ were also powerful, although the $Q_m^{\text{MLB}}$ case showed a serious under-rejection frequency. It is interesting that our proposed tests, $Q_{m,\rho}^{\text{NEW}}$ and $\text{LM}_{m,\rho}^{\text{NEW}}$, give similar powers for $m = 6, 10, 14$. This similarity is explained by Hosking [7, Section 4]. The portmanteau tests examine the goodness of fit without particular alternatives. However, Hosking [7, Section 4] noted that portmanteau tests can be approximately interpreted as LM tests for a particular form of ARMA models.

To compare the potential power properties, we also computed the size-adjusted powers; the results are listed in Tables 5 and 6. The tests using $Q_m^{\text{MLB}}$ are most powerful for $m = 6, 10, 14$. We confirmed that tests using $\text{LM}_{2,64}^{\text{NEW}}$ are the most powerful, and that $Q_{2,64}^{\text{NEW}}$ have a comparatively greater power than $Q_2^{\text{MLB}}$. Our proposed portmanteau test $Q_{m,\rho}^{\text{NEW}}$ exhibited a superior power to Kuan and Lee's $\widetilde{Q}_{m,\rho}^{\text{KL}}$.

From these simulations, we can state that our proposed tests are sufficiently efficient in terms of their empirical size and power properties compared with existing portmanteau tests. Besides their empirical size and power, our proposed tests are also superior in terms of computational cost. As $m$ increases, $Q_m^{\text{MLB}}$, $Q_m^{\text{K}}$, and the LM test need a large sample size $n$, because these statistics have to estimate long-run variance matrices containing $\Gamma$. In summary, we recommend our proposed test for determining the goodness of fit for the ARMA model.

**Acknowledgements** We would like to thank the editors, and the referees, Prof. Kohtaro Hitomi, Prof. Yoshihiko Nishiyama, Prof. Eiji Kurozumi, and Prof. Katsuto Tanaka, Prof. Tim Vogelsang, Prof. Shiqing Ling, Prof. Wai Keung Li, and Prof. A. Ian McLeod for their valuable comments and suggestions. An earlier version of this paper was presented at the 23rd (EC)2-conference at Maastricht University, held on 14–15 December 2012, and Festschrift for Prof. A. I. McLeod at Western University, held on 2–3 June 2014. I would also like to thank all participants of these conferences. This work is supported by JSPS Kakenhi (Grant Nos. 26380278, 26380279), Kansai University's overseas research program for the academic year of 2015, and Ministry of Scienece and Technology, TAIWAN for the project from MOST 104-2811-H-006-005.

## References

1. Box, G. E. P., & Pierce, A. (1970). Distribution of residual autocorrelations in autoregressive-integrated moving average time series models. *Journal of the American Statistical Association, 65*, 1509–1526.
2. Davidson, J. (1994). *Stochastic limit theory*. Oxford: Oxford University Press.
3. Francq, C., Roy, R., & Zakoïan, J. M. (2005). Diagnostic checking in ARMA models with uncorrelated errors. *Journal of the American Statistical Association, 100*, 532–544.
4. Francq, C., & Zakoïan, J. M. (1998). Estimating linear representations of nonlinear processes. *Journal of Statistical Planning and Inference, 68*(1), 145–165.
5. Francq, C., & Zakoïan, J. M. (2010). *GARCH models: Structure, statistical inference and financial applications*. Chichester: Wiley.
6. Godfrey, L. G. (1979). Testing the adequacy of a time series model. *Biometrika, 66*, 67–72.
7. Hosking, J. R. M. (1980). Lagrange-multiplier tests of time-series models. *Journal of the Royal Statistical Society: Series B, 42*, 170–181.

8. Katayama, N. (2013). *Proposal of robust M tests and their applications*. Woking paper series F-65, Economic Society of Kansai University.
9. Katayama, N. (2012). Chi-squared portmanteau tests for structural VARMA models with uncorrelated errors. *Journal of Time Series Analysis, 33*(6), 863–872.
10. Kiefer, N. M., & Vogelsang, T. J. (2002). Heteroskedasticity-autocorrelation robust testing using bandwidth equal to sample size. *Journal of Time Series Analysis, 18*, 1350–1366.
11. Kiefer, N. M., Vogelsang, T. J., & Bunzel, H. (2000). Simple robust testing of regression hypotheses. *Journal of Time Series Analysis, 68*, 695–714.
12. Kuan, C. M., & Lee, W. M. (2006). Robust *M* tests without consistent estimation of the asymptotic covariance matrix. *Journal of Time Series Analysis, 101*, 1264–1275.
13. Lee, W. M. (2007). Robust *M* tests using kernel-based estimators with bandwidth equal to sample size. *Economic Letters, 96*, 295–300.
14. Li, W. K. (2003). *Diagnostic checks in time series*. Boca Raton, FL: CRC Press.
15. Ljung, G. M., & Box, G. E. P. (1978). On a measure of lack of fit in time series models. *Biometrika, 65*(2), 297–303.
16. Lobato, I. N. (2001). Testing that dependent process is uncorrelated. *Biometrika, 96*, 1066–1076.
17. Lobato, I. N., Nankervis, J. C., & Savin, N. E. (2002). Testing for zero autocorrelation in the presence of statistical dependence. *Biometrika, 18*(3), 730–743.
18. McLeod, A. I. (1978). On the distribution of residual autocorrelations in Box–Jenkins models. *Journal of the Royal Statistical Society: Series B, 40*, 296–302.
19. Newbold, P. (1980). The equivalence of two tests of time series model adequacy. *Biometrika, 67*(2), 463–465.
20. Phillips, P. C. B., Sun, Y., & Jin, S. (2003). Consistent HAC estimation and robust regression testing using sharp original kernels with no truncation. Cowles Foundation discussion paper no. 1407.
21. Phillips, P. C. B., Sun, Y., & Jin, S. (2007). Long run variance estimation and robust regression testing using sharp origin kernels with no truncation. *Biometrika, 137*(3), 837–894.
22. Schott, J. R. (1997). *Matrix analysis for statistics*. New York, NY: Wiley.
23. Su, J. J. (2005). On the size and power of testing for no autocorrelation under weak assumptions. *Biometrika, 15*, 247–257.
24. White, H. (1994). *Estimation, inference, and specification analysis*. Cambridge: Cambridge University Press.

# Generalized $C(\alpha)$ Tests for Estimating Functions with Serial Dependence

**Jean-Marie Dufour, Alain Trognon and Purevdorj Tuvaandorj**

**Abstract** We propose generalized $C(\alpha)$ tests for testing linear and nonlinear parameter restrictions in models specified by estimating functions. The proposed procedures allow for general forms of serial dependence and heteroskedasticity, and can be implemented using any root-$n$ consistent restricted estimator. The asymptotic distribution of the proposed statistic is established under weak regularity conditions. We show that earlier $C(\alpha)$-type statistics are included as special cases. The problem of testing hypotheses fixing a subvector of the complete parameter vector is discussed in detail as another special case. We also show that such tests provide a simple general solution to the problem of accounting for estimated parameters in the context of two-step procedures where a subvector of model parameters is estimated in a first step and then treated as fixed.

J.-M. Dufour (✉)
Department of Economics, McGill University, Leacock Building, Room 414, 855 Sherbrooke Street West, Montréal, Québec H3A 2T7, Canada
e-mail: jean-marie.dufour@mcgill.ca
URL: http://www.jeanmariedufour.com

J.-M. Dufour
Centre Interuniversitaire de Recherche en Économie quantitative (CIREQ) and Centre Interuniversitaire de Recherche en Analyse des Organisations (CIRANO), Montréal, Canada

A. Trognon
CREST-ENSAE (Centre de Recherche en Économie et Statistique), CREST-PARIS, Timbre J310, 15 Boulevard Gabriel Péri, 92254 Malakoff Cedex, France
e-mail: trognon@ensae.fr
URL: http://www.crest.fr/component/

A. Trognon
University Paris 1, Paris, France

P. Tuvaandorj
CREST-ENSAI, ENSAI-Campus de Ker-Lann, Rue Blaise Pascal, BP 37203, 35712 Bruz Cedex, France
e-mail: purevdorj.tuvaandorj@ensai.fr
URL: https://sites.google.com/site/purevdorjtuvaandorj/

 
W.K. Li et al. (eds.), *Advances in Time Series Methods and Applications*,
Fields Institute Communications 78, DOI 10.1007/978-1-4939-6568-7_7

**Keywords** Testing · $C(\alpha)$ test · Estimating function · Generalized method of moment (GMM) · Serial dependence · Pseudo-likelihood · $M$-estimator · Nonlinear model · Score test · Lagrange multiplier test · Heteroskedasticity

## 1 Introduction

The $C(\alpha)$ statistic introduced by Neyman [52] embodies a general mechanism for dealing with nuisance parameters in tests of composite hypotheses. The basic idea of the method can be conveniently explained by using parameter subvector testing as an example. One first considers a score-type function for the tested parameter. The score function is then orthogonalized with respect to directions associated with the nuisance parameters under the null hypothesis. This removes the impact of the estimation error on the nuisance parameter: the residual vector from the projection—the *effective score function*—evaluated at the auxiliary estimator of the nuisance parameter is asymptotically equivalent to the effective score function evaluated at the true parameter. It is easy to see that the latter is asymptotically normally distributed, and consequently its normalized form—the $C(\alpha)$ statistic—has an asymptotic chi-square distribution under the null hypothesis.

The $C(\alpha)$ test enjoys a local optimality property while being computationally attractive (a few artificial regressions would be enough in many circumstances) and uses only $\sqrt{n}$-consistent estimator for the nuisance parameters which may not be asymptotically normal or even may not have an asymptotic distribution. When the restricted maximum likelihood (ML) estimator is used, the statistic reduces to Rao's score statistic. It is also useful to stress that the objects projected on the space spanned by the nuisance parameter scores can be more general functions (called Cramér functions by Neyman [52]), not necessarily the score function associated with the parameters of interest. For further discussions of $C(\alpha)$ tests and references, see Le Cam [44], Bhat and Nagnur [14], Bühler and Puri [15], Bartoo and Puri [3], Moran [46, 47], Chibisov [18], Chant [16], Ray [59], Singh and Zhurbenko [61], Foutz [27], Vorob'ev and Zhurbenko [68], Bernshtein [9–13], Le Cam and Traxler [45], Neyman [53], Tarone [65, 66], Tarone and Gart [67], Wang [69, 70], Basawa [4], Ronchetti [60], Smith [63, 64], Berger and Wallenstein [8], Hall and Mathiason [34], Paul and Barnwal [57], Wooldridge [71], Dagenais and Dufour [20], Davidson and MacKinnon [21, 22], Kocherlakota and Kocherlakota [43], Dufour and Dagenais [23], Bera and Yoon [7], Jaggia and Trivedi [39], Rao [58], Bera and Bilias [6], Pal [56], Dufour and Valéry [24] and Chaudhuri and Zivot [17].

In spite of numerous generalizations and modifications in parametric models, extensions of the $C(\alpha)$ test to other types of estimation criteria, e.g. estimating equations [5, 25, 28, 29, 38, 62], minimum distance, or the generalized method of moments (GMM [33, 36]), appear to be scarce. In particular, work on such tests has focused on linear hypotheses (especially, hypothesis setting the value of a parameter subvector) and/or independent observations; see Basawa [4].

In this paper, we propose and study a general $C(\alpha)$-type statistic in estimating-function and GMM setups, with weakly specified temporal dependence and heteroskedasticity. The proposed generalized statistic is quite comprehensive and includes earlier $C(\alpha)$-type statistics as special cases, as well as a wide spectrum of new ones. The null hypothesis takes the form of a general constraint (linear or nonlinear) on model parameters. This extends the $C(\alpha)$ test proposed by Smith [63] for nonlinear restrictions in parametric likelihood models. The asymptotic distribution of the test statistic is derived under a set of weak regularity conditions, allowing for general forms of serial dependence and heteroskedasticity.

A number of important special cases of the extended test statistic are discussed in detail. These include testing whether a parameter subvector has a given value—for which we give a number of alternative forms and special cases—and accounting for parameter uncertainty in two-stage procedures. The latter problem has considerable practical importance. Due to the fact that nonlinear estimating functions are often difficult to estimate, it is convenient to estimate some parameters by an alternative simpler method, and then use these estimates as if they were known. Such procedures can however modify the distributions of test statistics and induce distortions in test levels; see Gong and Samaniego [30], Pagan [54, 55], Murphy and Topel [48] and Newey and McFadden [49]. So it is important to make corrections for such effects. We underscore that generalized $C(\alpha)$ tests can provide relatively simple solutions to such difficulties in the context of estimating functions and GMM estimation, again in presence of general forms of serial dependence and heteroskedasticity. We first discuss tests based on a general first-stage estimator, as well as tests based on a two-stage GMM estimation.

The paper is organized as follows. Section 2 lays out the general framework considered in the paper and introduces the $C(\alpha)$ statistic. The regularity conditions are stated and the asymptotic properties of the generalized $C(\alpha)$ statistic are studied in Sect. 3. We discuss the forms that the $C(\alpha)$ statistic takes in some special cases in Sect. 4. Section 5 considers the problem of testing the value of parameter subvector. We formulate the $C(\alpha)$ statistic for models estimated by two-step procedures in Sect. 6. We briefly conclude in Sect. 7.

## 2 Generalized $C(\alpha)$ Statistic

We consider an $m \times 1$ vector estimating (or score-type) function $D_n(\theta; Z_n)$ which depends on an $n \times k$ data matrix $Z_n = [z_1, z_2, \ldots, z_n]'$ and a parameter vector $\theta \in \Theta \subseteq \mathbb{R}^p$ such that

$$D_n(\theta; Z_n) \xrightarrow[n\to\infty]{\mathrm{p}} D_\infty(\theta; \theta_0) \tag{1}$$

where $D_n(\theta; Z_n)$ is typically the sample mean of an estimating function, such as $D_n(\theta; Z_n) = \frac{1}{n}\sum_{t=1}^{n} h(\theta; z_t)$, $D_\infty(\cdot; \theta_0)$ is a mapping from $\Theta$ to $\mathbb{R}^m$, and $\theta_0$ denotes the "true" parameter vector. The parameter $\theta$ is estimated by minimizing a criterion function of the form

$$M_n(\theta, W_n) = D_n(\theta; Z_n)' \, W_n \, D_n(\theta; Z_n) \tag{2}$$

where $W_n$ is a symmetric positive definite matrix. This setup comprises as special cases the method of estimating functions [5, 25, 28, 29, 38, 62], the generalized method of moments [33, 36], maximum likelihood, pseudo-maximum likelihood, $M$-estimation and instrumental-variable methods.

A common assumption in such contexts consists in assuming that

$$\mathsf{E}_{\theta_0}[D_n(\theta_0; Z_n)] = 0 \tag{3}$$

where $\mathsf{E}_\theta[\,\cdot\,]$ represents the expected value under any data distribution such that $\theta$ can be interpreted as the true parameter vector, along with a number of additional regularity assumptions which allow the application of central limit theorems and laws of large numbers, such as:

$$\sqrt{n}\, D_n(\theta_0; Z_n) \xrightarrow[n\to\infty]{L} \mathrm{N}[0, I(\theta_0)], \tag{4}$$

$$J_n(\theta_0; Z_n) = \frac{\partial D_n(\theta_0; Z_n)}{\partial \theta'} \xrightarrow[n\to\infty]{\mathsf{p}} J(\theta_0), \tag{5}$$

where $I(\theta_0)$ and $J(\theta_0)$ are $m \times m$ and $m \times p$ full-column rank matrices. In Sect. 3, we relax the Assumptions (3) and (5).

The hypothesis we wish to test has the form

$$H_0 : \psi(\theta) = 0 \tag{6}$$

where $\psi(\theta)$ is a $p_1 \times 1$ continuously differentiable function of $\theta$ with $1 \le p_1 \le p$, and the $p_1 \times p$ matrix

$$P(\theta) = \frac{\partial \psi}{\partial \theta'} \tag{7}$$

has full row-rank $p_1$ (at least in an open neighborhood of $\theta_0$).

Let $\hat{\theta}_n$ be the unrestricted estimator of $\theta$ obtained by minimizing $M_n(\theta, W_n)$, $\hat{\theta}_n^0$ the corresponding constrained estimator under $H_0$, and $\tilde{\theta}_n^0$ any other restricted estimator of $\theta$ under $H_0$. Let us also denote estimators of $I(\theta)$ and $J(\theta)$ by $\hat{I}_n(\theta)$ and $\hat{J}_n(\theta)$ respectively, where $\theta$ may be replaced by unrestricted and restricted estimators of $\theta$ to obtain estimators of $I(\theta_0)$ and $J(\theta_0)$. If

$$D_n(\theta; Z_n) = \frac{1}{n} \sum_{t=1}^{n} h(\theta; z_t), \tag{8}$$

we may use the standard formula

$$\hat{J}_n(\theta) = \frac{\partial D_n(\theta; Z_n)}{\partial \theta'} = J_n(\theta; Z_n). \tag{9}$$

Depending on the problem at hand, different forms of $\hat{I}_n(\theta)$ may be considered. The standard estimator appropriate for random sampling models is

$$\hat{I}_n(\theta) = \frac{1}{n} \sum_{t=1}^{n} h(\theta; z_t) h(\theta; z_t)'. \tag{10}$$

Some authors also argue that the centered version of (10) given by

$$\hat{I}_n(\theta) = \frac{1}{n} \sum_{t=1}^{n} \left[h(\theta; z_t) - \overline{h}(\theta)\right]\left[h(\theta; z_t) - \overline{h}(\theta)\right]' \tag{11}$$

where $\overline{h}(\theta) = \frac{1}{n} \sum_{t=1}^{n} h(\theta; z_t)$, can yield power improvements; see Hall [32].

In this paper, we stress applications to time series data where serial dependence is present. In view of this, we focus on "heteroskedasticity-autocorrelation consistent" (HAC) covariance matrix estimators which account for the potential serial correlation and heteroskedasticity in the sequence $\{h(\theta; z_t)\}_{t=1}^{\infty}$:

$$\hat{I}_n(\theta) = \sum_{j=-n+1}^{n-1} \bar{k}(j/B_n)\, \hat{\Gamma}_n(j, \theta) \tag{12}$$

where $\bar{k}(\cdot)$ is a kernel function, $B_n$ is a bandwidth parameter (which depends on the sample size and, possibly, on the data), and

$$\hat{\Gamma}_n(j, \theta) = \begin{cases} \frac{1}{n} \sum_{t=j+1}^{n} h(\theta; z_t) h(\theta; z_{t-j})', & \text{if } j \geq 0, \\ \frac{1}{n} \sum_{t=-j+1}^{n} h(\theta; z_{t+j}) h(\theta; z_t)', & \text{if } j < 0. \end{cases} \tag{13}$$

The reader is referred to Newey and West [51], Andrews [1], Andrews and Monahan [2], Hansen [35], Cushing and McGarvey [19], Kiefer et al. [40], and Kiefer and Vogelsang [41, 42] for further properties of covariance estimators of the form (12).

We now consider the problem of formulating a test statistic for $H_0$ using a general restricted estimator of $\theta_0$. This means that we wish to use statistics based on estimators which may not be obtained by minimizing the objective function $M_n$ in (2). This is motivated by the fact that minimizing $M_n$ often constitutes a difficult numerical problem plagued by instabilities. Similarly, while some local efficiency arguments suggest taking $W_n = \hat{I}_n^{-1}$ (see Hansen [36, Theorem 3.2], Davidson and MacKinnon [22, Section 17.3], Gouriéroux and Monfort [31, Section 9.5.2], Hall [33, Section 3.6]), ill-conditioning can make this choice infeasible or harmful. So we allow here for a general weighting matrix $W_n$.

In order to obtain a unified test criterion which includes several other score-type statistics, we consider the following general "score-type" function:

$$s(\tilde{\theta}_n^0;\, W_n) = \sqrt{n}\, \tilde{Q}[W_n] D_n(\tilde{\theta}_n^0;\, Z_n)$$

where $\tilde{\theta}_n^0$ is a consistent restricted estimate of $\theta_0$ such that $\psi(\tilde{\theta}_n^0) = 0$ and $\sqrt{n}(\tilde{\theta}_n^0 - \theta_0)$ is asymptotically bounded in probability,

$$\tilde{Q}[W_n] := \tilde{P}_n\big(\tilde{J}_n'\, W_n\, \tilde{J}_n\big)^{-1} \tilde{J}_n'\, W_n,$$

$\tilde{P}_n = P(\tilde{\theta}_n^0)$, $\tilde{J}_n = \hat{J}_n(\tilde{\theta}_n^0)$, and $W_n$ is a symmetric positive definite (possibly random) $m \times m$ matrix such that

$$\operatorname*{plim}_{n\to\infty} W_n = W_0, \quad \det(W_0) \neq 0.$$

Under general regularity conditions (see Sect. 3), the asymptotic distribution of the score-type function is normal as described by (15) in Proposition 1, with

$$Q(\theta_0) = \operatorname*{plim}_{n\to\infty} \tilde{Q}[W_n] = P(\theta_0)\big[J(\theta_0)' W_0\, J(\theta_0)\big]^{-1} J(\theta_0)' W_0$$

and $\text{rank}[Q(\theta_0)] = p_1$. This suggests the following generalized $C(\alpha)$ criterion:

$$PC(\tilde{\theta}_n^0;\, \psi, W_n) = n\, \tilde{D}_n'\, \tilde{Q}[W_n]' \left\{\tilde{Q}[W_n] \tilde{I}_n\, \tilde{Q}[W_n]'\right\}^{-1} \tilde{Q}[W_n]\, \tilde{D}_n \tag{14}$$

where $\tilde{D}_n = D_n(\tilde{\theta}_n^0;\, Z_n)$ and $\tilde{I}_n = \hat{I}_n(\tilde{\theta}_n^0)$. We show in Sect. 3 that the asymptotic distribution of $PC(\tilde{\theta}_n^0;\, \psi, W_n)$ is $\chi^2(p_1)$ under $H_0$. The proposed test statistic includes as a special case several statistics proposed in the statistical and econometric literatures. We discuss these as well as other special cases in Sects. 4, 5 and 6.

## 3 Distribution of the Generalized $C(\alpha)$ Statistic

In this section, we derive the asymptotic distribution of the generalized $C(\alpha)$ statistic defined in (14) under the following set of assumptions. $\|\cdot\|$ refers to the Euclidean distance, applied to either vectors or matrices.

**Assumption 1** (*Existence of score-type functions*)

$$D_n(\theta, \omega) = \big(D_{1n}(\theta, \omega), \ldots, D_{mn}(\theta, \omega)\big)', \quad \omega \in \mathscr{Z}, \quad n = 1, 2, \ldots$$

is a sequence of $m \times 1$ random vectors, defined on a common probability space $(\mathscr{Z}, \mathscr{A}_{\mathscr{Z}}, \mathsf{P})$, which are functions of a $p \times 1$ parameter vector $\theta$, where $\theta \in \Theta \subseteq \mathbb{R}^p$

and $\Theta$ is a non-empty open subset of $\mathbb{R}^p$. All the random variables considered here as well in the following assumptions are functions of $\omega$, so the symbol $\omega$ may be dropped to simplify notations (e.g., $D_n(\theta) := D_n(\theta, \omega)$). There is a unique vector $\theta_0 \in \Theta$ called the "true parameter value".

**Assumption 2** (*Score asymptotic normality*)

$$\sqrt{n}\, D_n(\theta_0) \xrightarrow[n\to\infty]{\mathsf{p}} \bar{D}_\infty(\theta_0) \quad \text{where} \quad \bar{D}_\infty(\theta_0) \sim \mathrm{N}\big[0, I(\theta_0)\big].$$

**Assumption 3** (*Non-singularity of the score variance*) $I(\theta)$ is nonsingular for any $\theta \in \Theta$ which satisfies the restriction $\psi(\theta) = 0$.

**Assumption 4** (*Score expansion*) For $\theta$ in a non-empty open neighborhood $N_0$ of $\theta_0$, $D_n(\theta)$ admits an expansion of the form

$$D_n(\theta, \omega) = D_n(\theta_0, \omega) + J(\theta_0)(\theta - \theta_0) + R_n(\theta, \theta_0, \omega)$$

for $\omega \in \mathscr{D}_J$, an event with probability one, where $J(\theta)$ is an $m \times p$ (nonrandom) matrix function of $\theta$ and the remainder $R_n(\theta, \theta_0, \omega)$ satisfies the following condition: for any $\varepsilon > 0$ and $\delta > 0$, we have

$$\limsup_{n\to\infty} \mathsf{P}\big[\{\omega : r_n(\delta, \theta_0, \omega) > \varepsilon\}\big] \le U_D(\delta, \varepsilon, \theta_0)$$

$$r_n(\delta, \theta_0, \omega) = \sup\left\{ \frac{\|R_n(\theta, \theta_0, \omega)\|}{\|\theta - \theta_0\|} : \theta \in N_0 \text{ and } 0 < \|\theta - \theta_0\| \le \delta \right\},$$

$U_D(\delta, \varepsilon, \theta_0) \ge 0$ and $\lim_{\delta \downarrow 0} U_D(\delta, \varepsilon, \theta_0) = 0$.

**Assumption 5** (*Consistent estimator of* $J(\theta_0)$) There is a sequence of $m \times p$ random matrices $J_n(\theta, \omega)$ and a non-empty open neighborhood $V_0$ of $\theta_0$ such that, for all $\varepsilon > 0$ and $\delta > 0$,

$$\limsup_{n\to\infty} \mathsf{P}\big[\{\omega : \Delta_n(\theta_0, \delta, \omega) > \varepsilon\}\big] \le U_J(\delta, \varepsilon, \theta_0)$$

where

$$\Delta_n(\theta_0, \delta, \omega) := \sup\big\{\|J_n(\theta, \omega) - J(\theta_0)\| : \theta \in V_0 \text{ and } 0 \le \|\theta - \theta_0\| \le \delta\big\}$$

and $U_J(\delta, \varepsilon, \theta_0)$ is a non-random function such that

$$U_J(\delta, \varepsilon, \theta_0) \ge 0 \quad \text{and} \quad \lim_{\delta \downarrow 0} U_J(\delta, \varepsilon, \theta_0) = 0.$$

**Assumption 6** (*Asymptotic score non-degeneracy*) $\mathrm{rank}\big[J(\theta)\big] = p$ *for any* $\theta \in \Theta$ *which satisfies the restriction* $\psi(\theta) = 0$.

**Assumption 7** (*Restriction differentiability*) $\psi(\theta)$ is a $p_1 \times 1$ continuously differentiable vector function of $\theta$ with derivative $P(\theta) := \frac{\partial \psi}{\partial \theta'}$.

**Assumption 8** (*Restriction rank*) $\text{rank}\big[P(\theta)\big] = p_1$ for any $\theta \in \Theta$ which satisfies the restriction $\psi(\theta) = 0$.

**Assumption 9** (*Estimator $\sqrt{n}$ convergence*) $\tilde{\theta}_n^0 := \tilde{\theta}_n^0(\omega)$ is a consistent estimator of $\theta_0$, i.e.,

$$\underset{n\to\infty}{\text{plim}}\,\big(\tilde{\theta}_n^0 - \theta_0\big) = 0,$$

such that $\sqrt{n}\big(\tilde{\theta}_n^0 - \theta_0\big)$ is asymptotically bounded in probability, i.e.,

$$\limsup_{n\to\infty}\ \mathsf{P}\big[\{\omega : \sqrt{n}\,\big\|\tilde{\theta}_n^0 - \theta_0\big\| \geq y\}\big] \leq U(y;\theta_0), \quad \forall y > 0,$$

where $U(y;\theta_0)$ is a function such that $\lim_{y\to\infty} U(y;\theta_0) = 0$.

The latter assumption requires that the auxiliary estimator $\tilde{\theta}_n^0$ be $\sqrt{n}$-consistent only under the null hypothesis $H_0$, and corresponds to Neyman's [52] local $\sqrt{n}$-consistency assumption. It may also be written $\sqrt{n}\big(\tilde{\theta}_n^0 - \theta_0\big) = O_p(1)$ under $H_0$.

**Assumption 10** (*Restricted estimator*) $\psi\big(\tilde{\theta}_n^0\big) = \psi(\theta_0) = 0$ with probability 1.

**Assumption 11** (*Consistent estimator of score covariance matrix*) $\tilde{I}_n,\ n \geq 1$, is a sequence of $m \times m$ symmetric nonsingular (random) matrices such that $\text{plim}_{n\to\infty}$ $\tilde{I}_n = I(\theta_0)$.

**Assumption 12** (*Weight matrix convergence*) $W_n,\ n \geq 1$, is a sequence of $m \times m$ symmetric nonsingular (random) matrices such that $\text{plim}_{n\to\infty} W_n = W_0$ where $W_0$ is nonsingular.

The following proposition establishes the asymptotic distribution of the generalized $C(\alpha)$ statistic $PC\big(\tilde{\theta}_n^0;\psi, W_n\big)$ in (14).

**Proposition 1** (Asymptotic distribution of generalized $C(\alpha)$ statistic) *Let* $\tilde{Q}_n := \tilde{Q}[W_n] = \tilde{P}_n\,[\tilde{J}_n' W_n \tilde{J}_n]^{-1}\tilde{J}_n' W_n$ *where* $\tilde{J}_n = J_n(\tilde{\theta}_n^0; Z_n)$, $\tilde{P}_n = P\big(\tilde{\theta}_n^0\big)$. *If the Assumptions* 1–12 *are satisfied, then, under* $H_0$,

$$\sqrt{n}\,\tilde{Q}_n\,D_n\big(\tilde{\theta}_n^0; Z_n\big) \xrightarrow[n\to\infty]{L} \mathrm{N}\big[0,\,Q(\theta_0)I(\theta_0)Q(\theta_0)'\big] \tag{15}$$

*where* $Q(\theta_0) = P(\theta_0)\big[J(\theta_0)'W_0 J(\theta_0)\big]^{-1} J(\theta_0)'W_0$, *and*

$$PC\big(\tilde{\theta}_n^0;\psi, W_n\big) = n\,D_n\big(\tilde{\theta}_n^0; Z_n\big)'\tilde{Q}_n'\,\big[\tilde{Q}_n\,\tilde{I}_n\,\tilde{Q}_n'\big]^{-1}\,\tilde{Q}_n\,D_n\big(\tilde{\theta}_n^0; Z_n\big) \xrightarrow[n\to\infty]{L} \chi^2(p_1). \tag{16}$$

It is of interest to note here that the Assumptions 4 and 5 do not require that $D_n(\theta, \omega)$ be differentiable with respect to $\theta$. This is allowed by making a direct assumption on the existence of a linear expansion of $D_n(\theta, \omega)$ around $\theta_0$ [Assumption 4]. For the same reason, $J_n(\theta, \omega)$ does not have to be continuous with respect to $\theta$.

Since the differentiability of $D_n(\theta, \omega)$ with respect to $\theta$ is a common assumption, we will now show that the high-level Assumptions 4 and 5 hold in the standard case where $D_n(\theta, \omega)$ is differentiable, with probability limit $J(\theta)$, and both $J_n(\theta, \omega)$ and $J(\theta)$ are continuous at least at every point in a neighborhood of $\theta_0$. More precisely, consider the following assumptions.

**Assumption 13** (*Score differentiability*) $D_n(\theta, \omega)$ is almost surely (a.s.) differentiable with respect to $\theta$, for all $n$, in a non-empty open neighborhood $N_1$ of $\theta_0$. The derivative matrix of $D_n(\theta, \omega)$ is denoted

$$J_n(\theta, \omega) = \frac{\partial D_n(\theta, \omega)}{\partial \theta'} \tag{17}$$

where the sequence of matrices $J_n(\theta, \omega)$, $n \geq 1$, is well-defined for $\omega \in \mathscr{D}_J$ and $\mathscr{D}_J$ is an event with probability one (i.e., $\mathsf{P}[\omega \in \mathscr{D}_J] = 1$).

**Assumption 14** (*Score derivative uniform convergence*) $D_n(\theta, \omega)$ satisfies the following conditions:

(a) $J_n(\theta, \omega)$ is continuous with respect to $\theta$ for all $\theta \in N_2$, $\omega \in \mathscr{D}_J$ and $n \geq 1$;
(b) $\sup_{\theta \in N_2} \| J_n(\theta, \omega) - J(\theta) \| \xrightarrow[n \to \infty]{\mathsf{p}} 0$.

We then have the following implication, which shows that Proposition 1 still holds if the Assumptions 4 and 5 are replaced by the (stronger) Assumptions 13 and 14. Another implication is that $J(\theta)$ is continuous at $\theta = \theta_0$ in this special case.

**Proposition 2** (Sufficiency of score Jacobian continuity and uniform convergence) *Suppose the Assumptions* 1–3 *hold. Then the Assumptions* 13 *and* 14 *entail that:*

*(a)* $J(\theta)$ *is continuous at* $\theta = \theta_0$;
*(b) both the Assumptions* 4 *and* 5 *also hold.*

## 4 Alternative $C(\alpha)$-Type Statistics

It will be of interest to examine a number of special forms of the general statistic proposed in Sect. 2. In particular, the statistic $PC(\tilde{\theta}_n^0; \psi, W_n)$ nests several $C(\alpha)$-type and score-based statistics proposed in the statistical and econometric literatures, as well as new ones.[1] It will be of interest to spell out some of these.

[1] For further discussion of $C(\alpha)$ tests, the reader may consult Basawa [4], Ronchetti [60], Smith [63], Berger and Wallenstein [8], Dagenais and Dufour [20], and Kocherlakota and Kocherlakota [43].

On taking $W_n = \tilde{I}_n^{-1}$, as suggested by efficiency arguments, $PC(\tilde{\theta}_n^0;\, \psi,\, W_n)$ reduces to

$$PC\big(\tilde{\theta}_n^0;\, \psi\big) = n\, D_n\big(\tilde{\theta}_n^0;\, Z_n\big)'\, \tilde{W}_n\, D_n\big(\tilde{\theta}_n^0;\, Z_n\big) \tag{18}$$

where $\tilde{\theta}_n^0$ is any root-$n$ consistent estimator of $\theta$ which satisfies $\psi\big(\tilde{\theta}_n^0\big) = 0$, and

$$\tilde{W}_n = \tilde{I}_n^{-1}\tilde{J}_n'\big(\tilde{J}_n'\,\tilde{I}_n^{-1}\tilde{J}_n\big)^{-1}\tilde{P}_n'\big[\tilde{P}_n\big(\tilde{J}_n'\,\tilde{I}_n^{-1}\tilde{J}_n\big)^{-1}\tilde{P}_n'\big]^{-1}\tilde{P}_n\big(\tilde{J}_n'\,\tilde{I}_n^{-1}\tilde{J}_n\big)^{-1}\tilde{J}_n'\,\tilde{I}_n^{-1}$$

with $\tilde{P}_n = P\big(\tilde{\theta}_n^0\big)$, $\tilde{I}_n = \hat{I}_n(\tilde{\theta}_n^0)$ and $\tilde{J}_n = \hat{J}_n(\tilde{\theta}_n^0)$.

When the number of equations equals the number of parameters $(m = p)$, we have $\tilde{Q}\big[W_n\big] = \tilde{P}_n\tilde{J}_n^{-1}$ and $PC(\tilde{\theta}_n^0;\, \psi,\, W_n)$ does not depend on the choice of $W_n$:

$$\begin{aligned} PC(\tilde{\theta}_n^0;\, \psi,\, W_n) &= PC(\tilde{\theta}_n^0;\, \psi) \\ &= n\, D_n(\tilde{\theta}_n^0;\, Z_n)'(\tilde{J}_n^{-1})'\tilde{P}_n'\big[\tilde{P}_n\big(\tilde{J}_n'\,\tilde{I}_n^{-1}\tilde{J}_n\big)^{-1}\tilde{P}_n'\big]^{-1}\tilde{P}_n\,\tilde{J}_n^{-1}D_n(\tilde{\theta}_n^0;\, Z_n). \end{aligned} \tag{19}$$

In particular, this will be the case if $D_n(\theta;\, Z_n)$ is the derivative of a (pseudo) log-likelihood function.

For $m \geq p$, when $\tilde{\theta}_n^0$ is obtained by minimizing $M_n(\theta) = D_n(\theta;\, Z_n)'\,\tilde{I}_n^{-1}D_n(\theta;\, Z_n)$ subject to $\psi(\theta) = 0$, where $\tilde{I}_n$ is an estimator of $I(\theta_0)$, we can write $\tilde{\theta}_n^0 = \hat{\theta}_n^0$ and $PC\big(\tilde{\theta}_n^0;\, \psi,\, W_n\big)$ is identical to the *score-type statistic* suggested by Newey and West [50]:

$$S\big(\psi\big) = n\, D_n\big(\hat{\theta}_n^0;\, Z_n\big)'\,\hat{I}_n^{-1}\hat{J}_n\big(\hat{J}_n'\,\hat{I}_n^{-1}\hat{J}_n\big)^{-1}\hat{J}_n'\,\hat{I}_n^{-1}D_n\big(\hat{\theta}_n^0;\, Z_n\big) \tag{20}$$

where $\hat{I}_n = \hat{I}_n(\hat{\theta}_n^0)$ and $\hat{J}_n = \hat{J}_n(\hat{\theta}_n^0)$. This statistic is closely related with the *Lagrange-multiplier-type* (LM-type) statistic

$$LM\big(\psi\big) = n\,\hat{\lambda}_n'\,\hat{P}_n\big(\hat{J}_n'\,\hat{I}_n^{-1}\hat{J}_n\big)^{-1}\hat{P}_n'\,\hat{\lambda}_n \tag{21}$$

where $\hat{P}_n = P(\hat{\theta}_n^0)$ and $\hat{\lambda}_n$ is the Lagrange multiplier in the corresponding constrained optimization problem. Indeed, upon using the first-order condition

$$J_n\big(\hat{\theta}_n^0;\, Z_n\big)'\,\tilde{I}_n^{-1}D_n\big(\hat{\theta}_n^0;\, Z_n\big) = P\big(\hat{\theta}_n^0\big)'\hat{\lambda}_n, \tag{22}$$

we see easily that

$$S\big(\psi\big) = LM\big(\psi\big). \tag{23}$$

In (correctly specified) parametric models, we have $I(\theta) = -J(\theta)$ and the $C(\alpha)$ statistic in (19) reduces to

$$PC(\tilde{\theta}_n^0; \psi) = n\, D_n(\tilde{\theta}_n^0; Z_n)' \tilde{I}_n^{-1} \tilde{P}_n' \big[\tilde{P}_n \tilde{I}_n^{-1} \tilde{P}_n'\big]^{-1} \tilde{P}_n \tilde{I}_n^{-1} D_n(\tilde{\theta}_n^0; Z_n) \tag{24}$$

where $D_n(\tilde{\theta}_n^0; Z_n)$ is the score of the log-likelihood function and $\tilde{I}_n$ is the Fisher information matrix or a consistent estimate, each evaluated at the auxiliary estimator $\tilde{\theta}_n^0$. The extension of $C(\alpha)$ statistics to a general parameter constraint given in (24) was first proposed by Smith [64] in a likelihood setting; see Dagenais and Dufour [20] for further discussion of this test statistic.

## 5 Testing a Subvector

A common problem in statistics consists in testing an hypothesis of the form

$$H_0 : \theta_1 = \bar{\theta}_{10} \tag{25}$$

where $\theta_1$ is a subvector of $\theta$, and $\bar{\theta}_{10}$ is a given possible value of $\theta_1$, i.e. we consider $\psi(\theta) = \theta_1 - \bar{\theta}_{10}$. Without loss of generality, we can assume that $\theta = (\theta_1', \theta_2')'$ where $\theta_1$ is a $p_1 \times 1$ vector and $\theta_2$ is a $p_2 \times 1$ vector, and denote $\theta_0 = (\theta_{10}', \theta_{20}')'$ the "true value" of $\theta$. In this case,

$$P(\theta) = \big[I_{p_1}, 0_{p_1 \times p_2}\big] \tag{26}$$

where $I_{p_1}$ is the identity matrix of order $p_1$ and $0_{p_1 \times p_2}$ is the $p_1 \times p_2$ zero matrix. Let $\tilde{\theta}_n^0$ be a restricted $\sqrt{n}$-consistent estimator of $\theta$. We can then write $\tilde{\theta}_n^0 = (\bar{\theta}_{10}', \tilde{\theta}_{2n}^{0\prime})'$ where $\tilde{\theta}_{2n}^0$ is a $\sqrt{n}$-consistent estimator of $\theta_2$.

Let us partition $J(\theta)$ and $\tilde{J}_n = J_n(\tilde{\theta}_n^0; Z_n)$ conformably with $\theta = (\theta_1', \theta_2')'$:

$$J(\theta) = \big[J_{\cdot 1}(\theta), J_{\cdot 2}(\theta)\big], \; \tilde{J}_n = \big[\tilde{J}_{n\cdot 1}, \tilde{J}_{n\cdot 2}\big] = \big[\tilde{J}_{n\cdot 1}(\tilde{\theta}_n^0; Z_n), \tilde{J}_{n\cdot 2}(\tilde{\theta}_n^0; Z_n)\big], \tag{27}$$

where $J_{\cdot i}(\theta)$ and $\tilde{J}_{n\cdot i} = \tilde{J}_{n\cdot i}(\tilde{\theta}_n^0; Z_n)$ are $m \times p_i$ matrices, $i = 1, 2$. Let also

$$\tilde{J}_n^* = W_n^{1/2} \tilde{J}_n = \big[\tilde{J}_{n\cdot 1}^*, \tilde{J}_{n\cdot 2}^*\big], \quad \tilde{J}_{n\cdot i}^* = W_n^{1/2} \tilde{J}_{n\cdot i} \quad i = 1, 2, \tag{28}$$

and conformably partition the matrix $\tilde{J}_n' W_n \tilde{J}_n$ and its inverse $(\tilde{J}_n' W_n \tilde{J}_n)^{-1}$:

$$\tilde{J}_n' W_n \tilde{J}_n = \begin{bmatrix} (\tilde{J}_n' W_n \tilde{J}_n)_{11} & (\tilde{J}_n' W_n \tilde{J}_n)_{12} \\ (\tilde{J}_n' W_n \tilde{J}_n)_{21} & (\tilde{J}_n' W_n \tilde{J}_n)_{22} \end{bmatrix} = \begin{bmatrix} \tilde{J}_{n\cdot 1}' W_n \tilde{J}_{n\cdot 1} & \tilde{J}_{n\cdot 1}' W_n \tilde{J}_{n\cdot 2} \\ \tilde{J}_{n\cdot 2}' W_n \tilde{J}_{n\cdot 1} & \tilde{J}_{n\cdot 2}' W_n \tilde{J}_{n\cdot 2} \end{bmatrix}, \tag{29}$$

$$(\tilde{J}_n' W_n \tilde{J}_n)^{-1} = \begin{bmatrix} (\tilde{J}_n' W_n \tilde{J}_n)^{11} & (\tilde{J}_n' W_n \tilde{J}_n)^{12} \\ (\tilde{J}_n' W_n \tilde{J}_n)^{21} & (\tilde{J}_n' W_n \tilde{J}_n)^{22} \end{bmatrix}, \tag{30}$$

where $(\tilde{J}_n' W_n \tilde{J}_n)_{ij}$ and $(\tilde{J}_n' W_n \tilde{J}_n)^{ij}$ are $p_i \times p_j$ matrices, $i, j = 1, 2$. We denote $P[Z] = Z(Z'Z)^{-1}Z'$ the projection matrix on the space spanned by the columns of a full-column rank matrix $Z$, and $M[Z] = I - Z(Z'Z)^{-1}Z'$.

Let us now assume that the matrix $(\tilde{J}_n' W_n \tilde{J}_n)_{22}$ is invertible. This entails that $(\tilde{J}_n' W_n \tilde{J}_n)^{11}$ is invertible and, on using standard rules for multiplying partitioned matrices,

$$\begin{aligned}\big[(\tilde{J}_n' W_n \tilde{J}_n)^{11}\big]^{-1}(\tilde{J}_n' W_n \tilde{J}_n)^{12} &= -(\tilde{J}_n' W_n \tilde{J}_n)_{12}\big[(\tilde{J}_n' W_n \tilde{J}_n)_{22}\big]^{-1} \\ &= -(\tilde{J}_{n\cdot 1}' W_n \tilde{J}_{n\cdot 2})(\tilde{J}_{n\cdot 2}' W_n \tilde{J}_{n\cdot 2})^{-1}, \qquad (31)\end{aligned}$$

$$(\tilde{J}_n' W_n \tilde{J}_n)^{11} = \big[(\tilde{J}_{n\cdot 1}' W_n \tilde{J}_{n\cdot 1}) - \tilde{J}_{n\cdot 1}' W_n \tilde{J}_{n\cdot 2}(\tilde{J}_{n\cdot 2}' W_n \tilde{J}_{n\cdot 2})^{-1}\tilde{J}_{n\cdot 2}' W_n \tilde{J}_{n\cdot 1}\big]^{-1}; \qquad (32)$$

see Harville [37, Theorem 8.5.11]. We can then rewrite $\tilde{Q}[W_n]$ as

$$\begin{aligned}\tilde{Q}[W_n] &= \tilde{P}_n(\tilde{J}_n' W_n \tilde{J}_n)^{-1}\tilde{J}_n' W_n \\ &= [I_{p_1}, 0_{p_1 \times p_2}]\begin{bmatrix}(\tilde{J}_n' W_n \tilde{J}_n)^{11} & (\tilde{J}_n' W_n \tilde{J}_n)^{12} \\ (\tilde{J}_n' W_n \tilde{J}_n)^{21} & (\tilde{J}_n' W_n \tilde{J}_n)^{22}\end{bmatrix}\begin{bmatrix}\tilde{J}_{n\cdot 1}' \\ \tilde{J}_{n\cdot 2}'\end{bmatrix} W_n \\ &= \big[(\tilde{J}_n' W_n \tilde{J}_n)^{11}\tilde{J}_{n\cdot 1}' + (\tilde{J}_n' W_n \tilde{J}_n)^{12}\tilde{J}_{n\cdot 2}'\big]W_n \\ &= (\tilde{J}_n' W_n \tilde{J}_n)^{11}\big[\tilde{J}_{n\cdot 1}' + ((\tilde{J}_n' W_n \tilde{J}_n)^{11})^{-1}(\tilde{J}_n' W_n \tilde{J}_n)^{12}\tilde{J}_{n\cdot 2}'\big]W_n \\ &= (\tilde{J}_n' W_n \tilde{J}_n)^{11}\big[\tilde{J}_{n\cdot 1}' - (\tilde{J}_{n\cdot 1}' W_n \tilde{J}_{n\cdot 2})(\tilde{J}_{n\cdot 2}' W_n \tilde{J}_{n\cdot 2})^{-1}\tilde{J}_{n\cdot 2}'\big]W_n \\ &= \tilde{V}_{n\cdot 1|2}^{-1}\tilde{J}_{n\cdot 1|2}' W_n \qquad (33)\end{aligned}$$

where

$$\tilde{J}_{n\cdot 1|2} = \tilde{J}_{n\cdot 1} - \tilde{J}_{n\cdot 2}(\tilde{J}_{n\cdot 2}' W_n \tilde{J}_{n\cdot 2})^{-1}\tilde{J}_{n\cdot 2}' W_n \tilde{J}_{n\cdot 1} = W_n^{-1/2}M[\tilde{J}_{n\cdot 2}^*]\tilde{J}_{n\cdot 1}^*, \qquad (34)$$

$$\begin{aligned}\tilde{V}_{n\cdot 1|2} &= (\tilde{J}_{n\cdot 1}' W_n \tilde{J}_{n\cdot 1}) - \tilde{J}_{n\cdot 1}' W_n \tilde{J}_{n\cdot 2}(\tilde{J}_{n\cdot 2}' W_n \tilde{J}_{n\cdot 2})^{-1}\tilde{J}_{n\cdot 2}' W_n \tilde{J}_{n\cdot 1} \\ &= \tilde{J}_{n\cdot 1}^{*\prime}M[\tilde{J}_{n\cdot 2}^*]\tilde{J}_{n\cdot 1}^*. \qquad (35)\end{aligned}$$

Using (33), we get:

$$\begin{aligned}\tilde{Q}[W_n]\tilde{D}_n &= \tilde{V}_{n\cdot 1|2}^{-1}\tilde{J}_{n\cdot 1|2}' W_n \tilde{D}_n \\ &= \tilde{V}_{n\cdot 1|2}^{-1}\big[\tilde{J}_{n\cdot 1}' W_n \tilde{D}_n - (\tilde{J}_{n\cdot 1}' W_n \tilde{J}_{n\cdot 2})(\tilde{J}_{n\cdot 2}' W_n \tilde{J}_{n\cdot 2})^{-1}\tilde{J}_{n\cdot 2}' W_n \tilde{D}_n\big], \qquad (36)\end{aligned}$$

$$\tilde{J}_{n\cdot 1|2}' W_n \tilde{D}_n = \tilde{J}_{n\cdot 1}' W_n \tilde{D}_n - (\tilde{J}_{n\cdot 1}' W_n \tilde{J}_{n\cdot 2})(\tilde{J}_{n\cdot 2}' W_n \tilde{J}_{n\cdot 2})^{-1}\tilde{J}_{n\cdot 2}' W_n \tilde{D}_n, \qquad (37)$$

$$\tilde{Q}[W_n]\tilde{I}_n\,\tilde{Q}[W_n]' = \tilde{V}_{n\cdot1|2}^{-1}\,\tilde{J}_{n\cdot1|2}'\,W_n\,\tilde{I}_n\,W_n\,\tilde{J}_{n\cdot1|2}\,\tilde{V}_{n\cdot1|2}^{-1}$$
$$= \tilde{V}_{n\cdot1|2}^{-1}\,\tilde{J}_{n\cdot1}^{*\prime}M\big[\tilde{J}_{n\cdot2}^{*}\big]W_n^{1/2}\,\tilde{I}_n\,W_n^{1/2}M\big[\tilde{J}_{n\cdot2}^{*}\big]\tilde{J}_{n\cdot1}^{*}\,\tilde{V}_{n\cdot1|2}^{-1}, \quad (38)$$

where $\tilde{D}_n = D_n\big(\tilde{\theta}_n^0;\,Z_n\big)$. The generalized $C(\alpha)$ statistic then takes the form:

$$PC_1\big(\tilde{\theta}_n^0;\bar{\theta}_{10},W_n\big) = PC\big(\tilde{\theta}_n^0;\,\psi,\,W_n\big)$$
$$= n\,\tilde{D}_n'\,W_n\,\tilde{J}_{n\cdot1|2}\big(\tilde{J}_{n\cdot1|2}'\,W_n\,\tilde{I}_n\,W_n\,\tilde{J}_{n\cdot1|2}\big)^{-1}\tilde{J}_{n\cdot1|2}'\,W_n\,\tilde{D}_n$$
$$= n\,\tilde{D}_n'\,W_n^{1/2}M\big[\tilde{J}_{n\cdot2}^{*}\big]\tilde{J}_{n\cdot1}^{*}\,\tilde{\Sigma}_n(W_n)^{-1}\,\tilde{J}_{n\cdot1}^{*\prime}M\big[\tilde{J}_{n\cdot2}^{*}\big]W_n^{1/2}\tilde{D}_n \quad (39)$$

where

$$\tilde{\Sigma}_n(W_n) = \tilde{J}_{n\cdot1}^{*\prime}M\big[\tilde{J}_{n\cdot2}^{*}\big]\big(W_n^{1/2}\tilde{I}_n\,W_n^{1/2}\big)M\big[\tilde{J}_{n\cdot2}^{*}\big]\tilde{J}_{n\cdot1}^{*}$$

and the matrix $\tilde{V}_{n\cdot1|2}^{-1}$ cancels out.

It is also of interest to note that the transformed score $\tilde{S}_{n\cdot1|2} = \tilde{J}_{n\cdot1|2}'W_n\tilde{D}_n$ in $PC_1\big(\tilde{\theta}_n^0;\bar{\theta}_{10},W_n\big)$ is by construction uncorrelated with $\tilde{S}_{n\cdot2} = \tilde{J}_{n\cdot2}'\tilde{I}_n^{-1}\tilde{D}_n$ asymptotically. This follows on observing that:

$$\sqrt{n}\begin{bmatrix}\tilde{S}_{n\cdot1|2}\\ \tilde{S}_{n\cdot2}\end{bmatrix} = \sqrt{n}\,\tilde{R}_n\,\tilde{D}_n \xrightarrow[n\to\infty]{L} \mathrm{N}\big[0,\,\bar{R}(\theta_0)I(\theta_0)\bar{R}(\theta_0)'\big] \quad (40)$$

where

$$\tilde{R}_n = \begin{bmatrix}\tilde{J}_{n\cdot1|2}'W_n\\ \tilde{J}_{n\cdot2}'\,\tilde{I}_n^{-1}\end{bmatrix} \xrightarrow[n\to\infty]{p} R(\theta_0) = \begin{bmatrix}J_{\cdot1|2}(\theta_0)'W_0\\ J_{\cdot2}(\theta_0)'I(\theta_0)^{-1}\end{bmatrix}, \quad (41)$$

$$J_{\cdot1|2}(\theta_0) = J_{\cdot1}(\theta_0) - J_{\cdot2}(\theta_0)\big[J_{\cdot2}(\theta_0)'W_0\,J_{\cdot2}(\theta_0)\big]^{-1}J_{\cdot2}(\theta_0)'W_0\,J_{\cdot1}(\theta_0), \quad (42)$$

and the asymptotic covariance matrix between $\sqrt{n}\,\tilde{S}_{n\cdot2}$ and $\sqrt{n}\,\tilde{S}_{n\cdot1|2}$ is

$$\big[J_{\cdot2}(\theta_0)'I(\theta_0)^{-1}\big]I(\theta_0)\big[W_0\,J_{\cdot1|2}(\theta_0)\big] = J_{\cdot2}(\theta_0)'\big[W_0\,J_{\cdot1|2}(\theta_0)\big] = 0. \quad (43)$$

Indeed, the above orthogonality can be viewed as the source of the evacuation of the distribution of $\sqrt{n}\big(\tilde{\theta}_n^0 - \theta_0\big)$ from the asymptotic distribution of the generalized $C(\alpha)$ statistic: using the Assumptions 4 and 9 [see (77)], we see easily that, under $H_0$,

$$J_{\cdot1|2}(\theta_0)'W_0\sqrt{n}\,[D_n(\tilde{\theta}_n^0) - D_n(\theta_0)] = J_{\cdot1|2}(\theta_0)'W_0\,J(\theta_0)\sqrt{n}\big(\tilde{\theta}_n^0 - \theta_0\big) + o_p(1)$$
$$= J_{\cdot1|2}(\theta_0)'W_0\,J_{\cdot2}(\theta_0)\sqrt{n}\big(\tilde{\theta}_{2n}^0 - \theta_{20}\big) + o_p(1) = o_p(1)\,. \quad (44)$$

Thus the asymptotic null distribution of the modified score used by the generalized $C(\alpha)$ statistic does not depend on the limit distribution of the nuisance parameter estimator $\tilde{\theta}_n^0$, and similarly for the generalized $C(\alpha)$ statistic.

When $W_n = \tilde{I}_n^{-1}$, the formula in (39) simplifies to:

$$
\begin{aligned}
PC_1(\tilde{\theta}_n^0; \bar{\theta}_{10}) &= n\, \tilde{D}_n'\, \tilde{I}_n^{-1} \tilde{J}_{n\cdot 1|2}\big[\tilde{J}_{n\cdot 1|2}' \tilde{I}_n^{-1} \tilde{J}_{n\cdot 1|2}\big]^{-1} \tilde{J}_{n\cdot 1|2}'\, \tilde{I}_n^{-1} \tilde{D}_n \\
&= n\, \tilde{D}_n'\, \tilde{I}_n^{-1/2} M\big[\tilde{J}_{n\cdot 2}^*\big]\tilde{J}_{n\cdot 1}^*\big[\tilde{J}_{n\cdot 1}^{*\prime}\, M\big[\tilde{J}_{n\cdot 2}^*\big]\tilde{J}_{n\cdot 1}^*\big]^{-1} \tilde{J}_{n\cdot 1}^{*\prime}\, M\big[\tilde{J}_{n\cdot 2}^*\big]\tilde{I}_n^{-1/2} \tilde{D}_n \\
&= n\, \tilde{D}_n'\, \tilde{I}_n^{-1/2} P\big[M\big[\tilde{J}_{n\cdot 2}^*\big]\tilde{J}_{n\cdot 1}^*\big]\tilde{I}_n^{-1/2} \tilde{D}_n
\end{aligned}
\tag{45}
$$

where

$$
\tilde{J}_{n\cdot 1|2} = \big[I_m - \tilde{J}_{n\cdot 2}\big(\tilde{J}_{n\cdot 2}'\tilde{I}_n^{-1}\tilde{J}_{n\cdot 2}\big)^{-1}\tilde{J}_{n\cdot 2}'\tilde{I}_n^{-1}\big]\tilde{J}_{n\cdot 1} = \tilde{I}_n^{1/2} M\big[\tilde{I}_n^{-1/2}\tilde{J}_{n\cdot 2}\big]\tilde{I}_n^{-1/2}\tilde{J}_{n\cdot 1}\,, \tag{46}
$$

$$
\tilde{J}_{n\cdot 1}^* = \tilde{I}_n^{-1/2}\tilde{J}_{n\cdot 1}\,,\ \tilde{J}_{n\cdot 2}^* = \tilde{I}_n^{-1/2}\tilde{J}_{n\cdot 2}\,. \tag{47}
$$

Upon using (40)–(43), we see that $\tilde{J}_{n\cdot 1|2}'\, \tilde{I}_n^{-1}\tilde{D}_n$ and $\tilde{J}_{n\cdot 2}'\, \tilde{I}_n^{-1}\tilde{D}_n$ are asymptotically uncorrelated, and

$$
\begin{aligned}
\tilde{J}_{n\cdot 1|2}'\tilde{I}_n^{-1}\tilde{D}_n &= \tilde{J}_{n\cdot 1}'\, \tilde{I}_n^{-1/2} M\big[\tilde{I}_n^{-1/2}\tilde{J}_{n\cdot 2}\big]\tilde{I}_n^{-1/2}\tilde{D}_n \\
&= \tilde{J}_{n\cdot 1}'\, \tilde{I}_n^{-1/2}\left\{I_m - P\big[\tilde{I}_n^{-1/2}\tilde{J}_{n\cdot 2}\big]\right\}\tilde{I}_n^{-1/2}\tilde{D}_n
\end{aligned}
\tag{48}
$$

where $M[\tilde{I}_n^{-1/2}\tilde{J}_{n\cdot 2}]\tilde{I}_n^{-1/2}\tilde{D}_n$ is the residual from the projection of $\tilde{I}_n^{-1/2}\tilde{D}_n$ on $\tilde{I}_n^{-1/2}\tilde{J}_{n\cdot 2}$. Further, on applying the Frisch–Waugh–Lovell theorem, we see that

$$
P\big[\tilde{J}_n^*\big] = P\big[\tilde{J}_{n\cdot 2}^*\big] + P\big[M\big[\tilde{J}_{n\cdot 2}^*\big]\tilde{J}_{n\cdot 1}^*\big], \tag{49}
$$

hence

$$
\begin{aligned}
&PC_1(\tilde{\theta}_n^0;\, \bar{\theta}_{10}) = n\, \tilde{D}_n'\, \tilde{I}_n^{-1/2}\left\{P\big[\tilde{J}_n^*\big] - P\big[\tilde{J}_{n\cdot 2}^*\big]\right\}\tilde{I}_n^{-1/2}\tilde{D}_n \\
&= n\,\big[\tilde{D}_n'\,\tilde{I}_n^{-1}\tilde{J}_n(\tilde{J}_n'\,\tilde{I}_n^{-1}\tilde{J}_n)^{-1}\tilde{J}_n'\,\tilde{I}_n^{-1}\tilde{D}_n - \tilde{D}_n'\,\tilde{I}_n^{-1}\tilde{J}_{n\cdot 2}(\tilde{J}_{n\cdot 2}'\,\tilde{I}_n^{-1}\tilde{J}_{n\cdot 2})^{-1}\tilde{J}_{n\cdot 2}'\,\tilde{I}_n^{-1}\tilde{D}_n\big].
\end{aligned}
\tag{50}
$$

Finally, let us consider parametric models where $m = p$ and $D_{ni}(\theta;\, Z_n)$ denotes the $p_i \times 1$ score function (the derivative of the log-likelihood function) corresponding to $\theta_i$, $i = 1, 2$, along with the corresponding partition of $\tilde{D}_n$ and $\tilde{I}_n$ :

$$
\tilde{D}_n = \begin{bmatrix} \tilde{D}_{n1} \\ \tilde{D}_{n2} \end{bmatrix} = \begin{bmatrix} D_{n1}(\tilde{\theta}_n^0;\, Z_n) \\ D_{n2}(\tilde{\theta}_n^0;\, Z_n) \end{bmatrix},\ \tilde{I}_n = \begin{bmatrix} \tilde{I}_{n11} & \tilde{I}_{n12} \\ \tilde{I}_{n21} & \tilde{I}_{n22} \end{bmatrix}, \tag{51}
$$

where $\tilde{D}_{ni} = D_{ni}(\tilde{\theta}_n^0;\, Z_n)$ is a $p_i \times 1$ vector and $\tilde{I}_{nij}$ is $p_i \times p_j$ matrix, $i, j = 1, 2$. In such cases, we have $J(\theta_0) = -I(\theta_0)$, and upon setting $\tilde{J}_n = -\tilde{I}_n$, the formulas in (45) and (50) reduce to a simple difference between two statistics:

$$
\begin{aligned}
PC_1(\tilde{\theta}_n^0; \bar{\theta}_{10}) &= n\,(\tilde{D}_{n1} - \tilde{I}_{n12}\,\tilde{I}_{n22}^{-1}\,\tilde{D}_{n2})'(\tilde{I}_{n11} - \tilde{I}_{n12}\,\tilde{I}_{n22}^{-1}\,\tilde{I}_{n21})^{-1}(\tilde{D}_{n1} - \tilde{I}_{n12}\,\tilde{I}_{n22}^{-1}\,\tilde{D}_{n2}) \\
&= n\,\big[\tilde{D}_n'\,\tilde{I}_n^{-1}\,\tilde{D}_n - \tilde{D}_{n2}'\,\tilde{I}_{n22}^{-1}\,\tilde{D}_{n2}\big].
\end{aligned}
\tag{52}
$$

# 6 Two-Stage Procedures

In this section, we formulate the $C(\alpha)$ statistic for estimating functions (or GMM-type) models estimated by two-step procedures. The $C(\alpha)$ test procedure applies in a natural way to moment condition models estimated by a two-step procedure, because a correction for the first-stage estimation error is readily built into the statistic. Models of this kind typically involve a parameter vector $\theta = (\theta_1', \theta_2')'$ where $\theta_1$ is the parameter vector of interest (on which inference focuses), and $\theta_2$ denotes a vector of nuisance parameters which is consistently estimated by an auxiliary estimate $\tilde{\theta}_{2n}^0$ obtained from the first-stage estimation. Gong and Samaniego [30], Pagan [54, 55], and Murphy and Topel [48] among others study the properties of two-step estimation and testing procedures in a likelihood framework. Newey and McFadden [49] deal with the problem in a GMM framework, but do not consider the $C(\alpha)$ statistic.

In this section, we describe how generalized $C(\alpha)$ tests can provide relatively simple solutions to such problems in the context of estimating functions and GMM estimation, with serial dependence. We first consider the generic case where the nuisance vector $\theta_2$ is estimated in a first stage, and then treated as known for the purpose of testing the value of another parameter vector $\theta_1$. Second, we examine the special case of a two-step GMM estimation, where the estimation of the nuisance parameter is based on a separate set of estimating functions (or moment conditions).

## 6.1 *Tests Based on General Two-Step Estimation*

Suppose we are interested in testing the restriction $H_0 : \theta_1 = \bar{\theta}_{10}$ based on data $X_n = [x_1, \ldots, x_n]$ and an $m_1 \times 1$ vector of estimating functions

$$D_{n1}(\theta; X_n) = D_{n1}(\theta_1, \theta_2; X_n). \tag{53}$$

In particular, we may assume $D_{n1}(\theta; X_n)$ is a subvector of a larger vector

$$D_n(\theta; X_n) = [D_{n1}(\theta; X_n)', D_{n2}(\theta; X_n)']'. \tag{54}$$

A typical setup is the one where

$$D_{n1}(\theta; X_n) = \frac{1}{n}\sum_{t=1}^{n} h_1(\theta_1, \theta_2; x_t), \tag{55}$$

$$\mathsf{E}_\theta\big[h_1(\theta_1, \theta_2; x_t)\big] = 0, \quad t = 1, \ldots, n, \tag{56}$$

and $h_1(\theta; x_t) = h_1(\theta_1, \theta_2; x_t)$ is a subvector of a higher-dimensional vector $h(\theta; x_t) = [h_1(\theta; x_t)', h_2(\theta; x_t)']'$ of estimating functions.

If the dimension of $D_{n1}(\theta_1, \theta_2; X_n)$ is large enough ($m_1 \geq p$) and the regularity conditions of Proposition 1 are satisfied when $D_n(\theta; X_n)$ is replaced by $D_{n1}(\theta; X_n)$, we can build general $C(\alpha)$-type tests of $H_0 : \theta_1 = \bar{\theta}_{10}$ based on $D_{n1}(\theta_1, \theta_2; X_n)$. No information on the (eventual) left-out estimating functions $D_{n2}(\theta; X_n)$ is required. These features underscore the remarkable versatility of estimating functions in conjunction with the generalized $C(\alpha)$ procedure described in this paper.

Let $\tilde{\theta}_{2n}^0$ be an estimator of the nuisance parameter vector $\theta_2$ obtained from the data $Y_n = [y_1, \ldots, y_n]$ which may be different from $X_n$.[2] For example, $\tilde{\theta}_{2n}^0$ may be based on an "auxiliary" estimating function $D_{n2}(\theta; X_n)$, but this is not required. Consider now the restricted estimator $\tilde{\theta}_n^0 = \left(\bar{\theta}_{10}', \tilde{\theta}_{2n}^{0\prime}\right)'$, and denote $\tilde{D}_{n1} = D_{n1}\left(\tilde{\theta}_n^0; X_n\right)$, $\tilde{I}_{n11}$, $\tilde{J}_{n1i} := \hat{J}_{n1i}\left(\tilde{\theta}_n^0\right)$ and $W_{n11}$, the matrices corresponding to $\tilde{D}_n = D_n(\tilde{\theta}_n^0; Z_n)$, $\tilde{I}_n$, $\tilde{J}_{n \cdot i}$ and $W_n$ respectively for the system based on the estimating function $D_{n1}(\theta; X_n)$; $\tilde{D}_{n1}$ has dimension $m_1 \times 1$, $\tilde{J}_{n1i}$ is $m_1 \times p_i$, and $W_{n11}$ is $m_1 \times m_1$. In the case where $D_{n1}(\theta; X_n)$ is a subvector of $D_n(\theta; X_n)$ as in (54), $\tilde{I}_{n11}$, $\tilde{J}_{n1i}$ and $W_{n11}$ are the corresponding submatrices of $\tilde{I}_n$, $\tilde{J}_{n \cdot i}$ and $W_n$ respectively, where

$$W_n = \begin{bmatrix} W_{n11} & W_{n12} \\ W_{n21} & W_{n22} \end{bmatrix} \tag{57}$$

and $W_{nij}$ is a $p_i \times p_j$ matrix, $i, j = 1, 2$.

Making the appropriate substitutions in (39), we then get the following $C(\alpha)$-type statistic for $H_0 : \theta_1 = \bar{\theta}_{10}$:

$$PC_1\left(\tilde{\theta}_n^0; \bar{\theta}_{10}, W_{n11}\right) = n\, \tilde{D}_{n1}'\, W_{n11}\, \tilde{J}_{n11|2}\, \tilde{\Sigma}_{n11|2}^{-1}\, \tilde{J}_{n11|2}'\, W_{n11}\, \tilde{D}_{n1} \tag{58}$$

where $\tilde{\Sigma}_{n11|2} = \tilde{J}_{n11|2}'\, W_{n11}\, \tilde{I}_{n11}\, W_{n11}\, \tilde{J}_{n11|2}$, and

$$\begin{aligned} \tilde{J}_{n11|2} &= \tilde{J}_{n11} - \tilde{J}_{n12}(\tilde{J}_{n12}'\, W_{n11}\, \tilde{J}_{n12})^{-1} \tilde{J}_{n12}'\, W_{n11}\, \tilde{J}_{n11} \\ &= W_{n11}^{-1/2}\, M[W_{n11}^{1/2}\, \tilde{J}_{n12}] W_{n11}^{1/2}\, \tilde{J}_{n11}, \end{aligned} \tag{59}$$

$$\begin{aligned} \tilde{\Sigma}_{n11|2} &= \tilde{J}_{n11|2}'\, W_{n11}\, \tilde{I}_{n11}\, W_{n11}\, \tilde{J}_{n11|2} \\ &= \tilde{J}_{n11}'\, W_{n11}^{1/2} M[W_{n11}^{1/2}\, \tilde{J}_{n12}] W_{n11}^{1/2}\, \tilde{I}_{n11} W_{n11}^{1/2} M[W_{n11}^{1/2}\, \tilde{J}_{n12}] W_{n11}^{1/2}\, \tilde{J}_{n11}. \end{aligned} \tag{60}$$

By Proposition 1, $PC_1\left(\tilde{\theta}_n^0; \bar{\theta}_{10}, W_{n11}\right)$ has a $\chi^2(p_1)$ asymptotic distribution under $H_0$. On taking $W_{n11} = \tilde{I}_{n11}^{-1}$, $PC_1$ takes the following simplified form:

---

[2] The number of observations in the dataset $Y$ could be different from $n$, say is equal to $n_2$, $n_2 \neq n$. If the auxiliary estimate $\tilde{\theta}_{2n_2}^0$ obtained from the second dataset satisfies $\sqrt{n_2}(\tilde{\theta}_{2n_2}^0 - \theta_{20}) = O_p(1)$, then $\sqrt{n}(\tilde{\theta}_{2n_2}^0 - \theta_{20}) = \sqrt{n/n_2}\sqrt{n_2}(\tilde{\theta}_{2n_2}^0 - \theta_{20}) = O_p(1)$ provided $n/n_2 = O(1)$, and the arguments that follow remain valid. When a set of estimating functions $D_{n_2 2}(\theta_2)$ for the second dataset is considered, the argument presented here remains valid provided $\sqrt{n_2} D_{n_2 2}(\theta_{20})$ obeys a central limit theorem in addition to the previous conditions on the auxiliary estimate and the sample sizes.

$$PC_1(\tilde{\theta}_n^0; \bar{\theta}_{10}, \tilde{I}_{n11}^{-1}) = n\, \tilde{D}_{n1}'\, \tilde{I}_{n11}^{-1/2} \tilde{M}_{12}\, \tilde{I}_{n11}^{-1/2}\, \tilde{J}_{n11}\, \tilde{\Sigma}_{n11|2}^{-1}\, \tilde{J}_{n11}' \tilde{I}_{n11}^{-1/2}\, \tilde{M}_{12}\, \tilde{I}_{n11}^{-1/2}\, \tilde{D}_{n1} \tag{61}$$

where $\tilde{M}_{12} = M[\tilde{I}_{n11}^{-1/2}\, \tilde{J}_{n12}]$ and $\tilde{\Sigma}_{n11|2} = \tilde{J}_{n11|2}'\, \tilde{I}_{n11}^{-1}\, \tilde{J}_{n11|2} = \tilde{J}_{n11}'\, \tilde{I}_{n11}^{-1/2}\, \tilde{M}_{12}\, \tilde{I}_{n11}^{-1/2}$ $\tilde{J}_{n11}$.

When calculating the standard error of the estimator of $\theta_1$, one needs to take into account the sampling error associated with the first-stage estimator of the parameter $\theta_2$; see Newey and McFadden [49]. This is achieved transparently by the $C(\alpha)$ statistic, because its asymptotic distribution does not depend on the asymptotic distribution of the first-stage estimator. Here, the invariance of the the asymptotic distribution of $PC_1(\tilde{\theta}_n^0; \bar{\theta}_{10}, W_{n11})$ with respect to the distribution of $\tilde{\theta}_n^0$ is entailed by the orthogonality relation

$$\begin{aligned} J_{12}(\theta_0)'[W_{011} J_{11|2}(\theta_0)] &= J_{12}(\theta_0)' W_{011}\, W_{011}^{-1/2} M\big[W_{011}^{1/2}\, J_{12}(\theta_0)\big] W_{011}^{1/2}\, J_{11}(\theta_0) \\ &= \big[W_{011}^{1/2}\, J_{12}(\theta_0)\big]' M\big[W_{011}^{1/2}\, J_{12}(\theta_0)\big] W_{011}^{1/2}\, J_{11}(\theta_0) = 0, \end{aligned} \tag{62}$$

where $\text{plim}_{n\to\infty} W_{n11} = W_{011}$. This in turn implies that $\sqrt{n}\, \tilde{J}_{n11|2}'\, W_{n11}\, \tilde{D}_{n1}$ is asymptotically uncorrelated with $\sqrt{n}\, \tilde{J}_{n12}' \tilde{I}_{n11}^{-1}\, \tilde{D}_{n1}$; see (40)–(44) for a similar argument.

## 6.2 *Tests Based on a Two-Step GMM Estimation*

We now consider the case where the condition $m_1 \geq p$ may not hold—so rank conditions for applying a $C(\alpha)$-type test only based on $h_1$ cannot hold (without other restrictions)—but we have $m_2$ estimating functions $D_{n2}(\theta; X_n)$ as in (54) which be used to draw inference on $\theta_2$ and account for the uncertainty of $\theta_2$ estimates, where $m_2 \geq p_2$. Further, we suppose here that $D_{n2}(\theta; X_n)$ only depends on $\theta_2$, i.e. $D_{n2}(\theta; X_n) = D_{n2}(\theta_2; X_n)$, with $m_1 \geq p_1$ and $m_2 \geq p_2$.

In particular, these assumptions may be based on a system of moment equations

$$\mathsf{E}_\theta \begin{bmatrix} h_1(\theta_1, \theta_2; x_t) \\ h_2(\theta_2; y_t) \end{bmatrix} = 0, \quad t = 1, \ldots, n, \tag{63}$$

where $h_2(\theta_2; y_t)$ is typically used to estimate the nuisance parameter $\theta_2$ and

$$D_{n2}(\theta_2; Y_n) = \frac{1}{n} \sum_{t=1}^{n} h_2(\theta_2; y_t). \tag{64}$$

In this context, the sample estimating function is

$$\tilde{D}_n = \begin{bmatrix} \tilde{D}_{n1} \\ \tilde{D}_{n2} \end{bmatrix} = \begin{bmatrix} D_{n1}(\tilde{\theta}_n^0; X_n) \\ D_{n2}(\tilde{\theta}_{2n}^0; Y_n) \end{bmatrix} \tag{65}$$

where

$$J(\theta) = [J_{\cdot 1}(\theta), J_{\cdot 2}(\theta)] = \begin{bmatrix} J_{11}(\theta) & J_{12}(\theta) \\ 0_{m_2 \times p_1} & J_{22}(\theta) \end{bmatrix}. \tag{66}$$

The partitioned Jacobian estimator is then given by

$$\tilde{J}_n = [\tilde{J}_{n\cdot 1}, \tilde{J}_{n\cdot 2}] = \begin{bmatrix} \tilde{J}_{n11} & \tilde{J}_{n12} \\ 0_{m_2 \times p_1} & \tilde{J}_{n22} \end{bmatrix}. \tag{67}$$

On assuming that the regularity conditions of Proposition 1 are satisfied, we can use here the statistic $PC_1(\tilde{\theta}_n^0; \bar{\theta}_{10}, W_n)$ defined in (39). Further, the form (67) yields useful restrictions on the test statistic. We then have

$$PC_1(\tilde{\theta}_n^0; \bar{\theta}_{10}, W_n) = n\, \tilde{D}_n'\, W_n\, \tilde{J}_{n\cdot 1|2} \big(\tilde{J}_{n\cdot 1|2}'\, W_n\, \tilde{I}_n\, W_n\, \tilde{J}_{n\cdot 1|2}\big)^{-1} \tilde{J}_{n\cdot 1|2}'\, W_n\, \tilde{D}_n \tag{68}$$

with

$$\begin{aligned} \tilde{J}_{n\cdot 1|2}'\, W_n\, \tilde{D}_n = {} & \tilde{J}_{n11}'\, W_{n11}\, \tilde{D}_{n1} \\ & + \big[\tilde{J}_{n11}'\, W_{n12}\, \tilde{D}_{n2} - \tilde{J}_{n\cdot 1}'\, W_n\, \tilde{J}_{n\cdot 2} \big(\tilde{J}_{n\cdot 2}'\, W_n\, \tilde{J}_{n\cdot 2}\big)^{-1} \tilde{J}_{n\cdot 2}'\, W_n\, \tilde{D}_n\big]. \end{aligned} \tag{69}$$

In this case, the correction for the estimation of $\theta_2$ is accounted by the two last terms in the above expression for $\tilde{J}_{n\cdot 1|2}' W_n \tilde{D}_n$.

For moment equations of the form (63), it is natural to consider separate weightings for $\tilde{D}_{n1}$ and $\tilde{D}_{n2}$, i.e.

$$W_{n12} = W_{n21}' = 0. \tag{70}$$

On using both conditions (67) and (70), we see that

$$\begin{aligned} \tilde{J}_{n\cdot 1|2}'\, W_n\, \tilde{D}_n = {} & \tilde{J}_{n11}'\, W_{n11}\, \tilde{D}_{n1} \\ & - \tilde{J}_{n11}' W_{n11} \tilde{J}_{n12} \big(\tilde{J}_{n\cdot 2}'\, W_n\, \tilde{J}_{n\cdot 2}\big)^{-1} \big[\tilde{J}_{n12}'\, W_{n11}\, \tilde{D}_{n1} + \tilde{J}_{n22}'\, W_{n22}\, \tilde{D}_{n2}\big], \end{aligned} \tag{71}$$

$$\tilde{J}_{n\cdot 2}'\, W_n\, \tilde{J}_{n\cdot 2} = \tilde{J}_{n12}'\, W_{n11}\, \tilde{J}_{n12} + \tilde{J}_{n22}'\, W_{n22}\, \tilde{J}_{n22}. \tag{72}$$

Again the asymptotic distribution of the test statistic $PC_1(\tilde{\theta}_n^0; \bar{\theta}_{10}, W_n)$ is $\chi^2(p_1)$ under the null hypothesis $H_0 : \theta_1 = \bar{\theta}_{10}$, irrespective of the asymptotic distribution of $\tilde{\theta}_{2n}^0$.

## 7 Conclusion

In this paper, we have introduced a comprehensive $C(\alpha)$ statistic based on estimating functions (or GMM setups). As in Smith [63], the null hypothesis is specified in terms of a general possibly nonlinear constraint, rather than a restriction fixing a parameter subvector. The proposed procedure allows for general forms of serial dependence and heteroskedasticity, and can be implemented using any root-$n$ consistent restricted estimator. A detailed derivation of the asymptotic null distribution of the statistic was provided under weak regularity conditions.

The proposed generalized $C(\alpha)$-type statistic includes earlier ones as special cases, as well as a wide spectrum of new ones. A number of important special cases of the extended test statistic were discussed in detail. These include testing whether a parameter subvector has a given value—for which we give a number of alternative forms and special cases—and the important problem of accounting for parameter uncertainty in two-stage procedures.

**Acknowledgements** The authors thank Marine Carrasco, Jean-Pierre Cotton, Russell Davidson, Abdeljelil Farhat, V. P. Godambe, Christian Genest, Christian Gouriéroux, Stéphane Grégoir, Tianyu He, Frank Kleibergen, Sophocles Mavroeidis, Hervé Mignon, Julien Neves, Denis Pelletier, Mohamed Taamouti, Masaya Takano, Pascale Valéry, two anonymous referees, and the Editor Wai Keung for several useful comments. Earlier versions of this paper were presented at the Canadian Statistical Society 1997 annual meeting and at INSEE (CREST, Paris). This work was supported by the William Dow Chair in Political Economy (McGill University), the Bank of Canada (Research Fellowship), the Toulouse School of Economics (Pierre-de-Fermat Chair of excellence), the Universitad Carlos III de Madrid (Banco Santander de Madrid Chair of excellence), a Guggenheim Fellowship, a Konrad-Adenauer Fellowship (Alexander-von-Humboldt Foundation, Germany), the Canadian Network of Centres of Excellence [program on *Mathematics of Information Technology and Complex Systems* (MITACS)], the Natural Sciences and Engineering Research Council of Canada, the Social Sciences and Humanities Research Council of Canada, and the Fonds de recherche sur la société et la culture (Québec).

## Appendix

*Proof of Proposition 1* To simplify notation, we shall assume throughout that $\omega \in \mathscr{D}_J$ (an event with probability 1) and drop the symbol $\omega$ from the random variables considered. In order to obtain the asymptotic null distribution of the generalized $C(\alpha)$ statistic defined in (14), we first need to show that $P(\tilde{\theta}_n^0)$ and $J_n(\tilde{\theta}_n^0)$ converge in probability to $P(\theta_0)$ and $J(\theta_0)$ respectively. The consistency of $P(\tilde{\theta}_n^0)$, i.e.

$$\operatorname*{plim}_{n\to\infty}\left[P\left(\tilde{\theta}_n^0\right) - P(\theta_0)\right] = 0, \tag{73}$$

follows simply from the consistency of $\tilde{\theta}_n^0$ [Assumption 9] and the continuity of $P(\theta)$ at $\theta_0$ [Assumption 7]. Further, by Assumption 8, since $P(\theta)$ is continuous in open neighborhood of $\theta_0$, we also have

$$\text{rank}\,[\tilde{P}_n] = \text{rank}\,[P(\theta_0)] = p_1. \tag{74}$$

Consider now $J_n(\tilde{\theta}_n^0)$. By Assumption 5, for any $\varepsilon > 0$ and $\varepsilon_1 > 0$, we can choose $\delta_1 := \delta(\varepsilon_1, \varepsilon) > 0$ and a positive integer $n_1(\varepsilon, \delta_1)$ such that: (i) $U_J(\delta_1, \varepsilon, \theta_0) \leq \varepsilon_1/2$, and (ii) $n > n_1(\varepsilon, \delta_1)$ entails

$$\mathsf{P}\big[\Delta_n(\theta_0, \delta) > \varepsilon\big] = \mathsf{P}\big[\{\omega : \Delta_n(\theta_0, \delta, \omega) > \varepsilon\}\big] \leq U_J(\delta_1, \varepsilon, \theta_0) \leq \varepsilon_1/2.$$

Further, by the consistency of $\tilde{\theta}_n^0$ [Assumption 9], we can choose $n_2(\varepsilon_1, \delta_1)$ such that $n > n_2(\varepsilon_1, \delta_1)$ entails $\mathsf{P}[\|\tilde{\theta}_n^0 - \theta_0\| \leq \delta_1] \geq 1 - (\varepsilon_1/2)$. Then, using the Boole-Bonferroni inequality, we have for $n > \max\{n_1(\varepsilon, \delta_1), n_2(\varepsilon_1, \delta_1)\}$:

$$\begin{aligned}
\mathsf{P}\big[\big\|J_n(\tilde{\theta}_n^0) - J(\theta_0)\big\| \leq \varepsilon\big] &\geq \mathsf{P}\big[\big\|\tilde{\theta}_n^0 - \theta_0\big\| \leq \delta_1 \text{ and } \big\|J_n(\tilde{\theta}_n^0) - J(\theta_0)\big\| \leq \varepsilon\big] \\
&\geq \mathsf{P}\big[\big\|\tilde{\theta}_n^0 - \theta_0\big\| \leq \delta_1 \text{ and } \Delta_n\big(\theta_0, \delta_1\big) \leq \varepsilon\big] \\
&\geq 1 - \mathsf{P}\big[\big\|\tilde{\theta}_n^0 - \theta_0\big\| > \delta_1\big] - \mathsf{P}\big[\Delta_n\big(\theta_0, \delta_1\big) > \varepsilon\big] \\
&\geq 1 - (\varepsilon_1/2) - (\varepsilon_1/2) = 1 - \varepsilon_1.
\end{aligned}$$

Thus,

$$\liminf_{n\to\infty} \mathsf{P}\big[\big\|J_n(\tilde{\theta}_n^0) - J(\theta_0)\big\| \leq \varepsilon\big] \geq 1 - \varepsilon_1, \quad \text{for all } \varepsilon > 0,\ \varepsilon_1 > 0,$$

hence

$$\lim_{n\to\infty} \mathsf{P}\big[\big\|J_n(\tilde{\theta}_n^0) - J(\theta_0)\big\| \leq \varepsilon\big] = 1, \quad \text{for all } \varepsilon > 0, \tag{75}$$

or, equivalently,

$$\operatorname*{plim}_{n\to\infty} \big[J_n(\tilde{\theta}_n^0) - J(\theta_0)\big] = 0. \tag{76}$$

By Assumption 4, we can write [setting $0/0 = 0$] :

$$\begin{aligned}
\big\|\sqrt{n}\,\big[D_n(\tilde{\theta}_n^0) - D_n(\theta_0)\big] - J(\theta_0)\sqrt{n}\,\big(\tilde{\theta}_n^0 - \theta_0\big)\big\| &= \sqrt{n}\,\big\|R_n(\tilde{\theta}_n^0, \theta_0)\big\| \\
&= \frac{\big\|R_n(\tilde{\theta}_n^0, \theta_0)\big\|}{\big\|\tilde{\theta}_n^0 - \theta_0\big\|}\sqrt{n}\,\big\|\tilde{\theta}_n^0 - \theta_0\big\|
\end{aligned}$$

where

$$\frac{\big\|R_n(\tilde{\theta}_n^0, \theta_0)\big\|}{\big\|\tilde{\theta}_n^0 - \theta_0\big\|} \leq r_n(\delta, \theta_0) \ \text{ when } \tilde{\theta}_n^0 \in N_0 \text{ and } \big\|\tilde{\theta}_n^0 - \theta_0\big\| \leq \delta$$

and $\limsup_{n\to\infty} \mathsf{P}\big[r_n(\delta, \theta_0) > \varepsilon\big] < U_D(\delta, \varepsilon, \theta_0)$. Thus, for any $\varepsilon > 0$ and $\delta > 0$, we have:

$$\mathsf{P}\Big[\frac{\|R_n(\tilde{\theta}_n^0,\theta_0)\|}{\|\tilde{\theta}_n^0-\theta_0\|}\le\varepsilon\Big]\ge\mathsf{P}\big[r_n(\delta,\theta_0)\le\varepsilon,\tilde{\theta}_n^0\in N_0\text{ and }\|\tilde{\theta}_n^0-\theta_0\|\le\delta\big]$$
$$\ge 1-\mathsf{P}\big[r_n(\delta,\theta_0)>\varepsilon\big]-\mathsf{P}\big[\tilde{\theta}_n^0\notin N_0\text{ or }\|\tilde{\theta}_n^0-\theta_0\|>\delta\big]$$

hence, using the consistency of $\tilde{\theta}_n^0$,

$$\begin{aligned}\liminf_{n\to\infty}\mathsf{P}\big[\|R_n(\tilde{\theta}_n^0,\theta_0)\|/\|\tilde{\theta}_n^0-\theta_0\|\le\varepsilon\big] &\ge 1-\limsup_{n\to\infty}\mathsf{P}\big[r_n(\delta,\theta_0)>\varepsilon\big]\\ &\quad-\limsup_{n\to\infty}\mathsf{P}\big[\tilde{\theta}_n^0\notin N_0\text{ or }\|\tilde{\theta}_n^0-\theta_0\|>\delta\big]\\ &\ge 1-U_D(\delta,\varepsilon,\theta_0).\end{aligned}$$

Since $\lim_{\delta\downarrow 0}U_D(\delta,\varepsilon,\theta_0)=0$, it follows that $\lim_{n\to\infty}\mathsf{P}[\|R_n(\tilde{\theta}_n^0,\theta_0)\|/\|\tilde{\theta}_n^0-\theta_0\|\le\varepsilon]=1$ for any $\varepsilon>0$, or equivalently,

$$\|R_n(\tilde{\theta}_n^0,\theta_0)\|/\|\tilde{\theta}_n^0-\theta_0\|\xrightarrow[n\to\infty]{\mathsf{p}}0.$$

Since $\sqrt{n}\,(\tilde{\theta}_n^0-\theta_0)$ is asymptotically bounded in probability (by Assumption 9), this entails:

$$\sqrt{n}\,\|R_n(\tilde{\theta}_n^0,\theta_0)\|=\frac{\|R_n(\tilde{\theta}_n^0,\theta_0)\|}{\|\tilde{\theta}_n^0-\theta_0\|}\sqrt{n}\,\|\tilde{\theta}_n^0-\theta_0\|\xrightarrow[n\to\infty]{\mathsf{p}}0 \tag{77}$$

and

$$\big\|\sqrt{n}\,[D_n(\tilde{\theta}_n^0)-D_n(\theta_0)]-J(\theta_0)\sqrt{n}\,(\tilde{\theta}_n^0-\theta_0)\big\|\xrightarrow[n\to\infty]{\mathsf{p}}0. \tag{78}$$

By Taylor's theorem and Assumptions 7 and 8, we also have the expansion:

$$\psi(\theta)=\psi(\theta_0)+P(\theta_0)(\theta-\theta_0)+R_2(\theta,\theta_0), \tag{79}$$

for $\theta\in N\subseteq N_0\cap V_0$, where $N$ is a non-empty open neighborhood of $\theta_0$ and

$$\lim_{\theta\to\theta_0}\|R_2(\theta,\theta_0)\|/\|\theta-\theta_0\|=0,$$

i.e., $R_2(\theta,\theta_0)=o(\|\theta-\theta_0\|)$, so that, using Assumption 10,

$$\begin{aligned}\sqrt{n}\,P(\theta_0)(\tilde{\theta}_n^0-\theta_0)&=\sqrt{n}\,[\psi(\tilde{\theta}_n^0)-\psi(\theta_0)]-\sqrt{n}\,R_2(\tilde{\theta}_n^0,\theta_0)\\ &=-\sqrt{n}\,R_2(\tilde{\theta}_n^0,\theta_0)\end{aligned} \tag{80}$$

for $\tilde{\theta}_n^0\in N$, and

$$\|\sqrt{n}\,P(\theta_0)(\tilde{\theta}_n^0-\theta_0)\|=\frac{\|R_2(\tilde{\theta}_n^0,\theta_0)\|}{\|\tilde{\theta}_n^0-\theta_0\|}\sqrt{n}\|\tilde{\theta}_n^0-\theta_0\|\xrightarrow[n\to\infty]{\mathsf{p}}0. \tag{81}$$

By (74) and (76) jointly with the Assumptions 3, 6, 7, 8, 11 and 12, we have:

$$\text{rank}\,[\tilde{P}_n] = p_1,\ \text{rank}\,[\tilde{J}_n] = p,\ \text{rank}\,[\tilde{I}_n] = m,\ \text{rank}\,[W_n] = m, \tag{82}$$

so the matrices $\tilde{J}_n$, $\tilde{I}_n$, and $W_n$ all have full column rank. Since $\underset{n\to\infty}{\text{plim}}\,\tilde{P}_n = P(\theta_0)$ and $\underset{n\to\infty}{\text{plim}}\,\tilde{J}_n = J(\theta_0)$, we can then write:

$$\underset{n\to\infty}{\text{plim}}\left[\tilde{J}_n'\,W_n\,\tilde{J}_n\right]^{-1} = \left[J(\theta_0)'W_0\,J(\theta_0)\right]^{-1},\ \underset{n\to\infty}{\text{plim}}\,\tilde{Q}_n = Q(\theta_0),$$
$$\underset{n\to\infty}{\text{plim}}\,\tilde{Q}_n\,\tilde{J}_n = \underset{n\to\infty}{\text{plim}}\,\tilde{Q}_n\,J(\theta_0) = Q(\theta_0)\,J(\theta_0) = P(\theta_0),$$

where $\tilde{Q}_n := \tilde{Q}[W_n] = \tilde{P}_n[\tilde{J}_n'W_n\tilde{J}_n]^{-1}\tilde{J}_n'W_n$. Then, using (78) and (81), it follows that:

$$\begin{aligned}
&\underset{n\to\infty}{\text{plim}}\left\{\sqrt{n}\,\tilde{Q}_n\,D_n(\tilde{\theta}_n^0) - \sqrt{n}Q(\theta_0)\,D_n(\theta_0)\right\}\\
&\quad = \underset{n\to\infty}{\text{plim}}\left\{\sqrt{n}\,\tilde{Q}_n\,D_n(\tilde{\theta}_n^0) - Q(\theta_0)\sqrt{n}\,D_n(\theta_0)\right\} - \underset{n\to\infty}{\text{plim}}\left\{P(\theta_0)\sqrt{n}\,(\tilde{\theta}_n^0 - \theta_0)\right\}\\
&\quad = \underset{n\to\infty}{\text{plim}}\quad\left\{\tilde{Q}_n\left[\sqrt{n}\,[D_n(\tilde{\theta}_n^0) - D_n(\theta_0)] - J(\theta_0)\sqrt{n}\,(\tilde{\theta}_n^0 - \theta_0)\right]\right\}\\
&\qquad + \underset{n\to\infty}{\text{plim}}\quad\left\{\left[\tilde{Q}_n - Q(\theta_0)\right]\sqrt{n}\,D_n(\theta_0) + \left[\tilde{Q}_n\,J(\theta_0) - P(\theta_0)\right]\sqrt{n}\,(\tilde{\theta}_n^0 - \theta_0)\right\}\\
&\quad = \underset{n\to\infty}{\text{plim}}\quad\left\{\tilde{Q}_n\left[\sqrt{n}\,[D_n(\tilde{\theta}_n^0) - D_n(\theta_0)] - J(\theta_0)\sqrt{n}\,(\tilde{\theta}_n^0 - \theta_0)\right]\right\} = 0.
\end{aligned}$$

We conclude that the asymptotic distribution of $\sqrt{n}\,\tilde{Q}_n D_n(\tilde{\theta}_n^0)$ is the same as the one of $Q(\theta_0)\sqrt{n}\,D_n(\theta_0)$, namely (by Assumption 2) a $\text{N}[0,\ V_\psi(\theta_0)]$ distribution where

$$V_\psi(\theta) = Q(\theta)\,I(\theta)\,Q(\theta)'$$

and $V_\psi(\theta_0)$ has rank $p_1 = \text{rank}\big[Q(\theta_0)\big] = \text{rank}[P(\theta_0)]$. Consequently, the estimator

$$\tilde{V}_\psi(\tilde{\theta}_n^0) = \tilde{Q}_n\,\tilde{I}_n\,\tilde{Q}_n' \tag{83}$$

converges to $V_\psi(\theta_0)$ in probability and, by (82),

$$\text{rank}\,\left[\tilde{V}_\psi(\tilde{\theta}_n^0)\right] = p_1. \tag{84}$$

Thus the test criterion

$$PC(\tilde{\theta}_n^0;\,\psi,\,W_n) = nD_n(\tilde{\theta}_n^0;\,Z_n)'\tilde{Q}[W_n]'\left\{\tilde{Q}[W_n]\tilde{I}_n\,\tilde{Q}[W_n]'\right\}^{-1}\tilde{Q}[W_n]D_n(\tilde{\theta}_n^0;\,Z_n)$$

has an asymptotic $\chi^2(p_1)$ distribution. □

*Proof of Proposition 2* Consider the (non-empty) open neighborhood $N = N_1 \cap N_2$ of $\theta_0$. For any $\theta \in N$ and $\omega \in \mathscr{X}$, we can write

$$\begin{aligned}\left\|J(\theta) - J(\theta_0)\right\| &\leq \left\|J_n(\theta, \omega) - J(\theta)\right\| + \left\|J_n(\theta_0, \omega) - J(\theta_0)\right\| \\ &\quad + \left\|J_n(\theta, \omega) - J_n(\theta_0, \omega)\right\| \\ &\leq 2 \sup_{\theta \in N}\left\|J_n(\theta, \omega) - J(\theta)\right\| + \left\|J_n(\theta, \omega) - J_n(\theta_0, \omega)\right\|\end{aligned}$$

By Assumption 14b, we have

$$\operatorname*{plim}_{n\to\infty} \left(\sup_{\theta\in N}\left\|J_n(\theta, \omega) - J(\theta)\right\|\right) \leq \operatorname*{plim}_{n\to\infty}\left(\sup_{\theta\in N_2}\left\|J_n(\theta, \omega) - J(\theta)\right\|\right) = 0$$

and we can find a subsequence $\left\{J_{n_t}(\theta, \omega) : t{=}1, 2, \ldots\right\}$ of $\{J_n(\theta, \omega) : n = 1, 2, \ldots\}$ such that

$$\sup_{\theta\in N}\left\{\left\|J_{n_t}(\theta, \omega) - J(\theta)\right\|\right\} \underset{t\to\infty}{\longrightarrow} 0 \quad a.s.$$

Let

$$CS = \left\{\omega \in \mathscr{X} : \lim_{t\to\infty}\left(\sup_{\theta\in N}\left\|J_{n_t}(\theta, \omega) - J(\theta)\right\|\right) = 0\right\}$$

and $\varepsilon > 0$. By definition, $\mathsf{P}\big[\omega \in CS\big] = 1$. For $\omega \in CS$, we can choose $t_0(\varepsilon, \omega)$ such that

$$t \geq t_0(\varepsilon, \omega) \Rightarrow 2 \sup_{\theta\in N}\left\{\left\|J_{n_t}(\theta, \omega) - J(\theta)\right\|\right\} < \varepsilon/2.$$

Further, since $J_n(\theta, \omega)$ is continuous in $\theta$ at $\theta_0$, we can find $\delta(n, \omega) > 0$ such that

$$\left\|\theta - \theta_0\right\| < \delta(n, \omega) \Rightarrow \left\|J_n(\theta, \omega) - J_n(\theta_0, \omega)\right\| < \varepsilon/2.$$

Thus, taking $t_0 = t_0(\varepsilon, \omega)$ and $n = n_{t_0}$, we find that $\left\|\theta - \theta_0\right\| < \delta(n_{t_0}, \omega)$ implies

$$\left\|J(\theta) - J(\theta_0)\right\| < \frac{\varepsilon}{2} + \frac{\varepsilon}{2} = \varepsilon.$$

In other words, for any $\varepsilon > 0$, we can choose $\delta = \delta(n_{t_0}, \varepsilon) > 0$ such that

$$\left\|\theta - \theta_0\right\| < \delta \Rightarrow \left\|J(\theta) - J(\theta_0)\right\| < \varepsilon,$$

and the function $J(\theta)$ must be continuous at $\theta_0$. Part (a) of the Proposition is established.

Set $\overline{\Delta}_n(N_2, \omega) := \sup\{\|J_n(\theta, \omega) - J(\theta)\| : \theta \in N_2\}$. To get Assumption 5, we note that

$$\begin{aligned}\Delta_n(\theta_0, \delta, \omega) &:= \sup\left\{\|J_n(\theta, \omega) - J(\theta_0)\| : \theta \in N_2 \text{ and } 0 \leq \left\|\theta - \theta_0\right\| \leq \delta\right\} \\ &\leq \overline{\Delta}_n(N_2, \omega)\end{aligned}$$

for any $\delta > 0$, hence, by Assumption 14b,

$$\begin{aligned}\limsup_{n\to\infty} \mathsf{P}\left[\{\omega : \Delta_n(\theta_0, \delta, \omega) > \varepsilon\}\right] &\leq \limsup_{n\to\infty} \mathsf{P}\left[\left\{\omega : \overline{\Delta}_n(N_2, \omega) > \varepsilon\right\}\right] \\ &\leq U_J(\delta, \varepsilon, \theta_0)\end{aligned}$$

for any function $U_J(\delta, \varepsilon, \theta_0)$ that satisfies the conditions of Assumption 5. The latter thus holds with $V_0$ any non-empty open neighborhood of $\theta_0$ such that $V_0 \subseteq N_2$.

To obtain Assumption 4, we note that Assumption 14 entails $D_n(\theta, \omega)$ is continuously differentiable in an open neighborhood of $\theta_0$ for all $\omega \in \mathscr{D}_J$, so that we can apply Taylor's formula for a function of several variables (see Edwards [26, Section II.7]) to each component of $D_n(\theta, \omega)$ : for all $\theta$ in an open neighborhood $U$ of $\theta_0$ (with $U \subseteq N_2$), we can write

$$\begin{aligned}D_{in}(\theta, \omega) &= D_{in}(\theta_0, \omega) + J_n\left(\bar{\theta}_n^i(\omega), \omega\right)_{i\cdot}(\theta - \theta_0) \\ &= D_{in}(\theta_0, \omega) + J(\theta_0)_{i\cdot}(\theta - \theta_0) + R_{in}\left(\bar{\theta}_n^i(\omega), \theta_0, \omega\right),\ i = 1, \ldots, m,\end{aligned}$$

where $J_n(\theta, \omega)_{i\cdot}$ and $J(\theta)_{i\cdot}$ are the $i$-th rows of $J_n(\theta, \omega)$ and $J(\theta)$ respectively,

$$R_{in}\left(\bar{\theta}_n^i(\omega), \theta_0, \omega\right) = \left[J_n\left(\bar{\theta}_n^i(\omega), \omega\right)_{i\cdot} - J(\theta_0)_{i\cdot}\right](\theta - \theta_0)$$

and $\bar{\theta}_n^i(\omega)$ belongs to the line joining $\theta$ and $\theta_0$. Further, for $\theta \in U$,

$$\begin{aligned}\left|R_{in}\left(\bar{\theta}_n^i(\omega), \theta_0, \omega\right)\right| &\leq \left\|J_n\left(\bar{\theta}_n^i(\omega), \omega\right)_{i\cdot} - J(\theta_0)_{i\cdot}\right\| \left\|\theta - \theta_0\right\| \\ &\leq \left\|J_n\left(\bar{\theta}_n^i(\omega), \omega\right) - J(\theta_0)\right\| \left\|\theta - \theta_0\right\| \\ &\leq \left\|\theta - \theta_0\right\| \sup\left\{\|J_n(\theta, \omega) - J(\theta)\| : \theta \in N_2\right\},\ i = 1, \ldots, m,\end{aligned}$$

hence, on defining $N_0 = U$,

$$R_n(\theta, \theta_0, \omega) = \left[R_{1n}\left(\bar{\theta}_n^1(\omega), \theta_0, \omega\right), \ldots, R_{mn}\left(\bar{\theta}_n^m(\omega), \theta_0, \omega\right)\right]',$$

we see that

$$\begin{aligned}\left\|R_n(\theta, \theta_0, \omega)\right\| &\leq \sum_{1=1}^{m}\left|R_{in}\left(\bar{\theta}_n^i(\omega), \theta_0, \omega\right)\right| \\ &\leq m\left\|\theta - \theta_0\right\| \sup_{\theta\in N_2}\left\{\|J_n(\theta, \omega) - J(\theta)\|\right\}\end{aligned}$$

and

$$r_n(\delta, \theta_0, \omega) \; := \sup \left\{ \frac{\| R_n(\theta, \theta_0, \omega) \|}{\| \theta - \theta_0 \|} : \theta \in N_0 \text{ and } 0 < \| \theta - \theta_0 \| \leq \delta \right\}$$
$$\leq m \sup \{ \| J_n(\theta, \omega) - J(\theta) \| : \theta \in N_2 \}$$

Thus $r_n(\delta, \theta_0, \omega) \xrightarrow[n \to \infty]{p} 0$ and

$$\limsup_{n \to \infty} \mathsf{P}\big[\{\omega : r_n(\delta, \theta_0, \omega) > \varepsilon\}\big] \leq U_D(\delta, \varepsilon, \theta_0) \tag{85}$$

must hold for any function that satisfies the conditions of Assumption 4. This completes the proof. □

## References


1. Andrews, D. W. K. (1991). Heteroskedasticity and autocorrelation consistent covariance matrix estimation. *Econometrica*, *59*, 817–858.
2. Andrews, D. W. K., & Monahan, J. C. (1992). An improved heteroskedasticity and autocorrelation consistent covariance matrix estimator. *Econometrica*, *60*, 953–966.
3. Bartoo, J. B., & Puri, P. S. (1967). On optimal asymptotic tests of composite statistical hypotheses. *The Annals of Mathematical Statistics*, *38*(6), 1845–1852.
4. Basawa, I. V. (1985). Neyman–Le Cam tests based on estimating functions. In L. Le Cam & R. A. Olshen (Eds.), *Berkeley Conference in Honor of Jerzy Neyman and Jack Kiefer* (pp. 811–825). Belmont, CA: Wadsworth.
5. Basawa, I. V., Godambe, V. P., & Taylor, R. L., (Eds.), (1997). *Selected proceedings of the symposium on estimating functions*, Vol. 32 of *IMS lecture notes monograph series*. Hayward, CA: Institute of Mathematical Statistics.
6. Bera, A., & Bilias, Y. (2001). Rao's score, Neyman's $C(\alpha)$ and Silvey's LM tests: An essay on historical developments and some new results. *Journal of Statistical Planning and Inference*, *97*, 9–44.
7. Bera, A. K., & Yoon, M. J. (1993). Specification testing with locally misspecified alternatives. *Econometric theory*, *9*(04), 649–658.
8. Berger, A., & Wallenstein, S. (1989). On the theory of $C_\alpha$-tests. *Statistics and Probability Letters*, *7*, 419–424.
9. Bernshtein, A. V. (1976). On optimal asymptotic tests for composite hypotheses under non-standard conditions. *Theory of Probability and its Applications*, *21*, 34–47.
10. Bernshtein, A. V. (1978). On optimal asymptotic tests of homogeneity. *Theory of Probability and Its Applications*, *22*, 377–383.
11. Bernshtein, A. V. (1980). On the construction of majorizing tests. *Theory of Probability and Its Applications*, *25*, 16–26.
12. Bernshtein, A. V. (1980). On verifying composite hypotheses with nuisance parameters in the multivariate case. *Theory of Probability and Its Applications*, *25*, 287–298.
13. Bernshtein, A. V. (1981). Asymptotically similar criteria. *Journal of Soviet Mathematics*, *17*(3), 1825–1857.
14. Bhat, B. R., & Nagnur, B. N. (1965). Locally asymptotically most stringent tests and Lagrangian multiplier tests of linear hypotheses. *Biometrika*, *52*(3–4), 459–468.
15. Bühler, W. J., & Puri, P. S. (1966). On optimal asymptotic tests of composite hypotheses with several constraints. *Zeitschrift für Wahrscheinlichkeitstheorie und Verwandte Gebiete*, *5*(1), 71–88.

16. Chant, D. (1974). On asymptotic tests of composite hypotheses in nonstandard conditions. *Biometrika, 61*(2), 291–298.
17. Chaudhuri, S., & Zivot, E. (2011). A new method of projection-based inference in gmm with weakly identified nuisance parameters. *Journal of Econometrics, 164*(2), 239–251.
18. Chibisov, D. M. (1973). Asymptotic expansions for Neyman's $C(\alpha)$ tests. *Proceedings of the second Japan-USSR symposium on probability theory* (pp. 16–45). Berlin: Springer.
19. Cushing, M. J., & McGarvey, M. G. (1999). Covariance matrix estimation. In L. Mátyás (Ed.), Generalized method of moments estimation, Chap. 3 (pp. 63–95). Cambridge: Cambridge University Press.
20. Dagenais, M. G., & Dufour, J.-M. (1991). Invariance, nonlinear models and asymptotic tests. *Econometrica, 59*, 1601–1615.
21. Davidson, R., & MacKinnon, J. G. (1991). Artificial regressions and $C(\alpha)$ tests. *Economics Letters, 35*, 149–153.
22. Davidson, R., & MacKinnon, J. G. (1993). *Estimation and inference in econometrics*. New York, NY: Oxford University Press.
23. Dufour, J.-M., & Dagenais, M. G. (1992). Nonlinear models, rescaling and test invariance. *Journal of Statistical Planning and Inference, 32*, 111–135.
24. Dufour, J.-M., & Valéry, P. (2009). Exact and asymptotic tests for possibly non-regular hypotheses on stochastic volatility models. *Journal of Econometrics, 150*, 193–206.
25. Durbin, J. (1960). Estimation of parameters in time series regression models. *Journal of the Royal Statistical Society, Series A, 22*, 139–153.
26. Edwards, C. H. (1973). *Advanced calculus of several variables*. New York, NY: Dover.
27. Foutz, R. V. (1976). On the consistency of locally asymptotically most stringent tests. *The Canadian Journal of Statistics/ La Revue Canadienne de Statistique, 4*(2), 211–219.
28. Godambe, V. P. (1960). An optimum property of regular maximum likelihood estimation. *The Annals of Mathematical Statistics, 31*, 1208–1212. (Ackowledgement 32 (1960), 1343).
29. Godambe, V. P. (Ed.). (1991). *Estimating functions*. Oxford: Clarendon Press.
30. Gong, G., & Samaniego, F. J. (1981). Pseudo maximum likelihood estimation: Theory and applications. *The Annals of Statistics, 13*, 861–869.
31. Gouriéroux, C., & Monfort, A. (1995). *Statistics and econometric models* (trans: Quang Vuong) (Vols. 1 & 2). Cambridge: Cambridge University Press.
32. Hall, A. R. (2000). Covariance matrix estimation and the power of the overidentifying restrictions test. *Econometrica, 68*, 1517–1528.
33. Hall, A. R. (2004). *Generalized method of moments, Advanced texts in econometrics*. Oxford: Oxford University Press.
34. Hall, W. J., & Mathiason, D. J. (1990). On large-sample estimation and testing in parametric models. *International Statistical Review, 58*(1), 77–97.
35. Hansen, B. E. (1992). Consistent covariance matrix estimation for dependent heterogeneous processes. *Econometrica, 60*, 967–972.
36. Hansen, L. (1982). Large sample properties of generalized method of moments estimators. *Econometrica, 50*, 1029–1054.
37. Harville, D. A. (1997). *Matrix algebra from a statistician's perspective*. New York: Springer.
38. Heyde, C. C. (1997). *Quasi-likelihood and its application: A general approach to optimal parameter estimation*. Springer series in statistics New York, NY: Springer.
39. Jaggia, S., & Trivedi, P. K. (1994). Joint and separate score tests for state dependence and unobserved heterogeneity. *Journal of Econometrics, 60*(1), 273–291.
40. Kiefer, N. M., Vogelsang, T., & Bunzel, H. (2000). Simple robust testing of regression hypotheses. *Econometrica, 68*, 695–714.
41. Kiefer, N. M., & Vogelsang, T. J. (2002). Heteroskedasticity-autocorrelation robust standard errors using the bartlett kernel without truncation. *Econometrica, 70*, 2093–2095.
42. Kiefer, N. M., & Vogelsang, T. J. (2005). A new asymptotic theory for heteroskedasticity-autocorrelation robust tests. *Econometric Theory, 21*(6), 1130–1164.
43. Kocherlakota, S., & Kocherlakota, K. (1991). Neyman's $C(\alpha)$ test and Rao's efficient score test for composite hypotheses. *Statistics and Probability Letters, 11*, 491–493.

44. Le Cam, L. (1956). On the asymptotic theory of estimation and testing hypotheses. In J. Neyman (Ed.), *Proceedings of the third Berkeley symposium on mathematical statistics and probability*. Berkeley, CA: The University of California Press.
45. Le Cam, L., & Traxler, R. (1978). On the asymptotic behavior of mixtures of Poisson distributions. *Probability Theory and Related Fields*, *44*(1), 1–45.
46. Moran, P. A. P. (1970). On asymptotically optimal tests of composite hypotheses. *Biometrika*, *57*(1), 47–55.
47. Moran, P. A. P. (1973). Asymptotic properties of homogeneity tests. *Biometrika*, *60*(1), 79–85.
48. Murphy, K. M., & Topel, R. H. (1985). Estimation and inference in two-step econometric models. *Journal of Business and Economic Statistics*, *3*, 370–379.
49. Newey, W. K., & McFadden, D. (1994). Large sample estimation and hypothesis testing. In R. F. Engle & D. L. McFadden (Eds.), *Handbook of econometrics*, chapter 36 (Vol. 4, pp. 2111–2245). Amsterdam: North-Holland.
50. Newey, W. K., & West, K. D. (1987). Hypothesis testing with efficient method of moments estimators. *International Economic Review*, *28*, 777–787.
51. Newey, W. K., & West, K. D. (1987). A simple, positive semi-definite, heteroskedasticity and autocorrelation consistent covariance matrix. *Econometrica*, *55*, 703–708.
52. Neyman, J. (1959). Optimal asymptotic tests of composite statistical hypotheses. In U. Grenander (Ed.), *Probability and statistics, the Harald Cramér Volume* (pp. 213–234). Uppsala: Almqvist and Wiksell.
53. Neyman, J. (1979). C($\alpha$) tests and their use. *Sankhyā: The Indian Journal of Statistics, Series A*, *41*, 1–21.
54. Pagan, A. (1984). Econometric issues in the analysis of regressions with generated regressors. *International Economic Review*, *25*, 221–247.
55. Pagan, A. (1986). Two stage and related estimators and their applications. *Review of Economic Studies*, *53*, 517–538.
56. Pal, C. (2003). Higher order $C(\alpha)$ tests with applications to mixture models. *Journal of Statistical Planning and Inference*, *113*(1), 179–187.
57. Paul, S. R., & Barnwal, R. K. (1990). Maximum likelihood estimation and a $C(\alpha)$ test for a common intraclass correlation. *Journal of the Royal Statistical Society, Series D (The Statistician)*, *39*(1), 19–24.
58. Rao, B. L. S. P. (1996). Optimal asymptotic tests of composite hypotheses for continuous time stochastic processes. *Sankhyā: The Indian Journal of Statistics, Series A*, *58*, 8–24.
59. Ray, R. M. (1974). Maxmin $C(\alpha)$ tests against two-sided alternatives. *The Annals of Statistics*, *2*(6), 1175–1188.
60. Ronchetti, E. (1987). Robust $C(\alpha)$-type tests for linear models. *Sankhyā: The Indian Journal of Statistics, Series A*, *49*, 1–16.
61. Singh, A. C., & Zhurbenko, I. G. (1975). The power of the optimal asymptotic tests of composite statistical hypotheses. *Proceedings of the National Academy of Sciences*, *72*(2), 577–580.
62. Small, C. G., & McLeish, D. L. (1994). *Hilbert space methods in probability and statistical inference*. New York, NY: Wiley.
63. Smith, R. J. (1987). Alternative asymptotically optimal tests and their application to dynamic specification. *Review of Economic Studies*, *LIV*, 665–680.
64. Smith, R. J. (1987). Testing the normality assumption in multivariate simultaneous limited dependent variable models. *Journal of Econometrics*, *34*, 105–123.
65. Tarone, R. E. (1979). Testing the goodness of fit of the binomial distribution. *Biometrika*, *66*(3), 585–590.
66. Tarone, R. E. (1985). On heterogeneity tests based on efficient scores. *Biometrika*, *72*(1), 91–95.
67. Tarone, R. E., & Gart, J. J. (1980). On the robustness of combined tests for trends in proportions. *Journal of the American Statistical Association*, *75*(369), 110–116.
68. Vorob'ev, L. S., & Zhurbenko, I. G. (1979). Bounds for $C(\alpha)$-tests and their applications. *Theory of Probability and its Applications*, *24*, 253–268.
69. Wang, P. C. C. (1981). Robust asymptotic tests of statistical hypotheses involving nuisance parameters. *The Annals of Statistics*, *9*(5), 1096–1106.

70. Wang, P. C. C. (1982). On the computation of a robust version of the optimal $C(\alpha)$ test. *Communications in Statistics-Simulation and Computation*, *11*(3), 273–284.
71. Wooldridge, J. M. (1990). A unified approach to robust, regression-based specification tests. *Econometric Theory*, *6*, 17–43.

# Regression Models for Ordinal Categorical Time Series Data

**Brajendra C. Sutradhar and R. Prabhakar Rao**

**Abstract** Regression analysis for multinomial/categorical time series is not adequately discussed in the literature. Furthermore, when categories of a multinomial response at a given time are ordinal, the regression analysis for such ordinal categorical time series becomes more complex. In this paper, we first develop a lag 1 transitional logit probabilities based correlation model for the multinomial responses recorded over time. This model is referred to as a multinomial dynamic logits (MDL) model. To accommodate the ordinal nature of the responses we then compute the binary distributions for the cumulative transitional responses with cumulative logits as the binary probabilities. These binary distributions are next used to construct a pseudo likelihood function for inferences for the repeated ordinal multinomial data. More specifically, for the purpose of model fitting, the likelihood estimation is developed for the regression and dynamic dependence parameters involved in the MDL model.

**Keywords** Category transition over time · Cumulative logits · Marginal multinomial logits · Multinomial dynamic logits · Pseudo binary likelihood

## 1 Introduction

There are situations in practice where a univariate multinomial response, for example, the economic profit status of a pharmaceutical industry such as poor, medium, or high, may be recorded over the years along with known covariates such as type of industry, yearly advertising cost, and other research and development expenditures. It is likely that the profit status of an industry in a given year is correlated with status of profits from the past years. It is of interest to know both (i) the effects of the time dependent

B.C. Sutradhar (✉)
Memorial University, St. John's A1C5S7, Canada
e-mail: bsutradh@mun.ca

R. Prabhakar Rao
Sri Sathya Sai Institute of Higher Learning, Prasanthi Nilayam, Anantapur, India
e-mail: rprabhakarrao@gmail.com



W.K. Li et al. (eds.), *Advances in Time Series Methods and Applications*,
Fields Institute Communications 78, DOI 10.1007/978-1-4939-6568-7_8

covariates, and (ii) the dynamic relationship among the responses over the years. This type of multinomial time series data has been analyzed by some authors such as Fahrmeir and Kaufmann [4], Kaufmann [8], Fokianos and Kedem [5–7], and Loredo-Osti and Sutradhar [10]. As far as the dynamic relationship is concerned, Loredo-Osti and Sutradhar [10] have considered a multinomial dynamic logit (MDL) model as a generalization of the binary time series model used in Tagore and Sutradhar [16] (see also Tong [17]).

Suppose that $y_t = (y_{t1}, \ldots, y_{tj}, \ldots, y_{t,J-1})'$ denotes the $(J-1)$-dimensional multinomial response variable and for $j = 1, \ldots, J-1$,

$$y_t^{(j)} = \left(y_{t1}^{(j)}, \ldots, y_{tj}^{(j)}, \ldots, y_{t,J-1}^{(j)}\right)' = \left(01'_{j-1}, 1, 01'_{J-1-j}\right)' \equiv \delta_{tj} \tag{1}$$

indicates that the multinomial response recorded at time t belongs to the $j$th category. For $j = J$, one writes $y_t^{(J)} = \delta_{tJ} = 01_{J-1}$. Here and also in (1), for a scalar constant $c$, we have used $c1_j$ for simplicity, to represent $c \otimes 1_j$, $\otimes$ being the well known Kronecker or direct product. This notation will also be used through out the rest of the paper when needed. Note that in the non-stationary case, that is, when covariates are time dependent, one uses the time dependent marginal probabilities. Specifically, suppose that at time point $t$ $(t = 1, \ldots, T)$, $x_t = (x_{t1}, \ldots, x_{t\ell}, \ldots, x_{t,p+1})'$ denotes the $(p+1)$-dimensional covariate vector and $\beta_j = (\beta_{j0}, \beta_{j1}, \ldots, \beta_{jp})'$ denotes the effect of $x_t$ on $y_t^{(j)}$ for $j = 1, \ldots, J-1$, and all $t = 1, \ldots, T$, $T$ being the length of the time series. In such cases, the multinomial probability at time $t$, has the form

$$P[y_t = y_t^{(j)}] = \pi_{(t)j} = \begin{cases} \dfrac{\exp(x_t'\beta_j)}{1 + \sum_{g=1}^{J-1} \exp(x_t'\beta_g)} & \text{for } j = 1, \ldots, J-1; \quad t = 1, \ldots, T \\ \dfrac{1}{1 + \sum_{g=1}^{J-1} \exp(x_t'\beta_g)} & \text{for } j = J; \quad t = 1, \ldots, T, \end{cases} \tag{2}$$

and the elements of $y_t = (y_{t1}, \ldots, y_{tj}, \ldots, y_{t,J-1})'$ at time $t$ follow the multinomial probability distribution given by

$$P[y_{t1}, \ldots, y_{tj}, \ldots, y_{t,J-1}] = \Pi_{j=1}^{J} \pi_{(t)j}^{y_{tj}}, \tag{3}$$

for all $t = 1, \ldots, T$. In (3), $y_{tJ} = 1 - \sum_{j=1}^{J-1} y_{tj}$, and $\pi_{tJ} = 1 - \sum_{j=1}^{J-1} \pi_{tj}$.

Next we define the transitional probability from the $g$th $(g = 1, \ldots, J)$ category at time $t-1$ to the $j$th category at time $t$, given by

$$\eta_{t|t-1}^{(j)}(g) = P\left(Y_t = y_t^{(j)} \,\middle|\, Y_{t-1} = y_{t-1}^{(g)}\right)$$
$$= \begin{cases} \dfrac{\exp\left[x_t'\beta_j + \gamma_j' y_{t-1}^{(g)}\right]}{1 + \sum_{v=1}^{J-1} \exp\left[x_t'\beta_v + \gamma_v' y_{t-1}^{(g)}\right]}, & \text{for } j = 1, \ldots, J-1 \\ \dfrac{1}{1 + \sum_{v=1}^{J-1} \exp\left[x_t'\beta_v + \gamma_v' y_{t-1}^{(g)}\right]}, & \text{for } j = J, \end{cases} \tag{4}$$

where $\gamma_j = (\gamma_{j1}, \ldots, \gamma_{jv}, \ldots, \gamma_{j,J-1})'$ denotes the dynamic dependence parameters. Note that this model in (4) is referred to as the multinomial dynamic logits (MDL) model. For the binary case ($J = 2$), this type of non-linear dynamic logit model has been studied by some econometricians. See, for example, Amemiya [3, p. 422] in time series setup, and the recent book by Sutradhar [13, Sect. 7.7] in the longitudinal setup. Now for further notational convenience, we re-express the conditional probabilities in (4) as

$$\eta_{t|t-1}^{(j)}(g) = \begin{cases} \frac{\exp\left[x_t'\beta_j + \gamma_j'\delta_{(t-1)g}\right]}{1+\sum_{v=1}^{J-1}\exp\left[x_t'\beta_v + \gamma_v'\delta_{(t-1)g}\right]}, & \text{for } j = 1, \ldots, J-1 \\ \frac{1}{1+\sum_{v=1}^{J-1}\exp\left[x_t'\beta_v + \gamma_v'\delta_{(t-1)g}\right]}, & \text{for } j = J, \end{cases} \tag{5}$$

where for $t = 2, \ldots, T$, $\delta_{(t-1)g}$, by (1), has the formula

$$\delta_{(t-1)g} = \begin{cases} [01_{g-1}', 1, 01_{J-1-g}']' & \text{for } g = 1, \ldots, J-1 \\ 01_{J-1} & \text{for } g = J. \end{cases}$$

Remark that in (5), the category $g$ occurred at time $t-1$. Thus the category $g$ depends on time $t-1$, and $\delta_{(t-1)g} \equiv \delta_{g_{t-1}}$. However for simplicity we have used $g$ for $g_{t-1}$.

Let $\beta = (\beta_1', \ldots, \beta_j', \ldots, \beta_{J-1}')' : (p+1)(J-1) \times 1$, and $\gamma = (\gamma_1', \ldots, \gamma_j', \ldots, \gamma_{J-1}')' : (J-1)^2 \times 1$. These parameters are involved in the unconditional mean, variance and covariances of the responses. More specifically one may show [10] that

$$\begin{aligned} E[Y_t] &= \tilde{\pi}_{(t)}(\beta,\gamma) = (\tilde{\pi}_{(t)1}, \ldots, \tilde{\pi}_{(t)j}, \ldots, \tilde{\pi}_{(t)(J-1)})' : (J-1)\times 1 \\ &= \begin{cases} [\pi_{(1)1}, \ldots, \pi_{(1)j}, \ldots, \pi_{(1)(J-1)}]' & \text{for } t = 1 \\ \eta_{(t|t-1)}(J) + \left[\eta_{(t|t-1),M} - \eta_{(t|t-1)}(J)1_{J-1}'\right]\tilde{\pi}_{(t-1)} & \text{for } t = 2, \ldots, T-1 \end{cases} \end{aligned} \tag{6}$$

$$\begin{aligned} \text{var}[Y_t] &= \text{diag}[\tilde{\pi}_{(t)1}, \ldots, \tilde{\pi}_{(t)j}, \ldots, \tilde{\pi}_{(t)(J-1)}] - \tilde{\pi}_{(t)}\tilde{\pi}_{(t)}' \\ &= (\text{cov}(Y_{tj}, Y_{tk})) = (\tilde{\sigma}_{(tt)jk}),\ j,k = 1, \ldots, J-1 \\ &= \tilde{\Sigma}_{(tt)}(\beta,\gamma), \text{ for } t = 1, \ldots, T \end{aligned} \tag{7}$$

$$\begin{aligned} \text{cov}[Y_u, Y_t] &= \Pi_{s=u+1}^{t}\left[\eta_{(s|s-1),M} - \eta_{(s|s-1)}(J)1_{J-1}'\right]\text{var}[Y_u], \text{ for } u < t, t = 2, \ldots, T \\ &= (\text{cov}(Y_{uj}, Y_{tk})) = (\tilde{\sigma}_{(ut)jk}),\ j,k = 1, \ldots, J-1 \\ &= \tilde{\Sigma}_{(ut)}(\beta,\gamma), \end{aligned} \tag{8}$$

where

$$\eta_{(s|s-1)}(J) = [\eta_{s|s-1}^{(1)}(J), \ldots, \eta_{s|s-1}^{(j)}(J) \ldots, \eta_{s|s-1}^{(J-1)}(J)]' = \pi_{(s)} : (J-1)\times 1$$

$$\eta_{(s|s-1),M} = \begin{pmatrix} \eta_{s|s-1}^{(1)}(1) & \cdots & \eta_{s|s-1}^{(1)}(g) & \cdots & \eta_{s|s-1}^{(1)}(J-1) \\ \vdots & \vdots & \vdots & \vdots & \vdots \\ \eta_{s|s-1}^{(j)}(1) & \cdots & \eta_{s|s-1}^{(j)}(g) & \cdots & \eta_{s|s-1}^{(j)}(J-1) \\ \vdots & \vdots & \vdots & \vdots & \vdots \\ \eta_{s|s-1}^{(J-1)}(1) & \cdots & \eta_{s|s-1}^{(J-1)}(g) & \cdots & \eta_{s|s-1}^{(J-1)}(J-1) \end{pmatrix} : (J-1)\times(J-1).$$

Notice that there is a relation between the vector $\eta_{(s|s-1)}(J)$ and the matrix $\eta_{(s|s-1),M}$. This is because the transition matrix $\eta_{(s|s-1),M}$ contains the transitional probabilities from any of the first $J-1$ states at time $s-1$ to any of the $J-1$ states at time $s$, whereas the transition vector $\eta_{(s|s-1)}(J)$ contains transitional probabilities from the $J$th state at time $s-1$ to any of the first $J-1$ states at time $s-1$. Consequently, once the transition matrix $\eta_{(s|s-1),M}$ is computed, the transition vector $\eta_{(s|s-1)}(J)$ becomes known.

It is of importance to estimate $\beta$ and $\gamma$ parameters mainly to understand the aforementioned basic properties including the pair-wise correlations of the responses.

Note however that the multinomial time series model (2)–(5) and its basic moment properties shown in (6)–(8) are derived without any order restrictions of the categories of the responses. The purpose of this paper is to estimate the parameters $\beta$ and $\gamma$ under an ordinal categorical response model which we describe in Sect. 2. In Sect. 3, we demonstrate the application of a pseudo likelihood approach for the estimation for these parameters. Some concluding remarks are made in Sect. 4.

## 2 Cumulative MDL Model for Ordinal Categorical Data

When categories for a response at a given time $t$ are ordinal, one may then collapse the $J > 2$ categories in a cumulative fashion into two ($J' = 2$) categories and use simpler binary model to fit such collapsed data. Note however that there will be various binary groups depending on which category in the middle is used as a cut point. For the transitional categorical response from time $t-1$ (say) to time $t$, cumulation of the categories at time $t$ has to be computed conditional on the cumulative categories at time $t-1$. This will also generate a binary model for cumulative transitional responses. These concepts of cumulative probabilities for a cumulative response are used in the next section to construct the desired cumulative MDL model.

### *2.1 Marginal Cumulative Model at Time t = 1*

Suppose that for a selected cut point $j (j = 1, \ldots, J-1)$, $F_{(1)j} = \sum_{c=1}^{j} \pi_{(1)c}$ represents the probability for a multinomial response to be in category $c$ between 1 and $j$, where $\pi_{(1)c}$ by (2) defines the probability for the response to be in category $c$ $(c = 1, \ldots, J)$ at time $t = 1$. Thus, $1 - F_{(1)j} = \sum_{c=j+1}^{J} \pi_{(1)c}$ would represent the probability for the multinomial response to be in category $c$ beyond $j$. To reflect this binary nature of the observed response in category $c$ with regard to cut point $j$, we define a binary variable $b_c^{(j)}(1)$ such that

$$P\big[b_c^{(j)}(1) = 1\big] = 1 - F_{(1)j} = \sum_{c=j+1}^{J} \pi_{(1)c}. \tag{9}$$

Notice that because there are $J-1$ possible cut points, if the categories are ordered and the response falls in $c$th category, by (11) below, we then obtain the cut points based observed vector at time $t=1$ as

$$\left[b_c^{(1)}(1)=1,\ldots,b_c^{(c-1)}(1)=1,b_c^{(c)}(1)=0,\ldots,b_c^{(J-1)}(1)=0\right].$$

For other values of $t$, the observed responses are constructed similarly depending on the response category.

## 2.2 Lag 1 Transitional Cumulative Model at Time $t = 2, \ldots, T$

In order to develop a transitional model, suppose we observe that the multinomial response at time $t-1(t=2,\ldots,T)$ was in $c_1$th category $(c_1=1,\ldots,J)$, whereas at time $t$ it is observed in $c_2(c_2=1,\ldots,J)$ category. Let $(g,j)$ denote a bivariate cut point which facilitates the binary variables [similar to (9)] given by

$$b_{c_1}^{(g)}(t-1)=\begin{cases}1 \text{ for the response in category } c_1>g \text{ at time } t-1\\ 0 \text{ for the response in category } c_1\le g \text{ at time } t-1,\end{cases} \tag{10}$$

and

$$b_{c_2}^{(j)}(t)=\begin{cases}1 \text{ for the response in category } c_2>j \text{ at time } t\\ 0 \text{ for the response in category } c_2\le j \text{ at time } t.\end{cases} \tag{11}$$

Consequently, a transitional probability model based on conditional probabilities (5) may be written as

$$P\left[b_{c_2}^{(j)}(t)=1|b_{c_1}^{(g)}(t-1)\right]=\tilde{\lambda}_{gj}^{(2)}\left(b_{c_1}^{(g)}(t-1)\right)$$
$$=\begin{cases}\tilde{\lambda}_{gj}^{(2)}(1) \text{ for } b_{c_1}^{(g)}(t-1)=0\\ \tilde{\lambda}_{gj}^{(2)}(2) \text{ for } b_{c_1}^{(g)}(t-1)=1,\end{cases} \tag{12}$$
$$=\begin{cases}\frac{1}{g}\sum_{c_1=1}^{g}\sum_{c_2=j+1}^{J}\lambda_{t|t-1}^{(c_2)}(c_1)\\ \frac{1}{J-g}\sum_{c_1=g+1}^{J}\sum_{c_2=j+1}^{J}\lambda_{t|t-1}^{(c_2)}(c_1),\end{cases} \tag{13}$$

where the conditional probability $\lambda_{t|t-1}^{(c_2)}(c_1)$, has the known multinomial dynamic logit (MDL) form given by (5). For convenience, following (12)–(13), we also write

$$P[b_{c_2}^{(j)}(t)=0|b_{c_1}^{(g)}(t-1)]=1-\tilde{\lambda}_{gj}^{(2)}(b_{c_1}^{(g)}(t-1))$$
$$=\begin{cases}\tilde{\lambda}_{gj}^{(1)}(1)=1-\tilde{\lambda}_{gj}^{(2)}(1) \text{ for } b_{c_1}^{(g)}(t-1)=0\\ \tilde{\lambda}_{gj}^{(1)}(2)=1-\tilde{\lambda}_{gj}^{(2)}(2) \text{ for } b_{c_1}^{(g)}(t-1)=1,\end{cases} \tag{14}$$

$$= \begin{cases} \frac{1}{g}\sum_{c_1=1}^{g}\left[1-\sum_{c_2=j+1}^{J}\lambda_{t|t-1}^{(c_2)}(c_1)\right] \\ \frac{1}{J-g}\sum_{c_1=g+1}^{J}\left[1-\sum_{c_2=j+1}^{J}\lambda_{t|t-1}^{(c_2)}(c_1)\right] \end{cases} \quad (15)$$

$$= \begin{cases} \frac{1}{g}\sum_{c_1=1}^{g}\sum_{c_2=1}^{j}\lambda_{t|t-1}^{(c_2)}(c_1) \\ \frac{1}{J-g}\sum_{c_1=g+1}^{J}\sum_{c_2=1}^{j}\lambda_{t|t-1}^{(c_2)}(c_1). \end{cases} \quad (16)$$

## 3 Pseudo Binary Likelihood Estimation for the Ordinal Model

In this section, we construct a binary data based likelihood function, where the binary data are obtained by collapsing the available ordinal multinomial observations. Consequently, we refer to this likelihood approach as the so-called pseudo likelihood approach. However, for convenience, we use the terminology ‘likelihood’ for the ‘pseudo likelihood’, through out the section.

At $t = 1$, the marginal likelihood for $\beta$ by (9) has the form

$$L_1(\beta) = \Pi_{j=1}^{J-1}\left[\{F_{(1)j}\}^{1-b_c^{(j)}(1)}\right]\left[\{1-F_{(1)j}\}^{b_c^{(j)}(1)}\right]$$

$$= \Pi_{j=1}^{J-1}\left[\left\{\sum_{c=1}^{j}\pi_{(1)c}\right\}^{1-b_c^{(j)}(1)}\right]\left[\left\{\sum_{c=j+1}^{J}\pi_{(1)c}\right\}^{b_c^{(j)}(1)}\right], \quad (17)$$

where

$$b_c^{(j)}(1) = \begin{cases} 1 & \text{for } c > j \\ 0 & \text{for } c \le j. \end{cases} \quad (18)$$

Next for the construction of the conditional likelihood at $t$ given the information from previous time point $t-1$, we first re-express the binary conditional probabilities in (12) and (14), as

$$\tilde{\lambda}_{gj}^{(2)}(g^*) = \begin{cases} \tilde{\lambda}_{gj}^{(2)}(1) & \text{for } b_{c_1}^{(g)}(t-1) = 0 \\ \tilde{\lambda}_{gj}^{(2)}(2) & \text{for } b_{c_1}^{(g)}(t-1) = 1, \end{cases} \quad (19)$$

$$\tilde{\lambda}_{gj}^{(1)}(g^*) = \begin{cases} \tilde{\lambda}_{gj}^{(1)}(1) & \text{for } b_{c_1}^{(g)}(t-1) = 0 \\ \tilde{\lambda}_{gj}^{(1)}(2) & \text{for } b_{c_1}^{(g)}(t-1) = 1. \end{cases} \quad (20)$$

One may then write the conditional likelihood for $\beta$ and $\gamma$, as

$$L_{t|t-1}(\beta,\gamma) = \Pi_{g=1}^{J-1}\Pi_{j=1}^{J-1}\Pi_{g^*=1}^{2}\left[\left\{\tilde{\lambda}_{gj}^{(2)}(g^*)\right\}^{b_{c_2}^{(j)}(t)}\left\{\tilde{\lambda}_{gj}^{(1)}(g^*)\right\}^{1-b_{c_2}^{(j)}(t)}\right], \quad (21)$$

where the binary data $b_{c_2}^{(j)}(t)$ for observed $c_2$ are obtained by (11), and similarly $b_{c_1}^{(g)}(t-1)$ to define $g^*$ for given $c_1$ are obtained from (10).

Next by combining (17) and (21), one obtains the likelihood function for $\beta$ and $\gamma$ as

$$\begin{aligned} L(\beta,\gamma) &= L_1(\beta)\Pi_{t=2}^{T} L_{t|t-1}(\beta,\gamma) \\ &= \Pi_{j=1}^{J-1}\left[\{F_{(1)j}\}^{1-b_c^{(j)}(1)}\right]\left[\{1-F_{(1)j}\}^{b_c^{(j)}(1)}\right] \\ &\quad \times \Pi_{t=2}^{T}\Pi_{g=1}^{J-1}\Pi_{j=1}^{J-1}\Pi_{g^*=1}^{2}\left[\left\{\tilde{\lambda}_{gj}^{(2)}(g^*)\right\}^{b_{c_2}^{(j)}(t)}\left\{\tilde{\lambda}_{gj}^{(1)}(g^*)\right\}^{1-b_{c_2}^{(j)}(t)}\right]. \end{aligned} \quad (22)$$

For the benefit of the practitioners, we now develop the likelihood estimating equations for these parameters $\beta$ and $\gamma$, as in the following sections. Remark that for the construction of similar likelihood estimating equations in the stationary longitudinal setup, one may be referred to Sutradhar [14, Sect. 3.6.2.2].

Note that the likelihood function in (22) is constructed by collapsing the ordinal multinomial responses to the binary responses at all suitable cut points. This likelihood function, therefore, can not be used for nominal multinomial time series data. When the categories are nominal, it is appropriate to construct the likelihood function by exploiting the marginal probability function $\pi_{(t)j}$ from (2) for $t = 1$, and the conditional multinomial logit probability function $\eta_{t|t-1}^{(j)}(g)$ from (4) for $t = 2, \ldots, T$ (see Loredo-Osti and Sutradhar [10]). Notice that in practice the time dependent covariates $x_t$ in (2) and (4) are fixed in general. However, by treating $x_t$ as a random covariate vector, Fokianos and Kedem [6] obtained parameter estimates by maximizing a partial likelihood function without requiring any extra characterization of the joint process $\{y_t, x_t\}$. Loredo-Osti and Sutradhar [10] have, however, argued that in Fokianos and Kedem's [6] approach, the conditional Fisher information matrix is not the same as the one obtained by conditioning on $\{x_t\}$, the observed covariates. In fact, when the estimation is carried out in a general linear models framework that uses the canonical link function, this conditional information matrix obtained by Fokianos and Kedem, is just the Hessian matrix multiplied by $-1$, i.e., the observed information matrix.

As far as the ordinal multinomial time series data are concerned, the construction of binary mapping based likelihood function in (22) is a new concept. The core idea comes from the cumulative binary property for the MDL (multinomial dynamic logit) model (4) because of the present ordinal nature of the data. In the cross sectional setup, that is, for the case with $t = 1$ only, the likelihood function for ordinal multinomial data has been used by many authors such as Agresti [1]. Note that the marginal multinomial probability in (2) has the multinomial logit form. In the cluster data setup, many existing studies use this multinomial logit model (2) as the marginal model at a given time $t$. As far as the correlations between repeated responses are concerned, some authors such as Agresti [1], Lipsitz et al. [9], Agresti and Natarajan [2] do not model them, rather they use 'working' correlations to construct the so-called generalized estimating equations and solve them to obtain the estimates for regression parameters involved in the marginal multinomial logits model (2). These estimates however may not be reliable as they can be inefficient as compared to the 'working' independence assumption based estimates (see Sutradhar and Das [15], Sutradhar [13, Chap. 7] in the context of binary longitudinal data analysis). Thus, their extension to the time series setup may be useless. Moreover, it is not clear how to model the ordinal data using this type of 'working' correlations approach.

### 3.1 Likelihood Estimating Equations for the Regression Effects $\beta$

Recall that $\beta = (\beta_1', \ldots, \beta_j', \ldots, \beta_{J-1}')' : (J-1)(p+1) \times 1$, with $\beta_j = (\beta_{j0}, \beta_{j1}, \ldots, \beta_{jp})'$. For known $\gamma$, in this section, we exploit the likelihood function (22) and develop the likelihood estimating equation for $\beta$. For convenience, we use log likelihood function, which, following the likelihood function in (22), is written as

$$\text{Log } L(\beta,\gamma) = \sum_{j=1}^{J-1} \left[\{1 - b_c^{(j)}(1)\} \log F_{(1)j} + \{b_c^{(j)}(1)\} \log\{1 - F_{(1)j}\}\right]$$
$$+ \sum_{t=2}^{T} \sum_{g=1}^{J-1} \sum_{j=1}^{J-1} \sum_{g^*=1}^{2} \left[b_{c_2}^{(j)}(t) \log\left\{\tilde{\lambda}_{gj}^{(2)}(g^*)\right\} + \{1 - b_{c_2}^{(j)}(t)\} \log\left\{\tilde{\lambda}_{gj}^{(1)}(g^*)\right\}\right], \quad (23)$$

yielding the likelihood estimating equation for $\beta$ as

$$\frac{\partial \text{Log } L(\beta,\gamma)}{\partial \beta} = \sum_{j=1}^{J-1} \left[\frac{\{1 - b_c^{(j)}(1)\}}{F_{(1)j}} - \frac{\{b_c^{(j)}(1)\}}{\{1 - F_{(1)j}\}}\right] \frac{\partial F_{(1)j}}{\partial \beta}$$
$$+ \sum_{t=2}^{T} \sum_{g=1}^{J-1} \sum_{j=1}^{J-1} \sum_{g^*=1}^{2} \left[\frac{b_{c_2}^{(j)}(t)}{\tilde{\lambda}_{gj}^{(2)}(g^*)} - \frac{\{1 - b_{c_2}^{(j)}(t)\}}{\{1 - \tilde{\lambda}_{gj}^{(2)}(g^*)\}}\right] \frac{\partial \tilde{\lambda}_{gj}^{(2)}(g^*)}{\partial \beta}$$
$$= 0, \quad (24)$$

where

$$\frac{\partial F_{(1)j}}{\partial \beta} = \sum_{c=1}^{j} \left[\pi_{(1)c}(\delta_{(1)c} - \pi_{(1)})\right] \otimes x_1; \quad (25)$$

and

$$\frac{\partial \tilde{\lambda}_{gj}^{(2)}(g^*)}{\partial \beta} = \begin{cases} \frac{1}{g} \sum_{c_1=1}^{g} \sum_{c_2=j+1}^{J} \left[\eta_{t|t-1}^{(c_2)}(c_1)(\delta_{(t-1)c_2} - \eta_{t|t-1}(c_1))\right] \otimes x_t & \text{for } g^* = 1 \\ \frac{1}{J-g} \sum_{c_1=g+1}^{J} \sum_{c_2=j+1}^{J} \left[\eta_{t|t-1}^{(c_2)}(c_1)(\delta_{(t-1)c_2} - \eta_{t|t-1}(c_1))\right] \otimes x_t & \text{for } g^* = 2, \end{cases} \quad (26)$$

with

$$\pi_{(1)} = \left[\pi_{(1)1}, \ldots, \pi_{(1)c}, \ldots, \pi_{(1)(J-1)}\right]'$$
$$\delta_{(t-1)c} = \begin{cases} [01'_{c-1}, 1, 01'_{J-1-c}]' & \text{for } c = 1, \ldots, J-1 \\ 01_{J-1} & \text{for } c = J, \end{cases}$$
$$\eta_{t|t-1}(c_1) = \left[\eta_{t|t-1}^{(1)}(c_1), \ldots, \eta_{t|t-1}^{(c_2)}(c_1), \ldots, \eta_{t|t-1}^{(J-1)}(c_1)\right]'. \tag{27}$$

The details for the derivatives in (25) and (26) are given in "Appendix".

For given $\gamma$, the likelihood equations in (24) may be solved iteratively by using the iterative equation for $\beta$ given by

$$\hat{\beta}(r+1) = \hat{\beta}(r) - \left[\left\{\frac{\partial^2 \text{Log } L(\beta, \gamma)}{\partial \beta' \partial \beta}\right\}^{-1} \frac{\partial \text{Log } L(\beta, \gamma)}{\partial \beta}\right]_{|\beta = \hat{\beta}(r)} ; \ (J-1)(p+1) \times 1, \tag{28}$$

where the formula for the second order derivative matrix $\frac{\partial^2 Log\ L(\beta, \gamma_M)}{\partial \beta' \partial \beta}$ may be derived by taking the derivative of the $(J-1)(p+1) \times 1$ vector with respect to $\beta'$. The exact second order derivative matrix has a complicated formula. We provide an approximation as follows.

**An approximation to** $\frac{\partial^2 Log\ L(\beta, \gamma_M)}{\partial \beta' \partial \beta}$ **:**

Re-express the likelihood estimating equation from (24) as

$$\frac{\partial \text{Log } L(\beta, \gamma)}{\partial \beta} = \sum_{j=1}^{J-1} \frac{\partial F_{(1)j}}{\partial \beta} \{(1 - F_{(1)j}) F_{(1)j}\}^{-1} \left[\{1 - b_c^{(j)}(1)\} - F_{(1)j}\right]$$
$$+ \sum_{t=2}^{T} \sum_{g=1}^{J-1} \sum_{j=1}^{J-1} \sum_{g^*=1}^{2} \frac{\partial \tilde{\lambda}_{gj}^{(2)}(g^*)}{\partial \beta} \left\{\tilde{\lambda}_{gj}^{(2)}(g^*)\left(1 - \tilde{\lambda}_{gj}^{(2)}(g^*)\right)\right\}^{-1} \left[b_{c_2}^{(j)}(t) - \tilde{\lambda}_{gj}^{(2)}(g^*)\right]$$
$$= 0. \tag{29}$$

Notice that in the first term in the left hand side of (29), $\{1 - b_c^{(j)}(1)\}$ is, by (9), a binary variable with

$$E\left\{1 - b_c^{(j)}(1)\right\} = F_{(1)j}$$
$$\text{var}\{1 - b_c^{(j)}(1)\} = F_{(1)j}\{1 - F_{(1)j}\}, \tag{30}$$

and similarly in the second term, by (12), $b_{c_2}^{(j)}(t)$ conditional on $b_{c_1}^{(g)}(t-1)$ is a binary variable with

$$E\left[b_{c_2}^{(j)}(t) | b_{c_1}^{(g)}(t-1)\right] = \tilde{\lambda}_{gj}^{(2)}(g^*)$$
$$\text{var}\left[b_{c_2}^{(j)}(t) | b_{c_1}^{(g)}(t-1)\right] = \tilde{\lambda}_{gj}^{(2)}(g^*)\left[1 - \tilde{\lambda}_{gj}^{(2)}(g^*)\right], \tag{31}$$

for $g^* \equiv b_{c_1}^{(g)}(t-1)$. Thus, the likelihood estimating function in (29) is equivalent to a conditional quasi-likelihood (CQL) function in $\beta$ for the cut points based binary data [e.g. see Tagore and Sutradhar [16, Eq. (27), p. 888]. Now because the variance of the binary data is a function of the mean, the variance and gradient functions in (29) may be treated to be known when mean is known. Thus, when a QL estimating equation is solved iteratively, the gradient and variance functions use $\beta$ from a previous iteration [11, 18]. Consequently, by (29), the second derivative matrix required to compute (28) has a simpler approximate formula

$$\frac{\partial^2 \mathrm{Log}\, L(\beta,\gamma)}{\partial\beta\partial\beta'} = -\sum_{j=1}^{J-1} \frac{\partial F_{(1)j}}{\partial\beta}\{(1-F_{(1)j})F_{(1)j}\}^{-1}\frac{\partial F_{(1)j}}{\partial\beta'}$$
$$-\sum_{t=2}^{T}\sum_{g=1}^{J-1}\sum_{j=1}^{J-1}\sum_{g^*=1}^{2} \frac{\partial\tilde{\lambda}_{gj}^{(2)}(g^*)}{\partial\beta}\left\{\tilde{\lambda}_{gj}^{(2)}(g^*)\left(1-\tilde{\lambda}_{gj}^{(2)}(g^*)\right)\right\}^{-1}\frac{\partial\tilde{\lambda}_{gj}^{(2)}(g^*)}{\partial\beta'}. \quad (32)$$

Furthermore for known $\gamma$, by (28) and (29), under some mild conditions it follows that the solution of (29), say $\hat{\beta}$, satisfies

$$\hat{\beta} \sim N(\beta, V(\beta,\gamma)), \quad (33)$$

(see Kaufmann [8, Sect. 5]) where the covariance matrix is estimated by

$$\hat{V}(\cdot) = \left[-\frac{\partial^2 \mathrm{Log}\, L(\beta,\gamma)}{\partial\beta\partial\beta'}\right]^{-1}_{\beta=\hat{\beta}}. \quad (34)$$

### 3.2 Likelihood Estimating Equations for the Dynamic Dependence Parameters $\gamma$

In Sect. 3.1, we have estimated $\beta$ for known $\gamma$, for example, initially by using $\gamma = 0$, where by (4)–(5),

$$\gamma = \left(\gamma_1',\ldots,\gamma_j',\ldots,\gamma_{J-1}'\right)', \text{ with } \gamma_j = \left(\gamma_{j1},\ldots,\gamma_{jv},\ldots,\gamma_{j,J-1}\right)'.$$

Note that $F_{(1)j}$ for all $j = 1,\ldots,J-1$, are free from $\gamma$. Hence, by exploiting the log likelihood function (23), similar to (24), we write the likelihood equation for $\gamma$ as

$$\frac{\partial \mathrm{Log}\, L(\beta,\gamma)}{\partial\gamma} = \sum_{t=2}^{T}\sum_{g=1}^{J-1}\sum_{j=1}^{J-1}\sum_{g^*=1}^{2}\left[\frac{b_{c_2}^{(j)}(t)}{\tilde{\lambda}_{gj}^{(2)}(g^*)} - \frac{\left\{1-b_{c_2}^{(j)}(t)\right\}}{\left\{1-\tilde{\lambda}_{gj}^{(2)}(g^*)\right\}}\right]\frac{\partial\tilde{\lambda}_{gj}^{(2)}(g^*)}{\partial\gamma} = 0, \quad (35)$$

where

$$\frac{\partial \tilde{\lambda}_{gj}^{(2)}(g^*)}{\partial \gamma} = \begin{cases} \frac{1}{g}\sum_{c_1=1}^{g}\sum_{c_2=j+1}^{J}\left[\eta_{t|t-1}^{(c_2)}(c_1)(\delta_{(t-1)c_2} - \eta_{t|t-1}(c_1))\right] \otimes \delta_{(t-1)c_1} & \text{for } g^* = 1 \\ \frac{1}{J-g}\sum_{c_1=g+1}^{J}\sum_{c_2=j+1}^{J}\left[\eta_{t|t-1}^{(c_2)}(c_1)(\delta_{(t-1)c_2} - \eta_{t|t-1}(c_1))\right] \otimes \delta_{(t-1)c_1} & \text{for } g^* = 2. \end{cases} \quad (36)$$

An outline for this derivative is given in the "Appendix".

By similar calculations as in (28), one may solve the likelihood estimating equation in (35) for $\gamma$ using the iterative equation

$$\hat{\gamma}(r+1) = \hat{\gamma}(r) - \left[\left\{\frac{\partial^2 \text{Log } L(\beta,\gamma)}{\partial\gamma\partial\gamma'}\right\}^{-1} \frac{\partial \text{Log } L(\beta,\gamma)}{\partial\gamma}\right]_{|\gamma=\hat{\gamma}(r)} ; \ (J-1)^2 \times 1, \quad (37)$$

where the second order derivative matrix, following (32), may be computed as

$$\frac{\partial^2 \text{Log } L(\beta,\gamma)}{\partial\gamma\partial\gamma'} = -\sum_{t=2}^{T}\sum_{g=1}^{J-1}\sum_{j=1}^{J-1}\sum_{g^*=1}^{2} \frac{\partial \tilde{\lambda}_{gj}^{(2)}(g^*)}{\partial\gamma}\left\{\tilde{\lambda}_{gj}^{(2)}(g^*)\left(1-\tilde{\lambda}_{gj}^{(2)}(g^*)\right)\right\}^{-1}\frac{\partial \tilde{\lambda}_{gj}^{(2)}(g^*)}{\partial\gamma'}. \quad (38)$$

Furthermore for known $\beta$, by (37) and (38), it follows under some mild conditions that the solution of (35), say $\hat{\gamma}$, satisfies

$$\hat{\gamma} \sim N(\gamma, V^*(\beta,\gamma)), \quad (39)$$

(see Kaufmann [8, Sect. 5]) where the covariance matrix is estimated by

$$\hat{V}^*(\cdot) = \left[-\frac{\partial^2 \text{Log } L(\beta,\gamma)}{\partial\gamma\partial\gamma'}\right]^{-1}_{\gamma=\hat{\gamma}}. \quad (40)$$

### 3.3 Joint Likelihood Estimating Equations for $\beta$ and $\gamma$

Let $\theta = (\beta', \gamma')'$. One may then combine (28) and (37) and solve the iterative equation

$$\hat{\theta}(r+1) = \hat{\theta}(r) - \left[\begin{pmatrix} \frac{\partial^2 \text{Log } L(\beta,\gamma)}{\partial\beta\partial\beta'} & \frac{\partial^2 \text{Log } L(\beta,\gamma)}{\partial\beta\partial\gamma'} \\ \frac{\partial^2 \text{Log } L(\beta,\gamma)}{\partial\gamma\partial\beta'} & \frac{\partial^2 \text{Log } L(\beta,\gamma)}{\partial\gamma\partial\gamma'} \end{pmatrix}^{-1} \begin{pmatrix} \frac{\partial \text{Log } L(\beta,\gamma)}{\partial\beta} \\ \frac{\partial \text{Log } L(\beta,\gamma)}{\partial\gamma} \end{pmatrix}\right]_{|\theta=\hat{\theta}(r)} \quad (41)$$

to obtain the joint likelihood estimates for $\beta$ and $\gamma$. In order to construct the iterative equation (41), we require the formula for the second order derivative matrix $\frac{\partial^2 \text{Log } L(\beta,\gamma)}{\partial\beta\partial\gamma'}$ which, using (29), may be approximately computed as

$$\frac{\partial^2 \mathrm{Log}\, L(\beta,\gamma)}{\partial\beta\partial\gamma'} = -\sum_{t=2}^{T}\sum_{g=1}^{J-1}\sum_{j=1}^{J-1}\sum_{g^*=1}^{2} \frac{\partial\tilde{\lambda}_{gj}^{(2)}(g^*)}{\partial\beta}\left\{\tilde{\lambda}_{gj}^{(2)}(g^*)\left(1-\tilde{\lambda}_{gj}^{(2)}(g^*)\right)\right\}^{-1}\frac{\partial\tilde{\lambda}_{gj}^{(2)}(g^*)}{\partial\gamma'}, \quad (42)$$

where the formulas for $\frac{\partial\tilde{\lambda}_{gj}^{(2)}(g^*)}{\partial\beta}$ and $\frac{\partial\tilde{\lambda}_{gj}^{(2)}(g^*)}{\partial\gamma}$ are given by (26) and (36), respectively.

Furthermore, by similar arguments to (33) and (39), under some mild conditions it follows that the solution of (41), say $\begin{pmatrix}\tilde{\beta}\\ \tilde{\gamma}\end{pmatrix}$ has the multivariate Gaussian distribution

$$\begin{pmatrix}\tilde{\beta}\\ \tilde{\gamma}\end{pmatrix} \sim N\left[\begin{pmatrix}\beta\\ \gamma\end{pmatrix}, \begin{pmatrix}\tilde{V}_{11}(\beta,\gamma) & \tilde{V}_{12}(\beta,\gamma)\\ \tilde{V}_{12}'(\beta,\gamma) & \tilde{V}_{22}(\beta,\gamma)\end{pmatrix}\right], \quad (43)$$

where $\mathrm{cov}(\tilde{\beta}) = \tilde{V}_{11}(\beta,\gamma)$ and $\mathrm{cov}(\tilde{\gamma}) = \tilde{V}_{22}(\beta,\gamma)$ are estimated as

$$\begin{aligned}\hat{\mathrm{cov}}(\tilde{\beta}) &= A^{-1} + FE^{-1}F'\\ \hat{\mathrm{cov}}(\tilde{\gamma}) &= E^{-1},\end{aligned} \quad (44)$$

Rao [12, p. 33] with $E = D - B'A^{-1}B$, and $F = A^{-1}B$, where by (41)

$$A = \frac{\partial^2 \mathrm{Log}\, L(\beta,\gamma)}{\partial\beta\partial\beta'};\quad B = \frac{\partial^2 \mathrm{Log}\, L(\beta,\gamma)}{\partial\beta\partial\gamma'};\ \text{and}\quad D = \frac{\partial^2 \mathrm{Log}\, L(\beta,\gamma)}{\partial\gamma\partial\gamma'}.$$

## 4 Concluding Remarks

Recently some authors such as Loredo-Osti and Sutradhar [10] (see also Fokianos and Kedem [6]) have developed a likelihood approach for the estimation of regression and dynamic dependence parameters involved in a multinomial dynamic logit (MDL) model used for categorical time series data. This inference issue becomes more complex when the categorical response collected at a given time point also exhibit an order. In this paper we have demonstrated that this type of ordinal categorical responses collected over time may be analyzed by collapsing a multinomial response to a binary response at a given possible cut point and fitting binary dynamic model to all such binary responses collected based on all possible cut points over all times. For simplicity, we have fitted a low order, namely lag 1 dynamic model among all possible cut points based binary responses. A pseudo likelihood method using binary responses (in stead of the multinomial observations) is then constructed for the estimation of the regression and dynamic dependence parameters. The authors plan to undertake an empirical study involving simulations and real life data analysis in order to investigate the performance of the proposed estimation approach both for moderate and large size time series. The empirical results will be published elsewhere.

**Acknowledgments** The authors are grateful to Bhagawan Sri Sathya Sai Baba for His love and blessings to carry out this research in Sri Sathya Institute of Higher Learning. The authors thank the editorial committee for the invitation to participate in preparing this Festschrift honoring Professor Ian McLeod. It has brought back many pleasant memories of Western in early 80's experienced by the first author during his PhD study. We have prepared this small contribution as a token of our love and respect to Professor Ian McLeod for his long and sustained contributions to the statistics community through teaching and research in time series analysis, among other areas. The authors thank two referees for their comments and suggestions on the earlier version of the paper.

# Appendix

*Derivation for* $\frac{\partial F_{(1)j}}{\partial \beta}$ :

Recall from Sect. 2.1 that $F_{(1)j} = \sum_{c=1}^{j} \pi_{(1)c}$, where $\pi_{(1)c}$ is given by (2). It then follows that

$$\frac{\partial F_{(1)j}}{\partial \beta} = \frac{\partial}{\partial \beta} \sum_{c=1}^{j} \pi_{(1)c} = \frac{\partial}{\partial \beta} \sum_{c=1}^{j} \frac{\exp\left(x_1' \beta_c\right)}{1 + \sum_{g=1}^{J-1} \exp\left(x_1' \beta_g\right)}. \tag{45}$$

Now because

$$\frac{\partial \pi_{(1)c}}{\partial \beta_c} = \pi_{(1)c}[1 - \pi_{(1)c}]x_1, \text{ and } \frac{\partial \pi_{(1)c}}{\partial \beta_k} = -[\pi_{(1)c}\pi_{(1)k}]x_1, \tag{46}$$

it follows that

$$\begin{aligned} \frac{\partial \pi_{(1)c}}{\partial \beta} &= \begin{pmatrix} -\pi_{(1)1}\pi_{(1)c} \\ \vdots \\ \pi_{(1)c}[1 - \pi_{(1)c}] \\ \vdots \\ -\pi_{(1)(J-1)}\pi_{(1)c} \end{pmatrix} \otimes x_1 : (J-1)(p+1) \times 1 \\ &= \left[\pi_{(1)c}(\delta_{(1)c} - \pi_{(1)})\right] \otimes x_1. \end{aligned} \tag{47}$$

The formula for $\frac{\partial F_{(1)j}}{\partial \beta}$ in (25) follows by using (47) and (45).

*Derivation for* $\frac{\partial \tilde{\lambda}_{gj}^{(2)}(g^*)}{\partial \beta}$ :

By using the formula for $\tilde{\lambda}_{gj}^{(2)}(g^*)$ from (13) we write

$$\frac{\partial \tilde{\lambda}_{gj}^{(2)}(g^*)}{\partial \beta} = \frac{\partial}{\partial \beta} \begin{cases} \frac{1}{g} \sum_{c_1=1}^{g} \sum_{c_2=j+1}^{J} \lambda_{t|t-1}^{(c_2)}(c_1) & \text{for } g^* = 1 \\ \frac{1}{J-g} \sum_{c_1=g+1}^{J} \sum_{c_2=j+1}^{J} \lambda_{t|t-1}^{(c_2)}(c_1) & \text{for } g^* = 2, \end{cases} \tag{48}$$

where $\lambda^{(c_2)}_{t|t-1}(c_1)$ is given in (5), that is,

$$\eta^{(c_2)}_{t|t-1}(c_1) = \begin{cases} \frac{\exp\left[x_t'\beta_{c_2}+\gamma_{c_2}'\delta_{(t-1)c_1}\right]}{1+\sum_{v=1}^{J-1}\exp\left[x_t'\beta_v+\gamma_v'\delta_{(t-1)c_1}\right]}, & \text{for } c_2 = 1,\ldots,J-1 \\ \frac{1}{1+\sum_{v=1}^{J-1}\exp\left[x_t'\beta_v+\gamma_v'\delta_{(t-1)c_1}\right]}, & \text{for } c_2 = J. \end{cases} \tag{49}$$

Now, for $t = 2, \ldots, T$, it follows from (49) that

$$\begin{aligned} \frac{\partial \eta^{(c_2)}_{t|t-1}(c_1)}{\partial \beta_{c_2}} &= \eta^{(c_2)}_{t|t-1}(c_1)\left[1 - \eta^{(c_2)}_{t|t-1}(c_1)\right]x_t \\ \frac{\partial \eta^{(c_2)}_{t|t-1}(c_1)}{\partial \beta_k} &= -\left[\eta^{(c_2)}_{t|t-1}(c_1)\eta^{(k)}_{t|t-1}(c_1)\right]x_t, \end{aligned} \tag{50}$$

yielding

$$\begin{aligned} \frac{\partial \eta^{(c_2)}_{t|t-1}(c_1)}{\partial \beta} &= \begin{pmatrix} -\eta^{(1)}_{t|t-1}(c_1)\eta^{(c_2)}_{t|t-1}(c_1) \\ \vdots \\ \eta^{(c_2)}_{t|t-1}(c_1)[1-\eta^{(c_2)}_{t|t-1}(c_1)] \\ \vdots \\ -\eta^{(J-1)}_{t|t-1}(c_1)\eta^{(c_2)}_{t|t-1}(c_1) \end{pmatrix} \otimes x_t : (J-1)(p+1)\times 1 \\ &= \left[\eta^{(c_2)}_{t|t-1}(c_1)(\delta_{(t-1)c_2} - \eta_{t|t-1}(c_1))\right] \otimes x_t. \end{aligned} \tag{51}$$

The formula for the derivative in (26) follows now by applying (50) into (48).

*Derivation for* $\frac{\partial \tilde{\lambda}^{(2)}_{gj}(g^*)}{\partial \gamma}$ :

By using the formula for $\tilde{\lambda}^{(2)}_{gj}(g^*)$ from (13) we write

$$\frac{\partial \tilde{\lambda}^{(2)}_{gj}(g^*)}{\partial \gamma} = \frac{\partial}{\partial \gamma}\begin{cases} \frac{1}{g}\sum_{c_1=1}^{g}\sum_{c_2=j+1}^{J}\lambda^{(c_2)}_{t|t-1}(c_1) & \text{for } g^*=1 \\ \frac{1}{J-g}\sum_{c_1=g+1}^{J}\sum_{c_2=j+1}^{J}\lambda^{(c_2)}_{t|t-1}(c_1) & \text{for } g^*=2, \end{cases} \tag{52}$$

where $\lambda^{(c_2)}_{t|t-1}(c_1)$ is given in (5) [see also (49)].

Next, for $t = 2, \ldots, T$, it follows from (49) that

$$\begin{aligned} \frac{\partial \eta^{(c_2)}_{t|t-1}(c_1)}{\partial \gamma_{c_2}} &= \eta^{(c_2)}_{t|t-1}(c_1)\left[1 - \eta^{(c_2)}_{t|t-1}(c_1)\right]\delta_{(t-1)c_1} \\ \frac{\partial \eta^{(c_2)}_{t|t-1}(c_1)}{\partial \gamma_k} &= -\left[\eta^{(c_2)}_{t|t-1}(c_1)\eta^{(k)}_{t|t-1}(c_1)\right]\delta_{(t-1)c_1}, \end{aligned} \tag{53}$$

where

$$\delta_{(t-1)c_1} = \begin{cases} [01'_{c_1-1}, 1, 01'_{J-1-c_1}]' & \text{for } c_1 = 1, \ldots, J-1 \\ 01_{J-1} & \text{for } c_1 = J. \end{cases}$$

These derivatives in (53) may further be re-expressed as

$$\frac{\partial \eta^{(c_2)}_{t|t-1}(c_1)}{\partial \gamma} = \begin{pmatrix} -\eta^{(1)}_{t|t-1}(c_1)\eta^{(c_2)}_{t|t-1}(c_1) \\ \vdots \\ \eta^{(c_2)}_{t|t-1}(c_1)[1 - \eta^{(c_2)}_{t|t-1}(c_1)] \\ \vdots \\ -\eta^{(J-1)}_{t|t-1}(c_1)\eta^{(c_2)}_{t|t-1}(c_1) \end{pmatrix} \otimes \delta_{(t-1)c_1} : (J-1)^2 \times 1$$

$$= \left[\eta^{(c_2)}_{t|t-1}(c_1)(\delta_{(t-1)c_2} - \eta_{t|t-1}(c_1))\right] \otimes \delta_{(t-1)c_1}. \tag{54}$$

The formula for the derivative in (36) now follows by applying (53) into (52).

## References


1. Agresti, A. (1989). A survey of models for repeated ordered categorical response data. *Statistics in Medicine*, *8*, 1209–1224.
2. Agresti, A., & Natarajan, R. (2001). Modeling clustered ordered categorical data: A survey. *International Statistical Review*, *69*, 345–371.
3. Amemiya, T. (1985). *Advanced econometrics*. Cambridge, MA: Harvard University Press.
4. Fahrmeir, L., & Kaufmann, H. (1987). Regression models for non-stationary categorical time series. *Journal of Time Series Analysis*, *8*, 147–160.
5. Fokianos, K., & Kedem, B. (1998). Prediction and classification of non-stationary categorical time series. *Journal of Multivariate Analysis*, *67*, 277–296.
6. Fokianos, K., & Kedem, B. (2003). Regression theory for categorical time series. *Statistical Science*, *18*, 357–376.
7. Fokianos, K., & Kedem, B. (2004). Partial likelihood inference for time series following generalized linear models. *Journal of Time Series Analysis*, *25*, 173–197.
8. Kaufmann, H. (1987). Regression models for nonstationary categorical time series: Asymptotic estimation theory. *Annals of Statistics*, *15*, 79–98.
9. Lipsitz, S. R., Kim, K., & Zhao, L. (1994). Analysis of repeated categorical data using generalized estimating equations. *Statistics in Medicine*, *13*, 1149–1163.
10. Loredo-Osti, J. C., & Sutradhar, B. C. (2012). Estimation of regression and dynamic dependence parameters for non-stationary multinomial time series. *Journal of Time Series Analysis*, *33*, 458–467.
11. McCullagh, P. (1983). Quasilikelihood functions. *Annals of Statistics*, *11*, 59–67.
12. Rao, C. R. (1973). *Linear statistical inference and its applications*. New York, NY: Wiley.
13. Sutradhar, B. C. (2011). *Dynamic mixed models for familial longitudinal data*. New York, NY: Springer.
14. Sutradhar, B. C. (2014). *Longitudinal categorical data analysis*. New York, NY: Springer.
15. Sutradhar, B. C., & Das, K. (1999). On the efficiency of regression estimators in generalized linear models for longitudinal data. *Biometrika*, *86*, 459–465.
16. Tagore, V., & Sutradhar, B. C. (2009). Conditional inference in linear versus nonlinear models for binary time series. *Journal of Statistical Computation and Simulation*, *79*, 881–897.

17. Tong, H. (1990). *Nonlinear time series: A dynamical system approach.* Oxford statistical science series (Vol. 6). New York, NY: Oxford University Press (1990)
18. Wedderburn, R. W. M. (1974). Quasi-likelihood functions, generalized linear models, and the Gauss–Newton method. *Biometrika, 61*, 439–447.

# Identification of Threshold Autoregressive Moving Average Models

**Qiang Xia and Heung Wong**

**Abstract** Due to the lack of a suitable modeling procedure and the difficulty to identify the threshold variable and estimate the threshold values, the threshold autoregressive moving average (TARMA) model with multi-regime has not attracted much attention in application. Therefore, the chief goal of our paper is to propose a simple and yet widely applicable modeling procedure for multi-regime TARMA models. Under no threshold case, we utilize extended least squares estimate (ELSE) and linear arranged regression to obtain a test statistic $\hat{F}$, which is proved to follow an approximate $F$ distribution. And then, based on the statistic $\hat{F}$, we employ some scatter plots to identify the number and locations of the potential thresholds. Finally, the procedures are considered to build a TARMA model by these statistics and the Akaike information criterion (AIC). Simulation experiments and the application to a real data example demonstrate that both the power of the test statistic and the model-building can work very well in the case of TARMA models.

**Keywords** Arranged regression · Nonlinearity test · TMA Model

## 1 Introduction

Since Tong [29], the threshold autoregressive (TAR) model has provided a much wider spectrum of possible dynamics for the economic and financial time series data. A time series $y_t$ is said to follow a self-excited TAR model, if it satisfies

$$y_t = \sum_{j=1}^{k} \left[ \phi_0^{(j)} + \sum_{i=1}^{p} \phi_i^{(j)} y_{t-i} + \varepsilon_t^{(j)} \right] I(r_{j-1} \leq y_{t-d} < r_j).$$

Q. Xia
College of Mathematics and Informatics,
South China Agricultural University, Guangzhou 510642, China
e-mail: xiaqiang@scau.edu.cn

H. Wong (✉)
Department of Applied Mathematics,
The Hong Kong Polytechnic University, Hung Hom, Kowloon, Hong Kong, China
e-mail: mathwong@polyu.edu.hk



W.K. Li et al. (eds.), *Advances in Time Series Methods and Applications*,
Fields Institute Communications 78, DOI 10.1007/978-1-4939-6568-7_9

In principle, it can be easily extended to a threshold autoregressive moving average model (SETARMA) model, if it satisfies

$$y_t = \sum_{j=1}^{k} \left[ \phi_0^{(j)} + \sum_{i=1}^{p} \phi_i^{(j)} y_{t-i} + \sum_{i=1}^{q} \theta_i^{(j)} \varepsilon_{t-i}^{(j)} + \varepsilon_t^{(j)} \right] I(r_{j-1} \le y_{t-d} < r_j) \quad (1)$$

where $k$ is number of regimes and $d$ is a positive integer commonly referred to as the threshold delay. The thresholds are $-\infty = r_0 < r_1 < \cdots < r_k = \infty$; $\{\varepsilon_t^{(j)}\}$ is a sequence of independent and identically distributed (i.i.d.) random variables with mean zero and variance $\sigma_j^2$, $0 < \sigma_j^2 < \infty$ in each $j$. Then the one-dimensional Euclidean space is partitioned into $k$ regimes by such a process and each regime follows a linear ARMA model. When there are at least two regimes with different ARMA models, the overall process $y_t$ is a nonlinear TARMA model. For TAR models, which have been widely used in applications, some fundamental results on the probabilistic structure of TAR models were given by Tong and Lim [30], Chan et al. [6], Chan and Tong [7], Chan [5], Tong [31], Tsay [32–35], Wong and Li [36, 37] and Hansen [15]. For TARMA models, not many theoretical results have been reported for a long while.

In recent years, people realized that TMA and TARMA models are as important as TAR models in practice, and more attention has been paid to TMA and TARMA models in the literature. For instance, Brockwell et al. [3], Liu and Susko [25], de Gooijer [10] and Ling [22, 24]. For testing problems of TMA models, Ling and Tong [23] proposed a likelihood ratio test for linear MA model against TMA models. For testing problems of TARMA models, Li and Li [18] developed likelihood ratio test for ARMA model against its extension with two regimes. For estimation problems of TARMA models, Li et al. [19] proposed least squares estimate (LSE) for TARMA models with two regimes. However, modelling multi-regime TARMA models has not attracted much attention in the literature. The structure of $y_t$ depends on the threshold parameter $r$ and the delay parameter $d$. Due to the difficulty to estimate these parameters and the related computational problems, there is no simple procedure to identify the threshold. Consequently, TARMA models have not been widely used in applications. From a practical point of view, TARMA models should have more advantages over pure TAR or TMA models because they can provide a parsimonious form just like linear ARMA models. Therefore, the main objective of this paper is to propose a simple procedure for testing and modeling the threshold nonlinearity of TARMA models with multi-regime.

In this paper, on one hand, to test for threshold nonlinearity for TARMA models, we combine the extended least-squares (ELS; [26]) with arranged regression method [34] to construct a test statistics $\hat{F}$, which follows $F$ distribution and is suitable for TARMA model identification. On the other hand, with the statistics $\hat{F}$ and AIC, we propose a simple procedure for modelling TARMA models. Simulation studies are carried out to assess the performance of the statistics $\hat{F}$ and modelling procedure in finite samples.

The paper is organized as follows. We give the testing statistics and its null asymptotic distribution using arranged regression in Sect. 2. Section 3 introduces the modeling procedure in detail. In Sect. 4, we evaluate finite-sample performance and efficiency of the procedure by simulation and a real data set. Section 5 is our conclusion.

Throughout the paper, $A'$ denotes the transpose of a vector or a matrix $A$, $\rightarrow^{P}$ and $\rightarrow^{a.s.}$ denote the convergence in probability and almost sure convergence respectively.

## 2 Test Statistic and Its Asymptotic Distribution

### 2.1 Consistency of Least Squares Estimates

Conveniently, the model (1) is referred to as a TARMA($k, p, q, d$), where $k$ is the number regimes, $p$ the AR order, $q$ the MA order and $d$ is called the threshold lag. The interval $r_{j-1} \leq y_{t-d} < r_j$ is the $j$th regime of $y_t$, which has $n_j$ observations of $y_t$. From model (1), we can see that each regime follows an ARMA model. Hannan and Rissanen [14] suggested a regression approach to estimate ARMA models. It is natural to utilize the ELS or Hannan–Rissanen algorithm in studying model (1). For the $j$th regime, we first compute the estimated residuals $\{\hat{\varepsilon}_t^{(j)}\}$ by the ELS or the Hannan–Rissanen procedure, and then we obtain $\varepsilon_t{}^{(j)}(\Theta) = y_t - [\phi_0^{(j)} + \sum_{i=1}^{p} \phi_i^{(j)} y_{t-i} + \sum_{i=1}^{q} \theta_i^{(j)} \hat{\varepsilon}_{t-i}^{(j)}(\Theta)]$. Also, we denote $\Phi^{(j)} = (\phi_0^{(j)}, \phi_1^{(j)}, \ldots, \phi_p^{(j)}, \theta_1^{(j)}, \ldots, \theta_q^{(j)})'$, and the least squares estimates $\hat{\Phi}^{(j)} = (\hat{\phi}_0^{(j)}, \hat{\phi}_1^{(j)}, \ldots, \hat{\phi}_p^{(j)}, \hat{\theta}_1^{(j)}, \ldots, \hat{\theta}_q^{(j)})'$ respectively. Let $\Theta = (\Phi'^{(1)}, \ldots, \Phi'^{(k)})'$, and $\hat{\Theta} = (\hat{\Phi}'^{(1)}, \ldots, \hat{\Phi}'^{(k)})'$. Then, the sum of square errors function $L_n(\Theta)$ is defined as

$$L_n(\Theta) = \sum_{t=1}^{n} \sum_{j=1}^{k} [\varepsilon_t{}^{(j)}(\Theta)]^2,$$

the minimizer $\hat{\Theta}$ of $L_n(\Theta)$ is called the least squares estimate, that is,

$$\hat{\Theta} = \underset{\Theta}{argmin}\, L_n(\Theta),$$

where $\hat{\Phi}^{(j)} = (X'X^{(j)})^{-1}(X'Y^{(j)})$, $X'X^{(j)}$ is the associate $X'X$ matrix.

Our basic assumptions are as follows.

**Assumption 1** All $\{\varepsilon_t^{(j)}\}$ are $i.i.d.$ and $E(\varepsilon_t^{(j)})^4 < \infty, \quad j = 1, \ldots, k.$

**Assumption 2** $\frac{n_j}{n} \rightarrow^P c_j$ for all $j = 1, \ldots, k$, where $n$ is the total sample size and $c_j$ is a positive fraction satisfying $\sum_{j=1}^{n} c_j = 1$.

**Assumption 3** $\lambda_{j,min} \rightarrow^{a.s.} \infty$, $ln(\lambda_{j,max}) = o(\lambda_{j,min})$ $a.s.$, as $n \rightarrow \infty$ for $j$ regime, where $\lambda_{j,min}$ and $\lambda_{j,max}$ denotes the minimum and maximum eigenvalues of $X'X^{(j)}$, respectively.

**Assumption 4** $\Phi^{(j)} \neq \Phi^{(i)}$ for $j \neq i$, and $\sum_{i=1}^{q} \theta_i^{(j)} < 1$, where $\theta_i^{(j)} = 0$ for $i > q, j = 1, \ldots, k$.

*Remark 1* As Chen et al. [8] noted that it remains difficult to find necessary and sufficient conditions for stationarity and invertibility of TARMA models. In this study, hence, our focus is on the modelling procedure.

**Theorem 1** *Suppose Assumptions 1, 2, 3 and 4 hold, then, for given $k, d$ and the threshold values $r_j$, the least squares estimates $\hat{\Phi}^{(j)}$ converge to $\Phi^{(j)}$ almost surely.*

**Corollary 1** *If $\Phi^{(1)} = \cdots = \Phi^{(k)} = \Phi$, then model (1) is the ARMA model. Under Assumptions 1 and 3,*

$$\hat{\Phi} \rightarrow^{a.s.} \Phi.$$

## 2.2 A Test Statistics for Threshold Nonlinearity

To separate the regimes effectively, a rearranged TARMA($k, p, q, d$) model is useful. The separation can assemble the observation in groups and does not require knowing the precise value of the thresholds $r_j$. In order to see this easily, we consider $k = 2$. For a given TARMA($2, p, q, d$) model with a sample $\{y_1, y_2, \ldots, y_n\}$. The values of the threshold variable $y_{t-d}$ are assumed $\{y_{h-d}, y_{h+1-d}, \ldots, y_{n-d}\}$, where $h = max\{1, p+1-d, q+1-d\}$. Let $\pi_i$ denote the $i$th smallest observation of $\{y_{h-d}, y_{h+1-d}, \ldots, y_{n-d}\}$. If the first regime has $s$ observations, i.e., $y_{\pi_i+d}, i = 1, \ldots, s$, then for $i > s$ the observations belong to the second regime. Therefore, the model can be written as a rearranged TARMA model, i.e.,

$$y_{\pi_i+d} = \begin{cases} \phi_0^{(1)} + \sum_{j=1}^{p} \phi_j^{(1)} y_{\pi_i+d-j} + \sum_{u=1}^{q} \theta_u^{(1)} \varepsilon_{\pi_i+d-k}^{(1)} + \varepsilon_{\pi_i+d}^{(1)}, i > s \\ \phi_0^{(2)} + \sum_{j=1}^{p} \phi_j^{(2)} y_{\pi_i+d-j} + \sum_{u=1}^{q} \theta_u^{(2)} \varepsilon_{\pi_i+d-k}^{(2)} + \varepsilon_{\pi_i+d}^{(2)}, i \leq s \end{cases} \quad (2)$$

where $s$ satisfies $y_{\pi_s} \leq r < y_{\pi_s+1}$. More specifically, the arranged TARMA model provides a way, which can group the data points, and makes all of the observations in a group follow the same 'regression' model.

The motivation of the proposed test is to illustrate the potential use of arranged regression in studying TARMA models by considering model (2). If the threshold value $r$ were known, then we could easily obtain consistent estimates of the parameters. We in general do not know the threshold value, hence, we must proceed sequentially starting from the linear ARMA model. If the first regime has sufficiently large numbers of observations, i.e., many $i \leq s$, the least squares estimates $\hat{\Phi}_i^{(1)}$ are consistent for $\Phi_i^{(1)}$. In this case, the predictive residuals $\varepsilon_{\pi_i+d}$ are assumed white noise, which are asymptotically and orthogonal to the regressors $\{y_{\pi_i+d-j}, \varepsilon_{\pi_i+d-u}, j = 1, \ldots, p, u = 1, \ldots, q\}$. When $i$ arrives at or exceeds $s$, the model will have a change at time $\pi_{s+1} + d$. It will destroy the orthogonality between the predictive residuals ($\varepsilon_{\pi_i+d}$) and the regressors, which brings bias to the predictive residual for the observation with time index $\pi_{s+1} + d$. Thus, the consistency of $\hat{\Phi}_i^{(1)}$ is also destroyed. In this process, we need not know the actual value of $r$ here, all that is needed is the existence of a nontrivial threshold. According to the aforementioned consideration, one method to test for threshold nonlinearity is to regress the predictive residuals from the arranged ARMA regression on the regressors $\{y_{\pi_i+d-j}, \varepsilon_{\pi_i+d-u}, j = 1, \ldots, p, u = 1, \ldots, q\}$. Based on the residuals of the relevant regression, we can construct an $F$ statistic.

For the arranged regression (2), based on the first $m$ cases, let $\hat{\Phi}_m$ denote the vector of least squares estimates, $P_m$ be the associated $X'X$ inverse matrix, and $x_{m+1}$ is the vector of regressors of the next observations $y_{\pi_{m+1}+d}$ and $\varepsilon_{\pi_{m+1}+d}$ to enter the regression. Note that the variables $\varepsilon_{\pi_{m+1}+d}$ entering the vector $x_{m+1}$ are not observable, and the recursive LSE [11, 12, 34] cannot be implemented as it stands. We have to substitute some estimate $\hat{\varepsilon}_t$ for the components $\varepsilon_t$, computed according to $\hat{\varepsilon}_t = y_t - x_t'\hat{\Phi}_t$, where $x_t = (y_{t-1}, \ldots, y_{t-p}, \ldots, \hat{\varepsilon}_{t-1}, \ldots, \hat{\varepsilon}_{t-q})'$. The estimates of $\hat{\varepsilon}_t$ can be obtained by extended least squares estimate (ElSE), see Ljung and Soderstrom [26]. Then, we can compute ELSE efficiently by

$$\hat{\Phi}_{m+1} = \hat{\Phi}_m + K_{m+1}\left[y_{\pi_{m+1}+d} - x_{m+1}'\hat{\Phi}_m\right]$$

where $D_{m+1} = 1 + x_{m+1}' P_m x_{m+1}$, $P_{m+1} = (I - P_m \frac{x_{m+1}x_{m+1}'}{D_{m+1}})P_m$, and $K_{m+1} = \frac{P_{m+1}x_{m+1}}{D_{m+1}}$.

Moreover, we can obtain the following predictive and standardized predictive residuals

$$\tilde{\varepsilon}_{\pi_{m+1}+d} = y_{\pi_{m+1}+d} - x_{m+1}'\hat{\Phi}_m$$

and

$$\hat{e}_{\pi_{m+1}+d} = \frac{\tilde{\varepsilon}_{\pi_{m+1}+d}}{\sqrt{D_{m+1}}}.$$

Next, the predictive residuals are used to locate the threshold. Here, they can be used to construct the $F$ statistics for testing nonlinearity. We will give the details of the proposed nonlinearity test now. For fixed $p, q$ and $d$, there are $n - d - h + 1$ effective observations in the arranged regressions, with $h$ defined just before. Assume that the recursive regressions begin with $b$ observations, hence, the available number of predictive residuals is $n-d-b-h+1$. Then, the least squares regression is done as follows.

$$\hat{e}_{\pi_i+d} = \beta_0 + \sum_{k=1}^{p} \beta_{1_k} y_{\pi_i+d-k} + \sum_{l=1}^{q} \beta_{2_l} \hat{\varepsilon}_{\pi_i+d-l} + a_{\pi_i+d} \tag{3}$$

for $i = b+1, \ldots, n-d-h+1$, and the associated $F$ statistic can be computed as

$$\hat{F}(p,q,d) = \frac{\left(\sum \hat{e}_t^2 - \sum \hat{a}_t^2\right)/(p+q+1)}{\sum \hat{a}_t^2/(n-d-b-h-p-q)}, \tag{4}$$

where the summations are all from $b+1$ to $n-d-h+1$, and $\hat{a}_t$ is the least squares residual of (3). We use the argument $(p, q, d)$ of $\hat{F}$ to signify the dependence of the $F$ ratio on $p, q$ and $d$.

**Theorem 2** *Suppose that $y_t$ is a linear invertible ARMA process of order $p$ and $q$, that is to say, $y_t$ follows model* (1) *with $k = 1$. Using the Hannan–Rissanen algorithm or ELSE, then, the statistic $\hat{F}(p, q, d)$ defined in* (4) *follows approximately an $F$ distribution with degrees of freedom $p+q+1$ and $n-d-b-p-q-h$ for large $n$. Furthermore, $(p+q+1)\hat{F}(p,q,d)$ follows asymptotically a chi-squared random variable with degrees of freedom $p+q+1$.*

In practice that we do not know the number and locations of the thresholds, and as we know there is no simple method for testing threshold nonlinearity. The major considerations for proposing the $\hat{F}(p, q, d)$ statistic are relative power, feasibility and simplicity of implementation. As it requires only a sorting routine and the linear regression method, it is extremely simple.

## 3 Building TARMA Models

### 3.1 Selecting the Delay Parameter d

For building TARMA models, a major difficulty is selection of the delay parameter $d$ as well as the specification of the threshold variable. Similar to Tsay [34], which proposed using magnitude of the test statistics to select $d$ before locating the threshold values for modeling TAR models. It is assumed that the AR order $p$ and MA order $q$ are given, we can select an estimate of the delay parameter, say $d_{p,q}$, such that

$$\hat{F}(p,q,d_{p,q}) = max_{\{v \in S\}}\{\hat{F}(p,q,v)\}, \tag{5}$$

where $\hat{F}(p, q, v)$ is value of the $F$ statistic of (4), the subscript $(p, q)$ signifies that the estimate of $d$ may depend on $p$ and $q$, and $S$ is a set of possible positive integers, i.e., $S = \{1, 2, \ldots, max\{p, q\}\}$.

Notice that it is somewhat heuristic for the choice of $d_{p,q}$ in (5), which is based on the idea that if TARMA models are needed, then we might choose a delay parameter that gives the most significant result in testing for threshold nonlinearity. Several values of $d$ may be tried by a cautious data analyst, especially including the maximum and the second maximum of $\hat{F}(p, q, d)$ in (5). Generally speaking, $d_{p,q}$ depend upon $p$ and $q$, which is usually unknown. In this case, a reasonable AR order $p$ and MA order $q$ should be started with, such as suggested by AIC.

## 3.2 Locating the Values of Thresholds

For a TARMA model, it is very important to specify the threshold variable. Therefore, estimating the threshold $r_j$'s needs special care. For TAR models, Tong and Lim [30] considered the empirical percentiles as candidates for the threshold values, with the specification of a set of finite numbers of sample percentiles to work with. But Tsay [34] searched through the percentiles to locate the threshold values. Therefore, sample percentile point estimates or an interval estimate for each of the threshold values may be provided for TARMA models. Based on the latter idea, we make a try to locate the values of thresholds for TARMA models. This is illustrated by model (4), that is, assume that $k = 2$ and the true value of $r_1$ satisfies $y_{\pi_s} < r_1 < y_{\pi_{s+1}}$. Then, any value in the interval $[y_{\pi_s}, y_{\pi_{s+1}}]$ is as good as the other in providing an estimate of $r_1$, because all of them will give the same fitting results for a specified TARMA model. The limitation of this method is only that a threshold is not too close to the $0th$ or $100th$ percentile, otherwise there are not enough observations to provide an efficient estimate for these extreme points.

The tools proposed to locate the thresholds are scatter plots, which are plots of the specified threshold variable versus various statistics. scatter plots method has been applied extensively in the literature, such as Haggan et al. [13] which adopted the scatter plots of recursive AR estimates in studying the state-dependent model of Priestley [27]. Tsay [34] considered scatter plots of recursive AR estimates for TAR model. They all obtained fine results. Although the graphics are not formal testing statistics, useful information is provided by them in locating the thresholds. The plots used are: the ordinary predictive residuals or the standardized predictive residuals versus $y_{t-d}$; and $t$ ratios of recursive estimates of an AR or MA coefficient versus $y_{t-d}$. We will discuss the rationale of each of the plots in the following, while some illustrative examples are deferred to the simulation and the application section.

In the framework of arranged regression, the TARMA model consists of some model changes, which occur at each threshold value $r_j$. The predictive residuals will be biased at the threshold values. Thus, a scatter plots of the threshold variable versus the (standardized) predictive residuals may reveal the locations of the threshold values of a TARMA model. On the other hand, the plot is random for a linear time series, excluding the beginning of the recursion. We use the scatter plots because it can tell the locations of the threshold values directly.

It is best to begin with a linear time series, which can motivate the use of a scatter plot of the threshold variable versus recursive $t$ ratios of an AR or MA coefficient. In this case, the $t$ ratios have two functions: they can show the particular AR or MA coefficient to be significant or not; and when the significance of that coefficient is remarkable, the t ratios converge to a fixed value gradually and smoothly as the recursion continues. As an example, we consider the simple TARMA model

$$y_t = \begin{cases} \phi_1^{(1)} y_{t-1} + \theta_1^{(1)} \varepsilon_{t-1}^{(1)} + \varepsilon_t^{(1)}, \; y_{t-d} \leq r_1 \\ \phi_1^{(2)} y_{t-1} + \theta_1^{(2)} \varepsilon_{t-1}^{(2)} + \varepsilon_t^{(2)}, \; y_{t-d} > r_1 \end{cases} \tag{6}$$

where $\phi_1^{(1)}$ and $\phi_1^{(2)}$ are different as well as $\theta_1^{(1)}$ and $\theta_1^{(2)}$. Then, using LSE method and written as arranged regression (2), where we let $\phi_1$ or $\theta_1$ be the recursive estimate of the lag-1 AR or MA coefficient. Before the recursion reaches the threshold value $r_1$, the $t$ ratios of $\phi_1$ or $\theta_1$ behave exactly as those of a linear time series by Theorem 2.1. Once $y_{t-d}$ reaches $r_1$, the estimate $\phi_1$ or $\theta_1$ begins to change and the $t$ ratio starts to deviate. Thus, it will destroy the pattern of gradual convergence of the $t$ ratios. In practice, the $t$ ratio may begin to turn and change direction at the threshold value. Therefore, for model (3.2), $\phi_1$ or $\theta_1$ starts to change when $r_1$ is reached, and finally it has a compromise between $\phi_1^{(1)}$ and $\phi_1^{(2)}$ or $\theta_1^{(1)}$ and $\theta_1^{(2)}$. When this behavior also appears in the associated $t$ ratios, it shows information on the value of $r_1$. Generally, it is effortless to see the change in $t$ ratio when the two AR or MA coefficients are substantially different.

Using the $t$-ratio plot to locate the values of thresholds of TARMA models, we should pay attention to the following points. First, the constant term $\phi_0$ is important, because it signifies level changes. Second, if the variance of errors in each regime is obviously distinct, the residual plot will change clearly in the vicinity of thresholds. Third, as long as the sample size in every regime is reasonable, the usefulness of the plot in the previous discussion can be employed to the case of multi-threshold value actually. Finally, since the ordered $y_{t-d}$ are not equally spaced, when some data points of $y_{t-d}$ have relatively large values, omitting them in a scatter plot is often helpful. The last $b = (n/5) + min\{p, q\}$ points in all of the scatter plots are not shown in this article.

## 3.3 *Modeling TARMA Models*

The procedures for modeling TAR models were outlined by Tong and Lim [30], and Tsay [34] respectively, but each step of Tsay [34] was relatively simple. Based on Tsay's ideas, we propose a procedure for building TARMA models, as long as the delay parameter $d$ is selected and the values of thresholds are also located. We hope that the features of TARMA models can be exploited by this procedure in simulation and application, which consists of several steps and is described as follows.

Step 1. Select the appropriate AR order $p$ and MA order $q$ of a single ARMA model by AIC. The set of possible threshold lags is $S$, and $S = \{1, 2, \ldots, max\{p, q\}\}$.

Step 2. For a given $p$ and $q$, find the fitted residuals of the model in Step 1 by ELSE or the Hannan–Rissanen algorithm.

Step 3. For a given $p, q$ and every element $d$ of $S$, we use $b = (n/5) + min\{p, q\}$ as data points to initiate a recursion by arranged regression method, and compute the value of the test-statistic $\hat{F}(p, q, d)$. If the nonlinearity of the model is detected, then employ the approach of Sect. 3.1 to select the delay parameter $d_{p,q}$.

Step 4. For given $p, q$ and $d_{p,q}$, make use of the scatter plots of Sect. 3.2 to locate the threshold values $r$'s.

Step 5. For given $d_{p,q}$ and $r$'s, utilize the ELSE or the Hannan–Rissanen algorithm to estimate coefficients in each regime, and refine the AR order, MA order and threshold values with AIC.

In Step 1, the $acf$ and $pacf$ are often instinctive tools and may provide guidance for reasonable starting values of $p$ and $q$ for AR and MA model respectively. But for an ARMA model, its AR order $p$ and MA order $q$ may better be selected by considering AIC. In addition, if desired, Step 5 can refine the AR order and MA order. For a given $p$ and $q$ at Step 3, the set $S$ of possible threshold lags may be $\{1, 2, \ldots, max\{p, q\}\}$. We provide a method of selection of the delay parameter $d_{p,q}$ in Sect. 3.1. In Step 4, because scatter plots of insignificant AR or MA coefficients are usually not informative, $t$ ratios of various AR or MA coefficients can be examined as long as the AR or MA coefficients are significant. The model refinement at Step 5 may rely on AIC because of the linear nature of the TARMA model in each regime. See Tong and Lim [30], who gave the details of using AIC in modeling TAR models.

*Remark 2* For each regime of specified TARMA model, AIC [30] is taken the form $\text{AIC}(k) = Nln(RSS/N) + 2k$, and $RSS$ is the residual sum of squares of the fitted model, based on ELSE of the defining parameters. $N$ is the "effective number of observations" and $k$ is the number of independent parameters of the models.

# 4 Simulation Experiments and a Real Example

In this section, firstly, we present simulation results to examine the performance of the statistic $\hat{F}(p, q, d)$ and build TARMA models in finite samples through Monte Carlo (MC) experiments. Secondly, we apply the test statistic and the procedure to a real data set of the exchange rate of Japanese Yen versus USA dollar.

## *4.1 Simulation Experiments*

The power of the $\hat{F}(p, q, d)$ statistic in detecting the threshold nonlinearity will be studied firstly. Sample sizes used are 200 and 400 in the experiments respectively, and the number of replications is 1000. To reduce the effect of the starting value in generating a TARMA model, we generate $n + 200$ observations and discard the first 200 values for each realization of sample size $n$. The null is the ARMA(1, 1) model with constant term $\phi_0 = 0$, $\phi_1 = 0.5$ and $\theta_1 = 0.5$, and the alternative is the TARMA(2, 1, 2) model with constant terms $\phi_0^{(j)} = 0$, $j = 1, 2$, $r_1 = 0$, $\phi_1^{(2)} = \theta_1^{(2)} = 0.5$ and $\phi_1^{(1)} = \theta_1^{(1)} = 0, 0.2, 0.4, 0.5, 0.6, 0.8$. We choose $b = (n/5) + min\{p, q\}$, with $p$ the fitted AR order and $q$ the fitted MA order, and take significance levels $\alpha = 0.05$ and 0.1. The corresponding critical value are 3.00 and 2.30, respectively. The results are recorded in Table 1, which shows that the powers are very close to the nominal values 0.05 and 0.1 in the case of ARMA(1, 1) model. In particular, the power increases when the alternative departs from the linear ARMA model or when the sample size increases.

To study the power of our method further, we conduct another investigation about TARMA models with nonzero constant terms. Based on 1000 realizations and 10 and 5 % critical values, Table 2 gives the powers of rejecting a linear ARMA process. In the simulation, the model used is a TARMA(2, 1, 1), with parameters $(\phi_0^{(1)}, \theta_1^{(2)}, \phi_0^{(2)}, r_1, \sigma_1^2, \sigma_2^2) = (0.5, 0.5, 0.5, 0.5, 4.0, 1.0)$ or $(0.5, 0.5, -0.5, 0.5, 4.0, 1.0)$ and $\theta_1^{(2)}$ given $-1, -0.5, 0, 0.5$ respectively. The sample sizes are also 200 and 400. In the test statistic, $p = q = d = 1$ were used, let $b = (n/5) + q$. From the Table 2, when the constant term $\phi_0$ are identical and not equal to zero with different variance of errors in every regime, the power of $F$ statistic increases with larger difference of $\theta_1^{(1)}$. But each regime has a distinct constant and a different variance of errors, the power of $F$ statistic becomes stronger. These results indicate that the test can give good performance on the nonlinear case and should be useful in practice.

Secondly, the proposed procedure will be applied to some simulated examples in the following. Consider the following four models with simulated data set of sample size $n = 400$, where $i.i.d.$ denotes independent and identically distributed.

**Table 1** The power of rejecting a linear ARMA Model with $\phi_0^{(1)} = \phi_0^{(2)} = 0$ based on 1000 Replications

| $\alpha$ | n = 200 | | n = 400 | |
|---|---|---|---|---|
| | 5 % | 10 % | 5 % | 10 % |
| $\theta_1^{(1)}$ | Powers with $\theta_1^{(2)} = 0.5$ | | | |
| 0 | 0.998 | 0.999 | 1.000 | 1.000 |
| 0.2 | 0.851 | 0.918 | 0.996 | 1.000 |
| 0.4 | 0.163 | 0.260 | 0.279 | 0.410 |
| 0.5 | 0.042 | 0.086 | 0.048 | 0.092 |
| 0.6 | 0.139 | 0.214 | 0.285 | 0.387 |
| 0.8 | 0.627 | 0.746 | 0.923 | 0.960 |

[a] In fact, as $\phi_1^{(1)} = \theta_1^{(1)} = 0.5$, the power is the size. Meanwhile, the errors $\varepsilon_t^{(1)}, \varepsilon_t^{(2)} \sim N(0, 1)$

**Table 2** The power of rejecting a linear ARMA Model with $\phi_0^{(1)} = 0.5$ based on 1000 replications

| $\alpha$ | n = 200 | | | | n = 400 | | | |
|---|---|---|---|---|---|---|---|---|
| | 5 % | | 10 % | | 5 % | | 10 % | |
| $\phi_0^{(2)}$ | 0.5 | −0.5 | 0.5 | −0.5 | 0.5 | −0.5 | 0.5 | −0.5 |
| $\phi_1^{(1)} = \theta_1^{(1)}$ | Powers with $\phi_1^{(2)} = \theta_1^{(2)} = 0.5$ | | | | | | | |
| −1 | 1.000 | 1.000 | 1.000 | 1.000 | 1.000 | 1.000 | 1.000 | 1.000 |
| −0.5 | 1.000 | 1.000 | 1.000 | 1.000 | 1.000 | 1.000 | 1.000 | 1.000 |
| 0 | 0.986 | 0.987 | 0.993 | 0.996 | 1.000 | 1.000 | 1.000 | 1.000 |
| 0.5 | 0.266 | 0.298 | 0.362 | 0.407 | 0.271 | 0.440 | 0.364 | 0.536 |

[a] The errors $\varepsilon_t^{(1)} \sim N(0,1)$, $\varepsilon_t^{(2)} \sim N(0,4)$

- Model 1: TARMA(2, 1, 1, 1)

$$y_t = \begin{cases} -0.5y_{t-1} - 0.5\varepsilon_{t-1}^{(1)} + \varepsilon_t^{(1)}, & y_{t-1} \le 1.0 \\ 0.5y_{t-1} + 0.5\varepsilon_{t-1}^{(2)} + \varepsilon_t^{(2)}, & y_{t-1} > 1.0 \\ \varepsilon_t^{(1)} i.i.d. \sim N(0,1), \varepsilon_t^{(2)} i.i.d. \sim N(0,4) \end{cases} \tag{3.1}$$

- Model 2: TARMA(3, 1, 1, 1)

$$y_t = \begin{cases} -0.5y_{t-1} - 0.5\varepsilon_{t-1}^{(1)} + \varepsilon_t^{(1)}, & y_{t-1} \le 1 \\ 0.2y_{t-1} + 0.2\varepsilon_{t-1}^{(2)} + \varepsilon_t^{(2)}, & 1 < y_{t-1} \le 4 \\ 0.8y_{t-1} + 0.8\varepsilon_{t-1}^{(3)} + \varepsilon_t^{(3)}, & y_{t-1} > 4 \\ \varepsilon_t^{(1)} i.i.d. \sim N(0,1), \varepsilon_t^{(2)} i.i.d. \sim N(0,9), \varepsilon_t^{(3)} i.i.d. \sim N(0,4) \end{cases} \tag{3.2}$$

- Model 3: TARMA(2, 2, 1, 2)

$$y_t = \begin{cases} -0.5y_{t-1} + 0.5y_{t-2} - 0.5\varepsilon_{t-1}^{(1)} + \varepsilon_t^{(1)}, & y_{t-2} \le 0. \\ 0.5y_{t-1} - 0.5y_{t-2} + 0.5\varepsilon_{t-1}^{(2)} + \varepsilon_t^{(2)}, & y_{t-2} > 0. \\ \varepsilon_t^{(1)} i.i.d. \sim N(0,1), \varepsilon_t^{(2)} i.i.d. \sim N(0,4) \end{cases} \tag{3.3}$$

- Model 4: TARMA(3, 2, 1, 2)

$$y_t = \begin{cases} -0.5y_{t-1} + 0.5y_{t-2} - 0.5\varepsilon_{t-1}^{(1)} + \varepsilon_t^{(1)}, & y_{t-2} \le -3 \\ 0.2y_{t-1} - 0.2y_{t-2} + 0.2\varepsilon_{t-1}^{(2)} + \varepsilon_t^{(2)}, & -3 < y_{t-2} \le 2 \\ 0.8y_{t-1} - 0.8y_{t-2} + 0.8\varepsilon_{t-1}^{(3)} + \varepsilon_t^{(3)}, & y_{t-2} > 2 \\ \varepsilon_t^{(1)} i.i.d. \sim N(0,1), \varepsilon_t^{(2)} i.i.d. \sim N(0,9), \varepsilon_t^{(3)} i.i.d. \sim N(0,4) \end{cases} \tag{3.4}$$

For the model (3.1)–(3.4) with one sample, the order $p = q = 3$, $p = 1, q = 3$, $p = q = 3$ and $p = q = 3$ are suggested by AIC, respectively. Then, we compute the values of test-statistic $\hat{F}(p, q, d)$ as recorded in Table 4, which indicate $d_{p,q} = 1$ for all models except for (3.3). By looking at scatter plots to locate the thresholds, we choose the second maximum of $\hat{F}(p, q, d)$, i.e., $d_{p,q}$ is adjusted 2 to model (3.4).

Figures 1, 2, 3 and 4 show $t$-ratio of the lag-1 AR coefficient $\phi_1^{(1)}$ and residuals versus ordered threshold respectively. From the $t$-ratio of Fig. 1, we can find the obvious change in the vicinity of 1.0, which suggests a threshold around 1.0 clearly and is exactly confirmed by the residuals of Fig. 1. The $t$-ratio of Fig. 2 changes its direction twice: once near $y_{t-1} = 1$ and again near $y_{t-1} = 4$, suggesting that there

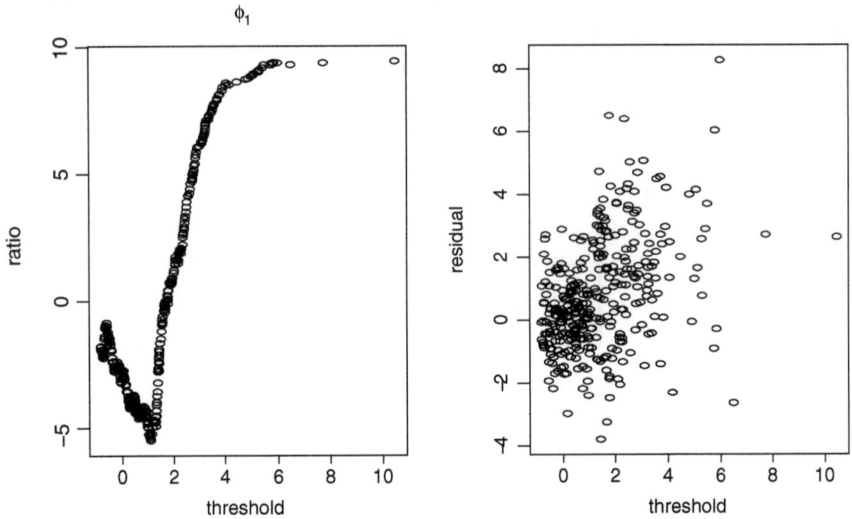


**Fig. 1** The $t$-ratio of $\phi_1^{(1)}$ and residuals versus threshold for model (3.1)

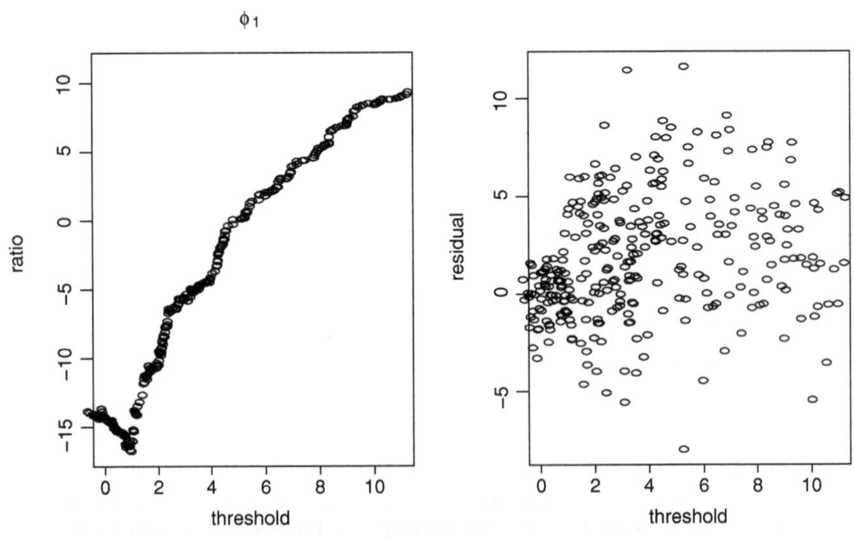


**Fig. 2** The $t$-ratio of $\phi_1^{(1)}$ and residuals versus threshold for model (3.2)

are two nontrivial thresholds, which are just ascertained by the residuals of Fig. 2. Similarly, we can find that Fig. 3 indicates a threshold around 0.0, and Fig. 4 implies two nontrivial thresholds, i.e., once near $y_{t-1} = -3$ and again near $y_{t-1} = 2$. Finally, M.L.E method is used to estimate coefficients in each regime by the given $d_{p,q}$ and

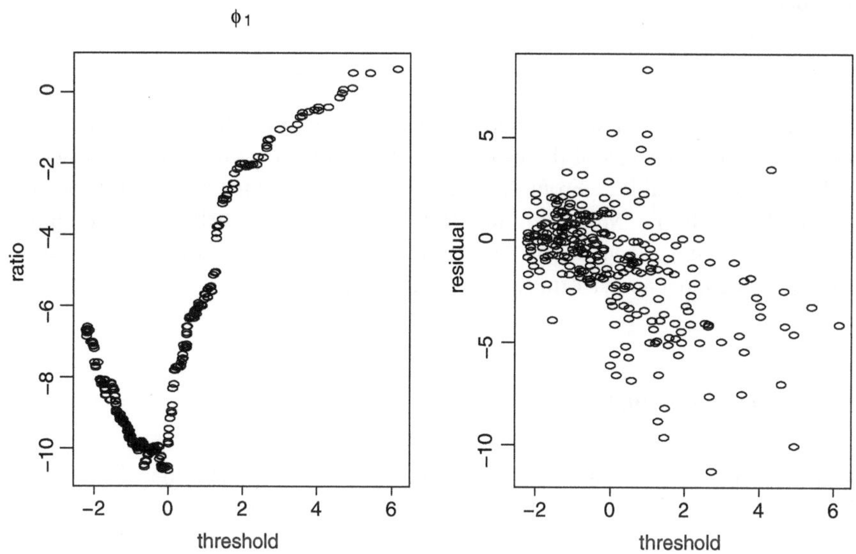


**Fig. 3** The $t$-ratio of $\phi_1^{(1)}$ and residuals versus threshold for model (3.3)

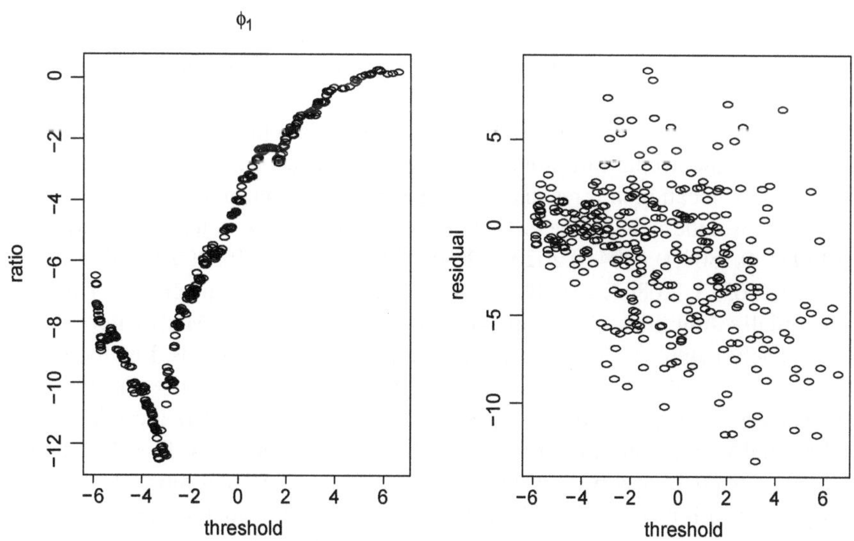


**Fig. 4** The $t$-ratio of $\phi_1^{(1)}$ and residuals versus threshold for model (3.4)

**Table 3** The estimate of $d$ in model (3.1)–(3.4)

| | $d$ | 1 | 2 | 3 |
|---|---|---|---|---|
| TARMA(2, 1, 1, 1) | $\hat{F}$ | 52.3278 | 3.6512 | 3.7048 |
| TARMA(3, 1, 1, 1) | $\hat{F}$ | 43.8513 | 2.2266 | 2.0466 |
| TARMA(2, 2, 1, 2) | $\hat{F}$ | 59.2395 | 63.8743 | 13.2984 |
| TARMA(3, 2, 1, 2) | $\hat{F}$ | 52.1521 | 50.8229 | 6.1295 |

$r$'s, and the AR and MA order and threshold values are refined with AIC. After completing the five steps of the procedure, the details of estimates in model (3.1)–(3.4) are found in Table 5. From it, we can see that the estimates of coefficients and thresholds are all fine. Moreover, all AR and MA orders in each regime are almost consistent with the true values by AIC. From the results of Tables 3 and 4, we believe that our procedure can perform satisfactorily in general.

## 4.2 A Real Example

Now we analyze the exchange rate of Japanese Yen versus USA dollar. The monthly data from Jan. 1971 to Dec. 2000 are used and there are 360 observations. This data set was analyzed by Ling and Tong [23]. $P_t$ denotes the exchange rate at $t$th month. Let $x_t = 100[\log(P_t) - \log(P_{t-1})]$ and $y_t = x_t - \sum_{i=2}^{360} x_i/359$ for $t \geq 2$.

Firstly, the AR and MA order are chosen to be $p = q = 5$ by the AIC, and then based on $b = n/10 + min\{p, q\} = 41$, the proposed $F$ statistic are computed in Table 5, which confirms that the process is nonlinear and selects $y_{t-3}$ as the threshold variable. Therefore, we go to Step 3. Figure 5 gives the scatter plots of the $t$ ratios of the lag-1 AR and MA coefficient and residuals versus ordered $y_{t-3}$. From the plot, it is clear that the $t$-ratio is significant and changes its direction twice, suggesting that there are two nontrivial thresholds. After examining the scatter plots carefully, we decide on the location of thresholds: one near $y_{t-3} = 0$ and again near $y_{t-3} = 2$. An examination of the actual values suggests that the possible estimates of $r_1's$ and $r_2's$, are $\{0.17925, 0.1809, 0.1904, 0.2020, 0.2195, 0.2344, 0.2347, 0.2360, 0.2656\}$, $\{2.0095, 2.0523, 2.0863, 2.0909, 2.1144, 2.1656, 2.1934\}$ respectively. This step substantially simplifies the complexity in modeling the TARMA model because it effectively identifies the number and locations of the thresholds. Finally, we use AIC to refine the threshold values, AR and MA orders in Step 5. The final threshold values are $r_1 = 0.2269$ and $r_2 = 2.0886$. The AR and MA orders are 1 in the first and second regime, and the third regime has 2 AR orders as well as MA orders, and the numbers of observations are 71, 114 and 171. The TMA model for this data was given by Ling and Tong [23] that has AIC$= 691.61$, whereas the AIC of ours is 678.74, which is substantially smaller. Details of the model are given in (5.1). Figure 6 gives the $acf$ of the standardized residuals of the model, as well as the

**Table 4** The estimate of parameters in model (3.1)–(3.4)

| Parameter | $\phi_1^{(1)}$ $\phi_1^{(2)}$ $\phi_1^{(3)}$ | $\phi_2^{(1)}$ $\phi_2^{(2)}$ $\phi_2^{(3)}$ | $\theta_1^{(1)}$ $\theta_1^{(2)}$ $\theta_1^{(3)}$ | $\theta_2^{(1)}$ $\theta_2^{(2)}$ $\theta_2^{(3)}$ | $\sigma_1^2$ $\sigma_2^2$ $\sigma_3^2$ | $r_1$ $r_2$ | $d$ |
|---|---|---|---|---|---|---|---|
| TARMA(2, 1, 1, 1) | −0.5041 | – | −0.4613 | – | 3.2917 | 1.0606 | 1 |
| | 0.4047 | – | 0.6231 | – | 0.9755 | – | |
| TARMA(3, 1, 1, 1) | −0.7178 | −0.3380 | – | – | 1.0364 | 0.9789 | 1 |
| | 0.1974 | 0.2035 | – | – | 8.5648 | 3.9452 | |
| | 0.8191 | 0.8182 | – | – | 4.7454 | | |
| TARMA(2, 2, 1, 2) | −0.5493 | 0.5134 | −0.5133 | – | 3.6981 | −0.0606 | 2 |
| | 0.3418 | −0.8358 | 0.8075 | – | 0.9577 | – | |
| TARMA(3, 2, 1, 2) | −0.5147 | 0.5131 | −0.4704 | – | 1.0147 | −3.0701 | 2 |
| | 0.3921 | −0.1510 | 0.0677 | – | 10.1832 | 2.0351 | |
| | 0.9215 | −0.5575 | 0.3169 | – | 5.4936 | | |

Table 5 The estimate of $d$ in modeling a real example

| $d$ | 1 | 2 | 3 | 4 | 5 |
|---|---|---|---|---|---|
| $\hat{F}$ | 2.3916 | 1.9004 | 3.0732 | 1.2764 | 1.3921 |

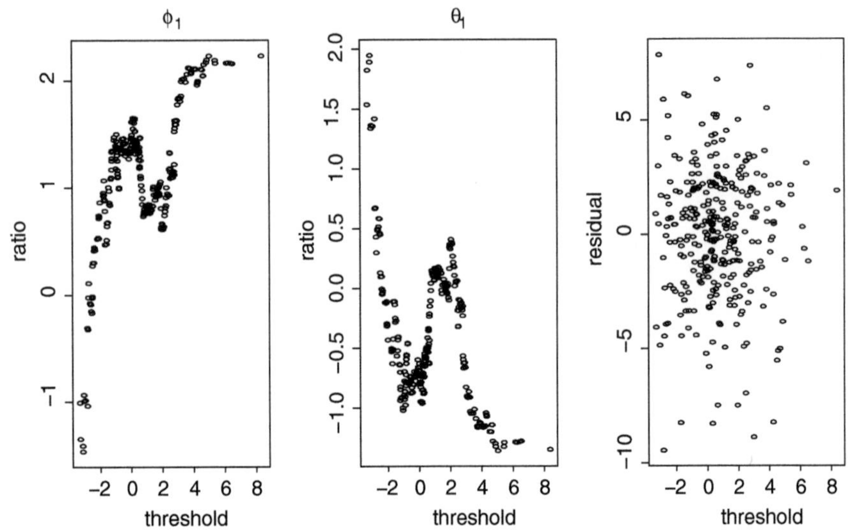


**Fig. 5** The $t$-ratio of $\phi_1^{(1)}$, $\theta_1^{(1)}$ and residuals versus threshold for model (3.5)

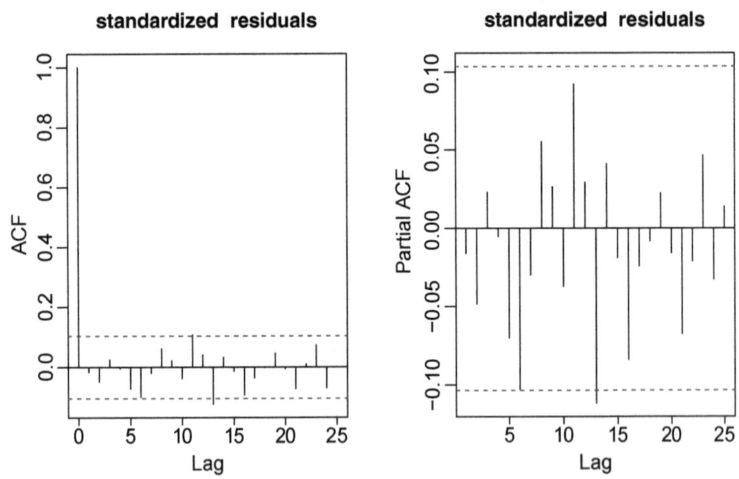


**Fig. 6** The *acf* and *pacf* of standardized residuals of model (3.5)

*pacf* of the standardized residuals. There is no rigorous diagnostic test for TARMA models yet. The use of *acf* and *pacf* are just crude methods. Both *acf* and *pacf* do not indicate any model inadequacy.

$$y_t = \begin{cases} 1.0587y_{t-1} + 0.0182y_{t-2} - 0.7018\varepsilon_{t-1}^{(1)} - 0.3695\varepsilon_{t-2}^{(2)} + \varepsilon_t^{(1)}, & y_{t-3} \leq 0.2269 \\ -0.0668y_{t-1} + 0.6869\varepsilon_{t-1}^{(2)} + \varepsilon_t^{(2)}, \quad 0.2269 < y_{t-3} \leq 2.0886 \\ 0.3804y_{t-1} - 0.3319\varepsilon_{t-1}^{(3)} + \varepsilon_t^{(3)}, \quad y_{t-3} > 2.0886 \\ \varepsilon_t^{(1)} i.i.d. \sim N(0, 5.86), \varepsilon_t^{(2)} i.i.d. \sim N(0, 6.25), \varepsilon_t^{(3)} i.i.d. \sim N(0, 8.44) \end{cases} \tag{3.5}$$

# 5 Conclusion

We propose a procedure for detection and modeling of TARMA models. The procedure is simple to implement and requires no pre-specification of the number of regimes of a TARMA model and its delay parameter. Using the proposed procedure, some simulation results and the application to a real example lend further support to our method. Firstly, through the MC experiments, we see the proposed $F$ test statistic gives good performance on detecting threshold nonlinearity. Secondly, our procedure obtain satisfactory results in the modeling of several simulated data sets. Finally, the application to a real data example confirms the above two aspects.

**Acknowledgements** We thank the two Referees for their criticisms and suggestions which have led to improvements of the paper. The research of Qiang Xia was supported by National Social Science Foundation of China (No:12CTJ019) and Ministry of Education in China Project of Humanities and Social Sciences (Project No.11YJCZH195).The research of Heung Wong was supported by a grant of the Research Committee of The Hong Kong Polytechnic University (Code: G-YBCV).

# Appendix: Proofs of Theorems

*Proof* (*Theorem* 1) For each regime of model (1.1), under the Assumption 4.1, we substitute the fitted residuals $\{\hat{\varepsilon}_{t-i}^{(j)}, i = 1, \ldots, q_j\}$ for $\{\varepsilon_{t-i}^{(j)}, i = 1, \ldots, q_j\}$ using ELSE. Then, in every regime, model (1.1) is the linear regression model, we can obtain least squares estimate $\hat{\Phi}^{(j)}$ of $j$th regime. Under the condition of Assumptions 1–4, they are fulfilled to the condition of Theorem 1 of Lai and Wei [17] and Theorem 2 of Liang et al. [21]. Therefore, for given $k$, $d$, and the threshold values $r_j$, the least squares estimates $\{\hat{\Phi}^{(j)}, j = 1, 2, \ldots, l\}$ converge to $\{\Phi^{(j)}, j = 1, 2, \ldots, l\}$ almost surely. □

*Proof* (*Theorem* 2) Consider the observation $\{y_t, t = 1, 2, \dots, n\}$ and define

$$X_t = (1, y_{t-1}, \dots, y_{t-p}, \hat{\varepsilon}_{t-1}, \dots, \hat{\varepsilon}_{t-q}),$$

with $\hat{\varepsilon}_{t-i}$'s being the residuals for model (1.1) fitted by the Hannan–Rissanen algorithm or ELSE.

Also define $\Phi$, $A_n$, $V_n$ by

$$\Phi' = (\phi_0, \phi_1, \dots, \phi_p, \theta_1, \dots, \theta_q),$$

$$A_n = (n-p-q)^{-1} \sum X_t' X_t, \quad V_n = (n-p-q)^{-1} \sum X_t' y_t,$$

Therefore, without loss of generality, the least squares estimate of $\Phi$ and the residuals are

$$\hat{\Phi} = A_n^{-1} V_n,$$

$$\hat{e}_t = y_t - \hat{y}_t = X_t \Phi + e_t - X_t \hat{\Phi} = X_t(\Phi - \hat{\Phi}) + e_t.$$

Also define $\Psi_n$ and $\hat{a}_t$ by

$$\Psi_n = (n-p-q)^{-1} \sum X_t' \hat{e}_t$$

$$= (n-p-q)^{-1} \sum X_t' X_t (\Phi - \hat{\Phi}) + (n-p-q)^{-1} \sum X_t' e_t$$

$$= A_n(\Phi - \hat{\Phi}) + (n-p-q)^{-1} \sum X_t' e_t,$$

$$\hat{a}_t = \hat{e}_t - X_t A_n^{-1} \Psi_n.$$

Hence,

$$\begin{aligned}
&\left(\sum \hat{e}_t^2 - \sum \hat{a}_t^2\right)/(n-p-q) \\
&= \left[\sum \hat{e}_t^2 - \sum (\hat{e}_t - X_t A_n^{-1} \Psi_n)^2\right]/(n-p-q) \\
&= \left[\sum \hat{e}_t' \hat{e}_t - \sum (\hat{e}_t - X_t A_n^{-1} \Psi_n)'(\hat{e}_t - X_t A_n^{-1} \Psi_n)\right]/(n-p-q) \\
&= \left[2\Psi_n' A_n^{-1} \sum X_t' \hat{e}_t - (n-p-q)\Psi_n' A_n^{-1} \Psi_n\right]/(n-p-q) \\
&= \Psi_n' A_n^{-1} \Psi_n \\
&= \left[A_n(\Phi - \hat{\Phi}) + (n-p-q)^{-1} \sum X_t' e_t\right]' A_n^{-1} \left[A_n(\Phi - \hat{\Phi}) + (n-p-q)^{-1} \sum X_t' e_t\right]
\end{aligned} \tag{3.6}$$

Because $X_t$ depends on $\{y_{t-k}; \hat{\varepsilon}_{t-l}, k = 1, \dots, p, l = 1, \dots, q\}$, which is independent of $e_t$. $(n-p-q)^{-\frac{1}{2}} \sum X_t' e_t$ forms a stationary and ergodic martingale difference process. Then $(n-p-q)^{-\frac{1}{2}} \sum X_t' e_t$ follows asymptotic normality according to a multivariate version of a martingale central limit theorem [2]. Theorem 2.1 shows $\Phi - \hat{\Phi} \to^{a.s.} o(1)$. $X_t$ is $p+q+1$ dimensional, therefore, (4) follows approximately an $F$ random variable with degrees of freedom $p+q+1$ and

$n - d - b - p - q - h$. As another point of view, it is obvious that $\frac{\sum \hat{a}_t^2}{(n-p-q)\sigma^2}$ is a chi-square random variable with degrees of freedom $n - d - b - h - p - q$. Also, the numerator and denominator of (4) have the same asymptotic variance $\sigma^2$. Then $(p + q + 1)\hat{F}(p, q, d)$ is asymptotically a chi-square random variable with degrees of freedom $p + q + 1$, which is a straightforward generalization of Corollary 3.1 of Keenan [16] or Tsay [32]. Theorem 2.2 is proved. □

## References


1. Akaike, H. (1974). A new look at statistical model identification. *IEEE Transactions on Automatic Control, 19*, 716–722.
2. Billingsley, P. (1961). The Lindeberg–Levy theorem for martingales. *Proceedings of the American Mathematical Society, 12*, 788–792.
3. Brockwell, P., Liu, J., & Tweedie, R. L. (1992). On the existence of stationary threshold autoregressive moving-average processes. *Journal of Time Series Analysis, 13*, 95–107.
4. Christopeit, N., & Helmes, K. (1980). Strong consistency of least squares estimators in linear regression models. *The Annals of Statistics, 4*, 778–788.
5. Chan, K. S. (1990). Testing for threshold autoregression. *Annals of Statistics, 18*, 1886–1894.
6. Chan, K. S., Petruccelli, J. D., Tong, H., & Woolford, S. W. (1985). A multiple-threshold AR(1) model. *Journal of Applied Probability, 22*, 267–279.
7. Chan, K. S., & Tong, H. (1986). On estimating thresholds in autoregressive models. *Journal of Time Series Analysis, 7*, 179–190.
8. Chen, W. S. C., So, K. P. M., & Liu, F. C. (2011). A review of threshold time series models in finance. *Statistics and Its Interface, 4*, 167–181.
9. Cryer, J. D., & Chan, K. S. (2008). *Time series analysis with applications in R. Springer texts in statistics* (2nd ed.). Berlin: Springer.
10. de Gooijer, J. G. (1998). On threshold moving-average models. *Journal of Time Series Analysis, 19*, 1–18.
11. Ertel, J. E., & Fowlkes, E. B. (1976). Some algorithms for linear spline and piecewise multiple linear regression. *Journal of the American Statistical Association, 71*, 640–648.
12. Goodwin, G. C., & Payne, R. L. (1977). *Dynamic system identification: Experiment design and data analysis*. New York, NY: Academic Press.
13. Haggan, V., Heravi, S. M., & Priestley, M. B. (1984). A study of the application of state-dependent models in nonlinear time series analysis. *Journal of Time Series Analysis, 5*, 69–102.
14. Hannan, E. J., & Rissanen, J. (1982). Recursive estimation of mixed autoregressive-moving average order. *Biometrika, 69*, 81–94.
15. Hansen, B. E. (2000). Sample splitting and threshold estimation. *Econometrica, 68*, 575–603.
16. Keenan, D. M. (1985). A Tukey nonadditivity-type test for time series nonlinearity. *Biometrika, 72*, 39–44.
17. Lai, T. L., & Wei, C. Z. (1982). Least squares estimates in stochastic regression models with applications to identification and control of dynamic systems. *The Annals of Statistics, 10*, 154–166.
18. Li, G., & Li, W. K. (2011). Testing a linear time series model against its threshold extension. *Biometrika, 98*, 243–250.
19. Li, D., Li, W. K., & Ling, S. (2011). On the least squares estimation of threshold autoregressive and moving-average models. *Statistics and Its Interface, 4*, 183–196.
20. Li, D., & Ling, S. (2012). On the least squares estimation of multiple-regime threshold autoregressive models. *Journal of Econometrics, 167*, 240–253.

21. Liang, R., Niu, C., Xia, Q., & Zhang, Z. (2015). Nonlinearity testing and modeling for threshold moving average models. *Journal of Applied Statistics*, *42*, 2614–2630.
22. Ling, S. (1999). On the probabilistic properties of a double threshold ARMA conditional heteroskedastic model. *Journal of Applied Probability*, *36*, 688–705.
23. Ling, S., & Tong, H. (2005). Testing a linear moving-average model against threshold moving-average models. *The Annals of Statistics*, *33*, 2529–2552.
24. Ling, S., Tong, H., & Li, D. (2007). The ergodicity and invertibility of threshold moving-average models. *Bernoulli*, *13*, 161–168.
25. Liu, J., & Susko, E. (1992). On strict stationarity and ergodicity of a nonlinear ARMA model. *Journal of Applied Probability*, *29*, 363–373.
26. Ljung, L., & Soderstrom, T. (1983). *Theory and practice of recursive identification*. Cambridge: MIT Press.
27. Priestley, M. B. (1980). State-dependent models: A general approach to nonlinear time series analysis. *Journal of Time Series Analysis*, *1*, 47–71.
28. Qian, L. (1998). On maximum likelihood estimators for a threshold autoregression. *Journal of Statistical Planning and Inference*, *75*, 21–46.
29. Tong, H. (1978). On a threshold model. In C. H. Chen (Ed.), *Pattern recognition and signal processing* (pp. 101–141). Amsterdam: Sijthoff and Noordhoff.
30. Tong, H., & Lim, K. S. (1980). Threshold autoregressions, limit cycles, and data. *Journal of the Royal Statistical Society B*, *42*, 245–292.
31. Tong, H. (1990). *Non-linear time series: A dynamical system approach*. Oxford: Oxford University Press.
32. Tsay, R. S. (1986). Nonlinearity tests for time series. *Biometrika*, *73*, 461–466.
33. Tsay, R. S. (1987). Conditional heteroscedastic time series models. *Journal of the American Statistical Association*, *82*, 590–604.
34. Tsay, R. S. (1989). Testing and modeling threshold autoregressive process. *Journal of the American Statistical Association*, *84*, 231–240.
35. Tsay, R. S. (2005). *Analysis of financial time series* (2nd ed.). London: Wiley.
36. Wong, C. S., & Li, W. K. (1997). Testing for threshold autoregression with conditional heteroscedasticity. *Biometrika*, *84*, 407–418.
37. Wong, C. S., & Li, W. K. (2000). Testing for double threshold autoregressive conditional heteroscedastic model. *Statistica Sinica*, *10*, 173–189.

# Improved Seasonal Mann–Kendall Tests for Trend Analysis in Water Resources Time Series

**Y. Zhang, P. Cabilio and K. Nadeem**

**Abstract** Nonparametric statistical procedures are commonly used in analyzing for trend in water resources time series (Chapter 23, Hipel and McLeod in Time series modelling of water resources and environmental systems. Elsevier, New York, 2005 [10]). One popular procedure is the seasonal Mann–Kendall tau test for detecting monotonic trend in seasonal time series data with serial dependence (Hirsch and Slack in Water Resour Res 20(6):727–732, 1984 [12]). However there is little rigorous discussion in the literature about its validity and alternatives. In this paper, the asymptotic normality of a seasonal Mann–Kendall test is determined for a large family of absolutely regular processes, a bootstrap sampling version of this test is proposed and its performance is studied through simulation. These simulations compare the performance of the traditional test, the bootstrapped version referred to above, as well as a bootstrapped version of Spearman's rho partial correlation. The simulation results indicate that both bootstrap tests perform comparably to the traditional test when the seasonal effect is deterministic, but the traditional test can fail to converge to the nominal levels when the seasonal effect is stochastic. Both bootstrapped tests perform similarly to each other in terms of accuracy and power.

**Keywords** Kendall correlation · Spearman partial correlation · Weakly dependent observations · Stationary ARMA process · Bootstrap · Hydrology

**Mathematical Subject Classification (2000)** Primary 62G10 · Secondary 62M10

Y. Zhang (✉) · P. Cabilio · K. Nadeem
Department of Mathematics and Statistics,
Acadia University, Wolfville, NS B4P 2R6, Canada
e-mail: ying.zhang@acadiau.ca

P. Cabilio
e-mail: paul.cabilio@acadiau.ca

K. Nadeem
e-mail: khurram.nadee@gmail.com



W.K. Li et al. (eds.), *Advances in Time Series Methods and Applications*,
Fields Institute Communications 78, DOI 10.1007/978-1-4939-6568-7_10

# 1 Introduction

Trend analysis is important in studying environmental time series data. Of these analyses, testing for the presence of a monotonic trend is of much interest in the study of water quality. "As a matter of fact, when only a small amount of data are available, the detection of the presence of trends is often all that one can realistically hope to achieve" [10]. Some of the characteristics commonly found in water resources time series data are non-normality, skewness, heavy tailed distributions, outliers, seasonal effects, missing values, censored data and serial dependence. In spite of the increasing computational power for some complex models, nonparametric statistical procedures remain popular since they are efficient and robust against non-normal underlying distributions. Chapter 23 of Hipel and McLeod [10] provides a comprehensive review of numerous publications concerned with testing for trend in hydrology time series data using rank based methods. As listed in Table 23.1.1 (p. 857, Chapter 23, Hipel and McLeod [10]), the tests for monotonic trend include (nonseasonal or univariate) Mann–Kendall tau, seasonal Mann–Kendall tau, Spearman's rho, Spearman's rho partial correlation, and aligned rank tests. When data are iid under the null hypothesis of no trend, the null distribution of such a statistic is usually free from the underlying distribution of the data and only depends on the sample size, however that no longer holds when the data are serially dependent. It has long been known that such rank based procedures are either liberal or conservative according to whether the data exhibit positive or negative autocorrelations. A number of remedial approaches exist in the literature. For the Mann–Kendall test, El-Shaarawi and Niculescu [8] derive expressions for the variance of the Mann–Kendall statistic in the case of $MA(1)$ and $MA(2)$, and then use the resulting exact variances to conduct the test using a normal approximation to the null distribution of the Mann–Kendall statistic. Yue et al. [27] consider various strategies for implementing the Mann–Kendall test for trend in the case that the serial dependency is known, such as an $AR(1)$ process. Under the assumption of a weakly dependent series, Cabilio et al. [3] study the asymptotic distribution properties of the Mann–Kendall trend test and of its bootstrap counterpart, and propose a bootstrap resampling test.

In this paper, we focus on seasonal trend tests. Assuming iid data, Hirsch et al. [11] apply a Jonckheere type statistic and introduce a seasonal Mann–Kendall trend test procedure for testing for a monotonic trend in monthly water quality data. Furthermore, Hirsch and Slack [12] improve on this seasonal Mann–Kendall tau trend test based on results by Dietz and Killeen [7] that considers serial correlation among different seasons (such as months) but assumes independence over longer time periods (such as years). We will explore the possibilities of further improvement of the seasonal Mann–Kendall tau trend procedures so as to relax this independence assumption. In Sect. 2, we will describe the models used for our seasonal trend tests, and then discuss the limiting behaviour of the test statistic as well as approximations to its null distribution. Simulation comparisons of the finite sample distribution behaviour and its bootstrap counterparts are detailed in Sect. 3. In Sect. 4 we illustrate the procedure for testing for trend in the average monthly water discharge at the Athabasca River downstream of Fort McMurray, followed by a discussion Sect. 5.

## 2 Null Distribution and Its Approximation

Consider the individual season as a block. The time series model may be written as

$$X_{i,j} = \tau(i, j) + e_{i,j}, \quad i = 1, \ldots, c; \; j = 1, \ldots, n, \tag{1}$$

where $n$ is the number of time points repeatedly measured within the $i$th season and $c$ is the number of seasons. Here $\tau(i, j)$ represents the overall deterministic trend that is due to seasonal or nonseasonal time effects, and $\{e_{i,j}\}$ are random noises. We assume that the random noise $\{e_{i,j}\}$ results from a zero mean/median weakly dependent stationary process. This is a realistic assumption in that the error terms exhibit autocorrelation which is strongest for observations contiguous in time and which weaken progressively with increasing lag times. The deterministic trend component in (1) may be simplified as

$$\tau(i, j) = s_i + f_j \tag{2}$$

where $f_j$ represents the time effect after controlling the periodically seasonal effects $s_i$. We further assume that the remaining season effect is stochastically stationary as a part of the random noise $\{e_{i,j}\}$. Under model (1), the seasonal effect could be either deterministic, stochastic, or both. For monthly data, $s_i$ may represent the monthly effect while $f_j$ would be the effect over years. To test for the presence of an increasing trend over time, the null hypothesis is

$$H_0 : f_j = constant \tag{3}$$

that is, without loss of generality, model (1) may be written as

$$X_{i,j} = s_i + e_{i,j} \tag{4}$$

and the alternative hypothesis is

$$H_1 : f_1 \leq f_2 \leq \cdots \leq f_n \tag{5}$$

with at least one inequality in the alternative strict.

This model mirrors the test of Jonckheere [13] in a randomized design for testing for an ordered alternative, so that Jonckheere's statistic is a natural choice as the test statistic. Denote the unstandardized Kendall's tau correlation between a season block's repeated responses and the alternative ordering as

$$\mathcal{A}_{\mathcal{K}}(i) = \sum_{l<k}^{n} sgn(R_i(k) - R_i(l)) \tag{6}$$

where data are ranked within the $i$th season block over time, $R_i(j)$ is the rank of the $j$th time data value in the $i$th season, and $sgn(R_i(k) - R_i(l))$ is either 1 or $-1$, depending on whether $R_i(k) > R_i(l)$ or $R_i(k) < R_i(l)$. The test statistic may be written as

$$J = \sum_{i=1}^{c} \mathcal{A}_{\mathcal{K}}(i). \tag{7}$$

Under $H_0$ and independent errors, for moderate sample sizes of $n$ and $c$, the exact distribution of $J$ can be readily calculated, and its large sample approximation is a normal distribution with zero mean and variance $cn(n-1)(2n+5)/18$. The standardized version of the Jonckheere statistic is used in this paper, that is,

$$J = \sum_{i=1}^{c} \mathcal{T}_{\mathcal{K}}(i) \tag{8}$$

where

$$\mathcal{T}_{\mathcal{K}}(i) = \binom{n}{2}^{-1} \mathcal{A}_{\mathcal{K}}(i), \tag{9}$$

is known as the Mann–Kendall statistic for testing for trend in a series of observations. Correspondingly the large sample approximation of the standardized version of Jonckheerre is a normal distribution with zero mean and variance $(4n+10)c/(9(n^2 - n))$ under $H_0$ and independent errors. With dependent errors under $H_0$, Lemma 1 in Zhang and Cabilio [28] shows that $J$ has asymptotically a normal distribution when data in each season follow a stationary ARMA process and the season blocks are mutually independent. A similar normality in the following more general Lemma 1 holds.

**Lemma 1** *In model (4), under the null hypothesis $H_0$, let $\{X_{i,j}\}$ form a c-dimensional strictly stationary sequences of stochastic vectors, $\{\mathbf{X}_j\}$ $(j = 1, \ldots, n)$, that is absolutely regular with a common absolutely continuous distribution function F satisfying condition i) of Theorem 1 in Yoshihara [26]. Then*

$$Var\,(J) = 4\sigma^2(n)^{-1} + O\left(n^{-2}\right) \tag{10}$$

*where*

$$\sigma^2 = \left[\sigma_1^2 + 2\sum_{s=1}^{\infty} \sigma_{1,s}\right] \tag{11}$$

*and* $\sigma_1^2 = Var(h_1(\mathbf{X}_1))$, $\sigma_{1,s} = Cov\,(h_1(\mathbf{X}_1), h_1(\mathbf{X}_{1+s}))$ *where*

$$h_1(\mathbf{x}_1) = \sum_{i=1}^{c} (1 - 2P(X_{i,1} < x_{i,1}))$$

*and if* $\sigma^2 > 0$, *then for* $n \to \infty$ *and fixed* $c$

$$\frac{\sqrt{n}}{2\sigma} J \xrightarrow{\mathcal{D}} N(0, 1). \tag{12}$$

An introduction to the absolutely regular process and the conditions of Yoshihara [26], together with a brief proof of Lemma 1 are provided in the "Appendix".

The exact null distribution of $J$ is unknown without knowledge of the explicit type of process underlying the correlation structure of $\{e_{i,j}\}$. Lemma 1 provides some justification for a possible approximation to the null distribution of the $J$ statistic, with one approach making use of the bootstrap sampling distribution. We consider two of the better known bootstrap methods, block and sieve. Politis [20] and Bühlmann [2] promote the block bootstrap for general stationary data generating processes including nonlinear models, while in the case of the linear time series model the expectation is that sieve bootstrap is superior (c.f. [2, 4].) As indicated by Kreiss et al. [14], validity of the different bootstrap procedures depends on the probabilistic structure of the underlying stochastic process and particular statistic considered. Cabilio et al. [3] discuss the validity of block and sieve bootstrap procedures to approximate the null distribution of the Mann–Kendall tau statistic, following the results in Dehling and Wendler [6] and Kreiss et al. [14]. Since $J$ is a sum of Mann–Kendall statistics, similar arguments may be applied to show the validity of these bootstrap procedures for approximating the null distribution of the Jonckheere statistic. The implementation would be equivalent to that of the bootstrap Mann–Kendall tau by individual seasons. Block or sieve bootstrap Jonckheere statistic samples are generated as follows.

Given time series samples, $X_{1,1}, \ldots, X_{1,c}; \ldots; X_{n,1}, \ldots, X_{n,c}$, we first decompose both the seasonal and time trends to obtain residuals,

$$\hat{e}_{i,j} = X_{i,j} - \hat{s}_i - \hat{f}_j \tag{13}$$

where $\hat{s}_i$ and $\hat{f}_j$ are consistent estimators of $s_i$ and $f_j$. The residuals are then ordered into a univariate time series sample by the natural time order denoted as $Y_1, Y_2, \ldots, Y_N$ where $Y_t = \hat{e}_{i,j}$ and $t = c(j-1) + i$ and $N = nc$. Moving block bootstrap (*MBB*) resampling is conducted on this univariate sample to reflect the dependence in both the short and/or the long term.

The *MBB* sample is generated as follows. For a given block size of length $l$, a total of $b = \lceil N/l \rceil$ blocks are randomly sampled so that each block is formed by $l$ consecutive observations with blocks starting with $Y_t, t = 1, \ldots, N - l + 1$, where $Y_t$ is selected at random from the sample $Y_1, Y_2, \ldots, Y_N$. The selected blocks are then combined to give the bootstrap sample $Y_1^*, Y_2^*, \ldots, Y_{bl}^*$, that is, $\{\hat{e}_{i,j}^*\}$.

To obtain the bootstrapped $J^*$, the bootstrap sample will be broken by season as $\hat{e}_{1,1}^*, \ldots, \hat{e}_{1,n}^*; \ldots; \hat{e}_{c,1}^*, \ldots \hat{e}_{c,n}^*$. The $i$th bootstrapped Mann–Kendall statistic is calculated as

$$\mathcal{T}_K(i)^* = \binom{n}{2}^{-1} \sum_{1 \le l < k \le n} sgn\left(\hat{e}_{i,k}^* - \hat{e}_{i,l}^*\right) \tag{14}$$

A bootstrapped $J^*$ may be written as

$$J^* = \sum_{i=1}^{c} \mathcal{T}_{\mathcal{K}}(i)^*. \tag{15}$$

The autoregressive sieve (*AR-sieve*) bootstrap approximates the data generating process by an autoregressive model with order $p = p(N)$, where $p(N) \to \infty$, $p(N) = o(N)$ as sample size $N \to \infty$. For given data, once the order is approximated by $\hat{p}$, the parameters of the AR($\hat{p}$) are estimated, and the estimated AR($\hat{p}$) process is used to generate a bootstrap sample by resampling from the AR residual process. The procedure may be detailed in terms of the steps in Kreiss et al. [14] as follows:

Step 1: select an order $\hat{p}$ by the Akaike Information Criterion (AIC) and fit a $\hat{p}$th order autoregressive model to $Y_1, Y_2, \ldots, Y_N$ obtaining either the Yule-Walker or Burg autoregressive parameter estimators, $(\hat{\phi}_1, \hat{\phi}_2, \ldots, \hat{\phi}_p)$.

Lyubchich et al. [15] claim that there is no significant difference between the Yule-Walker and Burg algorithms in their simulation results when they apply the sieve bootstrap to a F-ratio type statistic on detecting non-monotonic trends in environmental time series. However, based on superior finite sample properties of Burg (c.f. [25, 29]), the Burg algorithm is the one recommended for this step [21].

Step 2: compute the residuals as $\tilde{\epsilon}_i^0 = Y_i - \sum_{j=1}^{\hat{p}} \hat{\phi}_j Y_{i-j}$ where $i = \hat{p}+1, \hat{p}+2, \ldots, N$, and center the residuals as $\tilde{\epsilon}_i = \tilde{\epsilon}_i^0 - \bar{\epsilon}$ where $\bar{\epsilon} = (N-\hat{p})^{-1} \sum_{i=\hat{p}+1}^{N} \tilde{\epsilon}_i^0$. Denote the empirical distribution function of the centered residuals by $\hat{F}_N$.

Step 3: let $(Y_1^*, Y_2^*, \ldots, Y_N^*)$ be a set of observations from the time series generated from $Y_i^* = \sum_{j=1}^{\hat{p}} \hat{\phi}_j Y_{i-j}^* + \epsilon_i^*$ where $\epsilon_i^*$'s are *iid* variables having the identical distribution $\hat{F}_N$.

Step 4: Compute $J^*$ based on the sieve pseudo-time series $Y_1^*, Y_2^*, \ldots, Y_N^*$ as described above for the block bootstrap sample in Eqs. (14) and (15).

## 3 Simulation Results

In order to provide examples of the performance of the sampling distribution methods described in Sect. 2, this section details simulation results in the case of two seasonal models with monthly data $c = 12$. Both models considered for the trend and error processes were taken to be linear, so that the time trend decomposition is based on median slope estimation [24] and the seasonal decomposition is based on the sample median/mean across years in each season.

McLeod et al. [17] claim that the Spearman's rho partial rank correlation test has advantages over the original seasonal Mann–Kendall test [11] with the iid assumption. The simulation results in Yue et al. [27] show that for univariate time series the Mann–Kendall and Spearman's rho tests perform similarly in detecting trend to the point of being indistinguishable in practice. We compare the performance of two tests based on the sieve bootstrap sampling distribution, the sieve sampling Mann–Kendall tau (*SSMK*) and the sieve sampling Spearman's rho partial rank correlation (*SSP*). In addition we include a comparison with the seasonal Mann–Kendall test (*SMK*) [12] which is based on the asymptotic conditional covariance matrix [7]. The calculation of Spearman's rho partial rank correlation for the trend test is described in [17].

We consider two types of seasonality: deterministic and stochastic. For deterministic seasonality, the data were generated by

$$X_{i,j} = sin\left(\frac{\pi}{3} + \frac{\pi}{6}i\right) + \beta(12 * (j-1) + i) + e_{i,j} \tag{16}$$

where $i = 1, \ldots, 12$, $j = 1, \ldots, n$, and $e_{i,j}$ were generated by a stationary autoregressive model AR(1)

$$\mu_t = \phi\mu_{t-1} + \epsilon_t \tag{17}$$

where $\mu_{12*(j-1)+i} = e_{i,j}$, $t \geq 2$, and the innovation term $\epsilon_t$ is *iid* Student-$t(4)$. The seasonal component in model (16) was selected from Hirsch, Slack and Smith [11]. For the stochastic seasonality, the data were generated by

$$X_{i,j} = \beta(12 * (j-1) + i) + e_{i,j} \tag{18}$$

where $e_{i,j}$ were generated by a multiplicative seasonal ARMA model,

$$\mu_t = \Phi\mu_{t-12} + \epsilon_t - \theta\epsilon_{t-1} \tag{19}$$

or

$$\mu_t = \epsilon_t - \theta\epsilon_{t-1} - \Theta\epsilon_{t-12} + \theta\Theta\epsilon_{t-13} \tag{20}$$

where $\mu_{12*(j-1)+i} = e_{i,j}$, $t \geq 14$, and $\epsilon_t$ is *iid* $N(0, 1)$. These seasonal component models (19) and (20) seem to occur frequently in practice [5].

Using each of the three tests, Tables 1 and 2 provide the empirical significance levels for the deterministically seasonal model in (16) and for the stochastically seasonal model in (18) respectively, corresponding to the nominal sizes 0.10, 0.05, 0.01 and $n = 10, 20, 30$. Specifically, Table 1 is for the cases $\phi = -0.2, -0.5, 0, 0.2, 0.5$ with $t(4)$ innovation terms in (17), and Table 2 is for $\Phi = 0.5$ and $\theta = 0.4$ in (19), and $\Theta = -0.8$ and $\theta = 0.5$ in (20). In Table 3 we explore the empirical power ($\beta = 0.01, 0.05$) and again the empirical significance level ($\beta = 0$) of *SSMK* and *SSP* for the deterministic seasonal model in (16) with $t(4)$ distributed innovation terms and $n = 10, 15$.

All empirical levels and power of the tests were obtained by 2000 realizations. The simulation calculations were conducted using *R* [22]. The univariate Mann–Kendall statistics were calculated with *R* library *Kendall* [18]. The tests of *SMK* were conducted using *R* library *rkt* [16].

**Table 1** Empirical levels of 10, 5 and 1 % tests for the deterministic seasonal model in (16) with $t(4)$ distributed innovation terms

| $\phi$ | *Size* | n = 10 | | | n = 20 | | | n = 30 | | |
|---|---|---|---|---|---|---|---|---|---|---|
| | | *SSMK* | *SSP* | *SMK* | *SSMK* | *SSP* | *SMK* | *SSMK* | *SSP* | *SMK* |
| −0.5 | 0.10 | 0.103 | 0.112 | 0.091 | 0.095 | 0.101 | 0.101 | 0.099 | 0.115 | 0.099 |
| | 0.05 | 0.057 | 0.062 | 0.039 | 0.057 | 0.058 | 0.047 | 0.050 | 0.057 | 0.041 |
| | 0.01 | 0.015 | 0.013 | 0.003 | 0.014 | 0.019 | 0.006 | 0.012 | 0.017 | 0.005 |
| −0.2 | 0.10 | 0.099 | 0.094 | 0.095 | 0.100 | 0.096 | 0.092 | 0.096 | 0.097 | 0.092 |
| | 0.05 | 0.056 | 0.052 | 0.036 | 0.047 | 0.051 | 0.037 | 0.049 | 0.054 | 0.043 |
| | 0.01 | 0.013 | 0.015 | 0.003 | 0.013 | 0.013 | 0.006 | 0.009 | 0.014 | 0.008 |
| 0 | 0.10 | 0.116 | 0.114 | 0.109 | 0.093 | 0.097 | 0.100 | 0.097 | 0.098 | 0.101 |
| | 0.05 | 0.062 | 0.058 | 0.043 | 0.061 | 0.057 | 0.049 | 0.043 | 0.043 | 0.043 |
| | 0.01 | 0.015 | 0.020 | 0.001 | 0.017 | 0.017 | 0.007 | 0.012 | 0.010 | 0.007 |
| 0.2 | 0.10 | 0.114 | 0.118 | 0.099 | 0.104 | 0.102 | 0.097 | 0.099 | 0.098 | 0.102 |
| | 0.05 | 0.061 | 0.061 | 0.040 | 0.053 | 0.056 | 0.049 | 0.054 | 0.054 | 0.049 |
| | 0.01 | 0.022 | 0.025 | 0.003 | 0.015 | 0.014 | 0.006 | 0.016 | 0.016 | 0.007 |
| 0.5 | 0.10 | 0.122 | 0.126 | 0.118 | 0.120 | 0.112 | 0.119 | 0.124 | 0.121 | 0.136 |
| | 0.05 | 0.069 | 0.066 | 0.055 | 0.063 | 0.060 | 0.062 | 0.061 | 0.064 | 0.071 |
| | 0.01 | 0.018 | 0.021 | 0.002 | 0.019 | 0.018 | 0.010 | 0.013 | 0.015 | 0.014 |

**Table 2** Empirical levels of 10, 5 and 1 % tests for the stochastic seasonal model in (18). $s1$ is $\Phi = 0.5$ and $\theta = 0.4$ in (19) and $s2$ is $\Theta = -0.8$ and $\theta = 0.5$ in (20)

| $\phi$ | *Size* | n = 10 | | | n = 20 | | | n = 30 | | |
|---|---|---|---|---|---|---|---|---|---|---|
| | | *SSMK* | *SSP* | *SMK* | *SSMK* | *SSP* | *SMK* | *SSMK* | *SSP* | *SMK* |
| $s1$ | 0.10 | 0.147 | 0.170 | 0.217 | 0.135 | 0.157 | 0.257 | 0.139 | 0.151 | 0.279 |
| | 0.05 | 0.091 | 0.110 | 0.121 | 0.091 | 0.098 | 0.157 | 0.088 | 0.093 | 0.189 |
| | 0.01 | 0.041 | 0.056 | 0.020 | 0.030 | 0.037 | 0.046 | 0.028 | 0.033 | 0.066 |
| $s2$ | 0.10 | 0.113 | 0.124 | 0.167 | 0.089 | 0.094 | 0.174 | 0.098 | 0.093 | 0.188 |
| | 0.05 | 0.071 | 0.076 | 0.088 | 0.057 | 0.052 | 0.114 | 0.048 | 0.052 | 0.105 |
| | 0.01 | 0.024 | 0.034 | 0.009 | 0.012 | 0.016 | 0.028 | 0.014 | 0.014 | 0.031 |

It is seen from Table 1 that *SMK*, *SSMK* and *SSP* achieve empirical significance levels that, on the whole, are close to the nominal values, particularly for values of $\phi = -0.5, -0.2, 0, 0.2$. For negative correlations the empirical levels are generally accurate for all sample sizes. Notably for $\phi = 0, 0.2$, *SMK* is a little more accurate at $n = 10$, but at $n = 20, 30$ there is little difference between the three methods. All methods have less accurate empirical levels at $\phi = 0.5$, but curiously for $n = 30$ the empirical level of *SMK* decreases in accuracy. Turning to Table 2 for the first model it is seen that all three methods have inflated empirical levels and this is particularly true for *SMK*, which is extremely liberal at the 0.10 level in particular, and at the 0.05 level for $n = 20, 30$. In fact *SMK* becomes less accurate as $n$ increases at all

**Table 3** Empirical significance levels ($\beta = 0$) and powers ($\beta = 0.01$ or $\beta = 0.05$) of *SSMK* and *SSP* for the deterministic seasonal model in (16) with $t(4)$ distributed innovation terms

| | *Size* (%) | *SSMK* | | | *SSP* | | |
|---|---|---|---|---|---|---|---|
| | | $\beta = 0$ | $\beta = 0.01$ | $\beta = 0.05$ | $\beta = 0$ | $\beta = 0.01$ | $\beta = 0.05$ |
| $n = 10$ | | | | | | | |
| 0 | 10 | 0.111 | 0.948 | 1 | 0.110 | 0.952 | 1 |
| | 5 | 0.062 | 0.901 | 1 | 0.062 | 0.907 | 1 |
| | 1 | 0.016 | 0.717 | 1 | 0.015 | 0.741 | 1 |
| 0.2 | 10 | 0.098 | 0.853 | 1 | 0.100 | 0.855 | 1 |
| | 5 | 0.055 | 0.758 | 1 | 0.050 | 0.769 | 1 |
| | 1 | 0.019 | 0.542 | 1 | 0.021 | 0.552 | 1 |
| 0.4 | 10 | 0.116 | 0.719 | 1 | 0.115 | 0.716 | 1 |
| | 5 | 0.068 | 0.585 | 1 | 0.072 | 0.586 | 1 |
| | 1 | 0.019 | 0.341 | 1 | 0.020 | 0.330 | 1 |
| 0.5 | 10 | 0.124 | 0.588 | 1 | 0.129 | 0.583 | 1 |
| | 5 | 0.076 | 0.452 | 1 | 0.078 | 0.457 | 1 |
| | 1 | 0.027 | 0.241 | 1 | 0.029 | 0.245 | 1 |
| $n = 15$ | | | | | | | |
| 0 | 10 | 0.112 | 1 | 1 | 0.111 | 1 | 1 |
| | 5 | 0.056 | 1 | 1 | 0.058 | 1 | 1 |
| | 1 | 0.012 | 0.998 | 1 | 0.013 | 0.998 | 1 |
| 0.2 | 10 | 0.110 | 0.998 | 1 | 0.109 | 0.999 | 1 |
| | 5 | 0.060 | 0.995 | 1 | 0.057 | 0.994 | 1 |
| | 1 | 0.018 | 0.975 | 1 | 0.014 | 0.969 | 1 |
| 0.4 | 10 | 0.106 | 0.972 | 1 | 0.106 | 0.970 | 1 |
| | 5 | 0.056 | 0.932 | 1 | 0.062 | 0.928 | 1 |
| | 1 | 0.018 | 0.823 | 1 | 0.018 | 0.817 | 1 |
| 0.5 | 10 | 0.109 | 0.911 | 1 | 0.107 | 0.908 | 1 |
| | 5 | 0.068 | 0.837 | 1 | 0.065 | 0.832 | 1 |
| | 1 | 0.020 | 0.641 | 1 | 0.023 | 0.636 | 1 |

levels. Overall, the empirical level of *SSMK* is most often the closest to the nominal value. For the second model in Table 2, *SMK* is again the least accurate, becoming worse with increasing sample size. On the other hand, *SSMK* and *SSP* have similar good performances, with accuracy increasing with sample size. Finally, the power simulations in Table 3 indicate that there is little difference between *SSMK* and *SSP* for this model. Both achieve high values of power even for small values of $\beta$, even for $\phi$ large, and these power values increase dramatically with a modest increase in sample size.

## 4 Application

To illustrate the procedures discussed above, we apply them for testing trend in the average monthly water discharges ($m^3$/sec) at the Athabasca River downstream of Fort McMurray, Alberta from January, 1958 to December, 2008. The data were downloaded from the Water Survey of Canada archived database [9]. Schindler and Donahue [23] claim that climate warming and human modifications to catchments have significantly reduced the flows of major rivers of the Canadian western prairie provinces during the summer months (May–August). For the Athabasca River, using descriptive statistics and simple regression tools, they analyzed the annual mean records of average summer month water discharges during the similar time period at the same station. We are interested in confirming/determining whether there has been a significant downward trend using the all seasons monthly water discharges. Figure 1 plots the logarithm scaled data with a *lowess* smoothing line. The plot shows a weak downward trend mixed with a strong seasonality. Figure 2 is a plot of the sample autocorrelation function (ACF) of first (lag-1) differences of the log-scaled data, which indicates the presence of seasonality, thus providing a rationale for seasonal adjustment when testing for the trend over time. Further, Fig. 3 plots the sample ACF of first and seasonal (lag-12) differences of log-scaled water discharge levels. This plot suggests that the differenced data are still serially dependent, indicating the presence of stochastic seasonality.

We applied the original seasonal Mann–Kendall test (SMK) for testing for a decreasing trend and obtained a $p$ value of 0.006. The simulation results in the previous section show that this test would be very liberal when stochastic seasonality appears in the data. We further applied our seasonal Mann–Kendall as well as the Spearman partial correlation tests (SSMK and SSP) and obtained $p$ values 0.049 and 0.051 respectively, leading to a conclusion that there is still significant evidence for the presence of declining trend, but at a more reliable level.

## 5 Discussion

The seasonal Mann–Kendall tau test by Hirsch and Slack [12] is commonly used in detecting statistically significant trends in environmental time series analysis. Its popularity is due to its high detection power and robustness in terms of the underlying distributions and serial autocorrelations, as shown in a number of studies. Previous simulation studies that have been conducted to test its robustness against serial autocorrelation have assumed simple linear time series models with deterministic seasonal component. There is little discussion in the literature on what its range of validity is in terms of the data dependence structure. A different rank correlation based test, Spearman's rho partial correlation test, has not received the same attention as *SMK* in spite of the results in [17]. In this paper, following the ideas in Cabilio et al. [3], we focus on introducing a bootstrap sampling test, *SSMK*, and on comparing

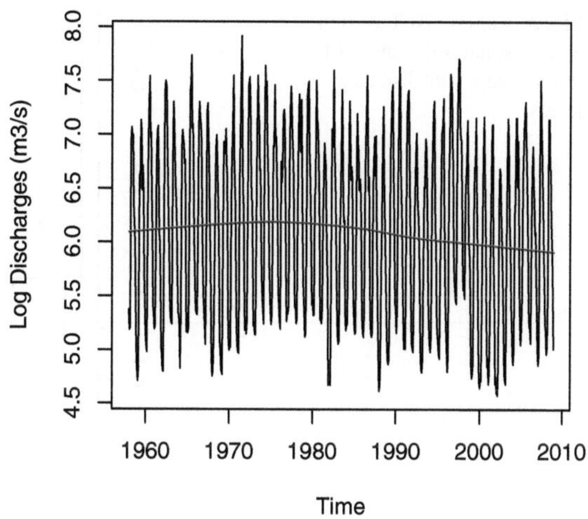


**Fig. 1** Monthly mean water discharges (logarithm), Athabasca River, 1958–2008

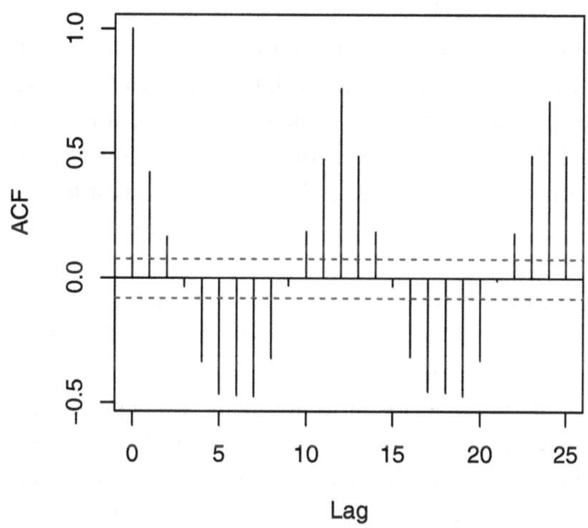


**Fig. 2** Sample ACF of first differences of log-scaled water discharge levels

its performance to both the *SMK*, and the bootstrap version of Spearman's rho partial correlation (*SSP*). Our simulation results show that both *SSMK* and *SSP* can improve the accuracy of the test for trend when stochastic seasonality is present in the data. As indicated by our simulations, the bootstrap sampling tests *SSMK* and *SSP* achieved empirical significance levels comparable to those produced by *SMK* for data generated from a simple AR(1) model with deterministic seasonal component. With regards to power of *SMK*, preliminary simulations that we have conducted for the model and parameters in Table 3, indicate that the tests *SSMK* and *SSP* appear to be generally at least as powerful as *SMK*, and for certain significance levels and values of $\phi$, more powerful. Hirsch and Slack [12] show that *SMK* is valid with autocorrelation less than 0.6 using an ARMA(1,1) simulation model, and claim that *SMK* could be accurate for monthly data since the monthly serial autocorrelation had been shown

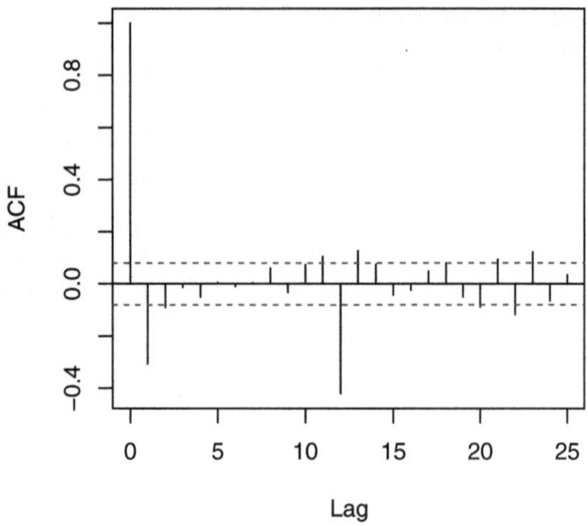


**Fig. 3** Sample ACF of first and seasonal differences of log-scaled water discharge levels

to be weak in many existing water resources time series data. However, this claim will not hold when the serial autocorrelation is seasonal since their method assumes independence within season. In fact, as noted in the previous section, the inaccuracy of the empirical level for *SMK* does not improve as sample size is increased in the presence of stochastic seasonality. Interestingly our simulation results indicate that the two bootstrap sampling tests *SSMK* and *SSP* perform similarly in terms of their accuracy and power. Our study is encouraging but not without limitations. In order to implement our sieve bootstrap approach, in our simulation study we assumed linear time trends and linearly serial autocorrelations. More extensive simulations will be needed to give a more complete picture of the robustness of the bootstrap approach in terms of trend forms and dependence structures. For this reason, a block bootstrap implementation procedure is provided in Sect. 2. The evidence presented here of the possibility of significant improvements, will hopefully encourage additional research aimed at providing further enhancement of nonparametric rank-based correlation methods for testing for trend in seasonal time series data.

**Acknowledgments** This research was supported in part by discovery grants from the Natural Sciences and Engineering Council of Canada and Acadia University Article 25.55. The authors wish to thank the Referees for their constructive comments and suggestions, and the program committee for organizing the event, *Time Series Methods and Applications: the A. Ian McLeod Festschrift*.

## Appendix

Weakly dependent stationary processes, loosely speaking, are characterized by the fact that the dependence between observations which are very far apart becomes very small, so that events which are functions of such far-flung observations behave almost as if they were independent. Background material and more precise definitions may be found in Bradley [1]. There are many ways of defining such weak dependence. For our purposes we consider the following coefficient. Let $\{A_i\}$ be a stationary vector sequence defined on a probability space $(\Omega, \mathcal{F}, P)$, and for $m < n$, let $\mathcal{F}_m^n$ be the $\sigma-$algebra generated by $A_m, \ldots, A_n$. For $m \geq 1$, define

$$\beta(n) = E\left\{\sup_{B \in \mathcal{F}_{n+m}^{\infty}} \left|P\left(B|\mathcal{F}_0^m\right) - P(B)\right|\right\}. \tag{21}$$

The process is said to be *absolutely regular (or $\beta$ mixing)* if $\beta(n) \to 0$. The rate of convergence to 0 of such a coefficient will determine the limiting behaviour of $U$-statistics based on such sequences. A *CLT* for $U$-statistics based on absolutely regular vector processes, derived in Yoshihara [26]. A general $U$-statistic with a degree $k$ kernel $h(A_{i_1}, \ldots, A_{i_k})$ may be defined as

$$U_n = \binom{n}{k}^{-1} \sum_{(n,k)} h\left(A_{i_1}, \ldots, A_{i_k}\right)$$

where $\{A_{i_l}\}$ is a sequence of $n$ random vectors from a common distribution $F$ and the sum extends over all subsets $1 \leq i_1 < \cdots < i_k \leq n$ of $(1, \ldots, n)$.

The results in Yoshihara [26] when specialized to a $U$-statistic with a degree $k = 2$ kernel based on a stationary absolutely regular sequence $\{A_i\}$ with a common d.f. $F(A_i)$ show that if the following conditions are satisfied for some $\delta > \delta' > 0$, and for some $M > 0$:

(i) $\beta(n) = O\left(n^{-(2+\delta')/\delta'}\right)$
(ii) $E\left|h\left(A_i, A_j\right)\right|^{2+\delta} \leq M$
(iii) $\int |h(a_1, a_2)|^{2+\delta} dF(a_1) dF(a_2) < M$,

then $Var(U_n)$ has the form given by the right side of Eq. (11) with the projection function defined as

$$h_1(a_1) = E(h(a_1, A_2))$$

and $U_n$ has an asymptotic normal distribution similar to Eq. (12).

Mokkadem [19] establishes the form of weak dependence of stationary vector *ARMA* processes as well as the rate of convergence to 0. His Theorem 1 on p. 310 states that if the sequence of independent identically distributed errors $\{\epsilon_i\}$ are absolutely continuous, the stationary *ARMA* process is absolutely regular with

$$\beta(n) = O\left(r^n\right) \text{ for some } 0 < r < 1. \tag{22}$$

so that condition (i) is satisfied in this case.
Outline of Proof for Lemma 1:

The proof of Lemma 1 relies on showing that $J$ is a U-statistic satisfying the Yoshihara conditions described above. In fact, $J$ in Eq. (8) can be written as

$$\begin{aligned} J &= \sum_{i=1}^{c} \binom{n}{2}^{-1} \sum_{l<k}^{n} sgn(X_{i,k} - X_{i,l}) \\ &= \binom{n}{2}^{-1} \sum_{l<k}^{n} \sum_{i=1}^{c} sgn(X_{i,k} - X_{i,l}) \end{aligned}$$

Thus $J$ is a U-statistic with a kernel function,

$$h(\mathbf{x}_l, \mathbf{x}_k) = \sum_{i=1}^{c} sng(k-l)sgn(x_{i,k} - x_{i,l}).$$

and

$$\begin{aligned} E(J) &= \sum_{i=1}^{c} E(sgn(X_{i,2} - X_{i,1})) \\ &= 0. \end{aligned}$$

Since $|h| \le c$, so that conditions (ii) and (iii) are immediately satisfied for all $\delta' < \delta$. When the condition (i) is satisfied, by Yoshihara [26], we obtain Eqs. (11) and (12).

## References


1. Bradley, R. C. (2005). Basic properties of strong mixing conditions. A survey and some open questions. *Probability Surveys, 2*, 107–144.
2. Bühlmann, P. (2002). Bootstraps for time series. *Statistical Science, 17*, 52–72.
3. Cabilio, P., Zhang, Y., & Chen, X. (2013). Bootstrap rank test for trend in time series. *Environmentrics, 24*(8), 537–549.
4. Choi, E., & Hall, P. (2000). Bootstrap confidence regions computed from autoregressions of arbitrary order. *Journal Royal Statistical Society Series B, 62*, 461–477.
5. Cryer, J. D., & Chan, K. S. (2008). *Time series analysis: With applications in R* (2nd ed.). New York: Springer.
6. Dehling, H., & Wendler, M. (2010). Central limit theorem and the bootstrap for U-statistics of strongly mixing data. *Journal of Multivariate Analysis, 101*, 126–137.
7. Dietz, E. J., & Killeen, T. J. (1981). A nonparametric multivariate test for monotone trend with pharmaceutical applications. *Journal of the American Statistical Association, 76*(373), 169–174.
8. El-Shaarawi, A. H., & Niculescu, S. (1992). On Kendall's tau as a test of trend in time series data. *Environmetrics, 3*, 385–411.
9. Environment Canada. (2011). Water survey of Canada archived hydrometric data. http://wateroffice.ec.gc.ca/
10. Hipel, K. W., & McLeod, A. I. (2005). *Time series modelling of water resources and environmental systems*. New York: Elsevier.

11. Hirsch, R. M., Slack, J. R., & Smith, R. A. (1982). Techniques of trend analysis for monthly water quality data. *Water Resources Research*, *18*(1), 107–121.
12. Hirsch, R. M., & Slack, J. R. (1984). A Nonparametric trend test for seasonal data with serial dependence. *Water Resources Research*, *20*(6), 727–732.
13. Jonckheere, A. R. (1954). A test of significance for the relation between m rankings and k ranked categories. *British Journal of Statistical Psychology*, *7*, 93–100.
14. Kreiss, J. K., Paparoditis, E., & Politis, D. N. (2011). On the range of validity of the autoregressive sieve bootstrap. *The Annals of Statistics*, *39*, 2103–2130.
15. Lyubchich, V., Gel, Y. R., & El-Shaarawi, A. (2013). On detecting non-monotonic trends in environmental time series: A fusion of local regression and bootstrap. *Environmetrics*, *24*, 209–226.
16. Marchetto, A. (2015). *rkt: Mann–Kendall test, seasonal and regional Kendall tests*. R package version 1.4. http://CRAN.R-project.org/package=rkt
17. McLeod, A. I., Hipel, K. W., & Bodo, B. A. (1991). Trend analysis methodology for water quality time series. *Environmetrics*, *2*(2), 169–200.
18. McLeod, A. I. (2011). *Kendall: Kendall rank correlation and Mann–Kendall trend test*. R package version 2.2. http://CRAN.R-project.org/package=Kendall
19. Mokkadem, A. (1988). Mixing properties of ARMA processes. *Stochastic Processes and their Applications*, *29*, 309–315.
20. Politis, D. N. (2003). The impact of bootstrap methods on time series analysis. *Statistical Science*, *18*, 219–230.
21. Poskitt, D. S. (2008). Properties of the Sieve bootstrap for fractionally integrated and non-invertible processes. *Journal of Time Series Analysis*, *29*, 224–250.
22. R Core Team. (2014). *R: A language and environment for statistical computing*. R Foundation for Statistical Computing, Vienna, Austria. http://www.R-project.org/
23. Schindler, D. W., & Donahue, W. F. (2006). An impending water crisis in Canada's western prairie provinces. *Proceedings of the National Academy of Sciences of the United States of America*, *103*(19), 7210–7216.
24. Sen, P. K. (1968). Estimates of the regression coefficient based on Kendall's tau. *Journal of the American Statistical Association*, *63*, 1379–1389.
25. Tjøstheim, D., & Paulsen, J. (1983). Bias of some commonly-used time series estimators. *Biometrika*, *70*, 389–400.
26. Yoshihara, K. (1976). Limiting behavior of U-statistics for stationary absolutely regular processes. *Zeitschrift für Wahrscheinlichkeitstheorie und verwandte Gebiete*, *35*, 237–252.
27. Yue, S., Pilon, P., Phinney, B., & Cavadias, G. (2002). The influence of autocorrelation on the ability to detect trend in hydrological series. *Hydrological Processes*, *16*, 1807–1829.
28. Zhang, Y., & Cabilio, P. (2013). A generalized Jonckheere test against ordered alternatives for repeated measures in randomized blocks. *Statistics in Medicine*, *32*, 1635–1645.
29. Zhang, Y., & McLeod, A. I. (2006). Computer algebra derivation of the bias of linear estimators of autoregressive models. *Journal of Time Series Analysis*, *27*, 157–165.

# A Brief Derivation of the Asymptotic Distribution of Pearson's Statistic and an Accurate Approximation to Its Exact Distribution

**Serge B. Provost**

**Abstract** A brief and accessible derivation of the asymptotic distribution of Pearson's goodness-of-fit statistic is proposed. Additionally, a shifted gamma distribution is introduced as an accurate approximation to be utilized when the chi-squared distribution proves to be inadequate. It is also explained that the exact probability mass function of this test statistic can be readily determined from its moment-generating function via symbolic computations. Two illustrative numerical examples are included.

**Keywords** Pearson's statistic · Asymptotic distribution · Goodness-of-fit tests · Shifted gamma distribution

**AMS Mathematics Subject Classification (2010)** 62E20 · 60E10 · 62E15 · 62E17

## 1 Introduction

The chi-squared goodness-of-fit statistic was initially proposed by [11]. Its main applications consist in assessing the extent to which a categorical data set is distributed according to certain specified probabilities or a given distribution, see [9], testing for the homogeneity of two multinomial populations, see [1, 3], and determining whether two attributes are independently distributed, see [10, 14].

Consider an experiment having $r$ mutually exclusive and exhaustive outcomes denoted by $\mathscr{O}_1, \ldots, \mathscr{O}_r$, whose respective probabilities of occurrence are hypothesized to be $p_1, \ldots, p_r$, so that $\sum_{j=1}^{r} p_j = 1$. Assuming that the experiment is replicated $n$ independent times and letting $Y_j$ denote the number of times the experiment

S.B. Provost (✉)
Department of Statistical and Actuarial Sciences, The University of Western Ontario, Western Science Centre Room 262, London, ON N6A 5B7, Canada
e-mail: sp@uwo.ca



W.K. Li et al. (eds.), *Advances in Time Series Methods and Applications*, Fields Institute Communications 78, DOI 10.1007/978-1-4939-6568-7_11

results in outcome $\mathcal{O}_j$, $j = 1, \ldots, r$, the random variables $Y_1, \ldots, Y_{r-1}$ jointly have the multinomial probability mass function

$$n! \prod_{j=1}^{r} \frac{p_j^{y_j}}{y_j!},$$

with $y_r = n - \sum_{j=1}^{r-1} y_j$ and $p_r = 1 - \sum_{j=1}^{r-1} p_j$. Pearson argued that, asymptotically, the statistic

$$\mathcal{P} = \sum_{j=1}^{r} \frac{(Y_j - np_j)^2}{np_j} \tag{1}$$

has a chi-squared distribution on $r - 1$ degrees of freedom. Several derivations of this result are available in the literature, including those provided by [4, 5, 8, 11]. However, these proofs can be somewhat lengthy and/or require certain specialized results such as Slutsky's theorem, some properties of projection or idempotent matrices, the singular value decomposition theorem or series expansions for certain functions of matrices. A short proof of the asymptotic distribution of $\mathcal{P}$, which is essentially based on the multivariate central limit theorem, is proposed in Sect. 2. When $s$ parameters have to be estimated, Watson [15] showed that the asymptotic distribution of $\mathcal{P}$ becomes chi-squared on $r - s - 1$ degrees of freedom.

The chi-squared approximation may turn out to be unreliable if, for instance, some of the expected values, $n\,p_j$, are too small. Although there is no consensus on the conditions that precludes its application, various criteria such as no cell count equal to zero, expected values, $n\,p_j$, greater than five for a certain proportion of the cells, a minimum sample size, a minimum number of classes and a sample size at least equal to a certain multiple of the number of classes, have been suggested, see for instance [2, 7, 12, 13, 16]. When such conditions are not satisfied, the chi-squared approximation may prove inaccurate, in which case one would have to forego utilizing Pearson's test. As a viable alternative, another continuous approximation, namely, the shifted gamma distribution is introduced in Sect. 3. Additionally, it is explained that section that the exact distribution of Pearson's statistic can be readily obtained from its moment-generating function by means of symbolic computations and that the parameters of the shifted gamma approximation can then easily be determined. As well, two numerical examples are provided in Sect. 4.

## 2 A Short Proof of the Asymptotic Distribution of Pearson's $\chi^2$ statistic

Let the random vector $\mathbf{Y} = (Y_1, \ldots, Y_r)'$ have a Multinomial $(n;\, p_1, \ldots, p_r)$ distribution with $p_j > 0,\ j = 1, \ldots, r$, and $\sum_{j=1}^{r} p_j = 1$,

$$\Sigma \equiv \mathrm{Cov}(Y_1, \dots, Y_{r-1}) = n \begin{pmatrix} -p_1^2 + p_1 & -p_1\, p_2 & \dots & -p_1\, p_{r-1} \\ -p_2\, p_1 & -p_2^2 + p_2 & \dots & -p_2\, p_{r-1} \\ \vdots & \vdots & \ddots & \vdots \\ -p_{r-1}\, p_1 & -p_{r-1}\, p_2 & \dots & -p_{r-1}^2 + p_{r-1} \end{pmatrix} \tag{2}$$

and $\mathbf{U} = (Y_1 - np_1, \dots, Y_{r-1} - np_{r-1})'$. Letting $\Sigma^{-1/2}$ denote the inverse of the symmetric square root of $\Sigma$, it follows from the multivariate central limit theorem that $\mathbf{Z} \equiv \Sigma^{-1/2}\mathbf{U} \to \mathcal{N}_{r-1}(\mathbf{0}, I)$, a standard normal distribution, which implies that $\mathbf{U}'\,\Sigma^{-1}\mathbf{U} = \mathbf{Z}'\,\mathbf{Z} \to \chi^2_{r-1}$ as $n \to +\infty$. On letting $Y_r = n - \sum_{j=1}^{r-1} Y_j$ and $p_r = 1 - \sum_{j=1}^{r-1} p_j$, Pearson's $\chi^2$ statistic denoted $\mathscr{P}$ can be expressed as follows:

$$\begin{aligned} \mathscr{P} = \sum_{j=1}^{r} \frac{(Y_j - np_j)^2}{np_j} &= \frac{1}{n}\left[\sum_{j=1}^{r-1} \frac{(Y_j - np_j)^2}{p_j} + \frac{(n - \sum_{j=1}^{r-1} Y_j - n(1 - \sum_{j=1}^{r-1} p_j))^2}{p_r}\right] \\ &= \frac{1}{n}\left[\sum_{j=1}^{r-1} \frac{(Y_j - np_j)^2}{p_j} + \frac{\left(\sum_{j=1}^{r-1}(Y_j - np_j)\right)^2}{p_r}\right]. \end{aligned} \tag{3}$$

It can then be verified that the inverse of $\Sigma = n\,(\mathscr{D}iag\,(\boldsymbol{p}) - \boldsymbol{p}\boldsymbol{p}')$ where $\boldsymbol{p} = (p_1, \dots, p_{r-1})$, is

$$\Sigma^{-1} = \frac{1}{n}\,\mathscr{D}iag\left(\frac{1}{p_1}, \dots, \frac{1}{p_{r-1}}\right) + \frac{\mathbf{1}\mathbf{1}'}{n\,p_r} \tag{4}$$

where $\mathbf{1} = (1, \dots, 1)'$, as $\Sigma\,(\mathscr{D}iag\,(1/p_1, \dots, 1/p_{r-1})/n) = I - \boldsymbol{p}\mathbf{1}'$ and $\Sigma\,(\mathbf{1}\mathbf{1}'/(n\,p_r)) = \boldsymbol{p}\mathbf{1}'/p_r - \boldsymbol{p}\,(\sum_{j=1}^{r-1} p_j)\,\mathbf{1}')/p_r = (1 - (1 - p_r))\boldsymbol{p}\,\mathbf{1}'/p_r = \boldsymbol{p}\mathbf{1}'$. Now, on noting that $\mathbf{1}'\,\mathbf{U} = \sum_{j=1}^{r-1}(Y_j - np_j)$, one has $\mathbf{U}'\,\Sigma^{-1}\mathbf{U} = \mathscr{P}$ as given in Eq. (3), so that $\mathscr{P} \to \chi^2_{r-1}$ as $n \to +\infty$. □

*Remarks* We observe that $\mathbf{Y} = (Y_1, \dots, Y_r)'$ can be expressed as the sum of $n$ independently distributed Multinomial$(1; p_1, \dots, p_r)$ random vectors $\mathbf{X}_i$ where each $\mathbf{X}_i$ is a vector of zeros except for its $k$th component, which is equal to 1 when the $k$th outcome occurs at the $i$th trial of the experiment and that, for instance, the last component of $\mathbf{Y}$ is fixed given its first $r-1$ components. As a result, the covariance matrix of $\mathbf{Y}$ is singular whereas that associated with its first $r-1$ components, that is, $\Sigma$ is invertible. Accordingly, it indeed follows from the multivariate central limit theorem that $\Sigma^{-1/2}\left(\sum_{i=1}^{n} \mathbf{X}_i^* - n\boldsymbol{p}\right) \to \mathcal{N}_{r-1}(\mathbf{0}, I)$, where $\mathbf{X}_i^*$ denotes the $(r-1)$-dimensional subvector of $\mathbf{X}_i$ consisting of its first $r-1$ components and $\boldsymbol{p} = (p_1, \dots, p_{r-1})'$. Note that, in the above notation, $\sum_{i=1}^{n} \mathbf{X}_i^* - n\boldsymbol{p} = \mathbf{U}$. Moreover, that $\Sigma^{-1}$ as specified by Eq. (4) is the inverse of $\Sigma$, can be deduced from Theorem 8.3.3 [6], for which, however, no explicit proof was provided.

## 3 An Accurate Approximation to the Exact Distribution of $\mathscr{P}$

The asymptotic distribution of Pearson's statistic can prove quite accurate as an approximation whenever certain conditions alluded to in the Introduction are satisfied. Otherwise, the chi-squared approximation may leave much to be desired and lead to invalid conclusions. As it turns out, in such instances, a reliable approximation can be obtained by replacing the chi-squared distribution by a shifted gamma distribution whose density and cumulative distribution functions are respectively given by

$$g(x) = \frac{e^{-\frac{x-\delta}{\theta}}(x-\delta)^{\alpha-1}}{\Gamma(\alpha)\,\theta^{\alpha}}, \quad x \in [\delta, \infty), \tag{5}$$

where $\alpha > 0$, $\theta > 0$ and $\delta$ is a real number (positive in this case), and

$$G(y) = 1 - \frac{\Gamma\left(\alpha, \frac{y-\delta}{\theta}\right)}{\Gamma(\alpha)}, \quad y \in [\delta, \infty), \tag{6}$$

where $\Gamma(\alpha, z) = \int_z^{\infty} x^{\alpha-1} e^{-x}\, \mathrm{d}x$ is the (upper) incomplete gamma function.

Consider the moment-generating function of $\mathscr{P}$, that is,

$$\mathscr{M}_{\mathscr{P}}(t) = \sum_{i=1}^{m} e^{t\sum_{j=1}^{r}(c_{i,j}-n\,p_j)^2/(n\,p_j)}\, n! \prod_{j=1}^{r} \frac{p_j^{c_{i,j}}}{c_{i,j}!} \tag{7}$$

where $n$, $r$ and $p_j$ are as previously defined and $m = \binom{n+r-1}{r-1}$ is the number of compositions of $n$ into $r$ ordered nonnegative integers $(y_1, \ldots, y_r)$, which can be obtained for example by utilizing the *Mathematica* command *Compositions*$[n, k]$, $c_{i,j}$ denoting the $j$th element of the $i$th composition.

Upon expanding the right-hand side of Eq. (7) and simplifying the resulting expression by making use of a symbolic computation software package, $\mathscr{M}_{\mathscr{P}}(t)$ can be represented as $\sum_{i=1}^{m^*} \beta_i\, \mathrm{e}^{b_i t}$ where $m^* \le m$, $(b_1, \ldots, b_{m^*})$ is the support of the distribution of $\mathscr{P}$, the $b_i$'s being listed in increasing order, and $\beta_i = \mathrm{Prob}(\mathscr{P} = b_i)$, $i = 1, \ldots, m^*$. This follows from standard results in connection with the moment-generating functions of discrete random variables. Accordingly, the exact cumulative distribution function of $\mathscr{P}$ at the point $b_i$ is $F_{\mathscr{P}}(b_i) = \sum_{\ell=1}^{i} \beta_\ell$.

The parameters of the shifted gamma approximation are then determined by minimizing

$$\sum_{i=1}^{m^*} (G(b_i) - F_{\mathscr{P}}(b_i))^2 \tag{8}$$

with respect to $\alpha$, $\theta$ and $\delta$, which can be achieved for instance with the *Mathematica* command *NMinimize*. Then, any required percentile of the distribution can be evaluated from the shifted gamma cumulative distribution function.

## 4 Numerical Examples

*Example 1* Letting $n = 2$, $r = 3$ and the null hypothesis be $\mathscr{H}_0 : (p_1, p_2, p_3) = (0.142857, 0.285714, 0.571429)$, the distribution of $\mathscr{P}$ is as specified in Table 1 (pmf and cdf respectively denoting the exact probability mass function and the exact cumulative distribution function of $\mathscr{P}$), which also includes the cumulative distribution functions obtained by making use of the chi-squared distribution on two degrees of freedom and the shifted gamma distribution with parameters $\alpha = 0.912125$, $\theta = 1.42627$ and $\delta = 0.147528$.

In this case, the compositions (that is, all the possible values of the observed vectors $(y_1, y_2, y_3)$) are $\{\{0, 0, 2\}, \{0, 1, 1\}, \{0, 2, 0\}, \{1, 0, 1\}, \{1, 1, 0\}, \{2, 0, 0\}\}$, and, under $\mathscr{H}_0$, the moment-generating function as determined from Eq. (7) is

$$\frac{16}{49}e^{5t/8} + \frac{16}{49}e^{3t/2} + \frac{8}{49}e^{19t/8} + \frac{4}{49}e^{13t/4} + \frac{4}{49}e^{5t} + \frac{e^{12t}}{49} .$$

Note that for an experimental value of $\mathscr{P}$ equal to 5 and a significance level of 5 %, the null hypothesis would correctly be rejected when the shifted gamma distribution is being utilized as an approximation, whereas it would mistakenly fail to be rejected under the usual chi-squared approximation. It can also be observed that the shifted gamma cumulative distribution function is in very close agreement with the exact one.

*Example 2* Letting $n = 5$, $r = 2$ and $(p_1, p_2) = (1/19, 18/19)$, the exact distribution of $\mathscr{P}$ is as specified in Table 2, which also includes the cumulative distribution functions obtained by making use of the chi-squared distribution on one degree of freedom and the shifted gamma distribution with parameters $\alpha = 0.177068, \theta = 1.51936$ and $\delta = 0.036122$.

**Table 1** Exact and approximate distributions of $\mathscr{P}$ as specified in Example 1

| $b_i$ | pmf | cdf | $G(b_i)$ | $\chi^2_2$ cdf |
|---|---|---|---|---|
| 0.625 | 0.326531 | 0.326531 | 0.326788 | 0.268384 |
| 1.5 | 0.326531 | 0.653061 | 0.651651 | 0.527633 |
| 2.375 | 0.163265 | 0.816327 | 0.816604 | 0.695017 |
| 3.25 | 0.0816327 | 0.897959 | 0.90274 | 0.803088 |
| 5 | 0.0816327 | 0.979592 | 0.97234 | 0.917915 |
| 12 | 0.0204082 | 1. | 0.999809 | 0.997521 |

**Table 2** Exact and approximate distributions of $\mathscr{P}$ as specified in Example 2

| $b_i$ | pmf | cdf | $G(b_i)$ | $\chi_1^2$ cdf |
|---|---|---|---|---|
| 0.277778 | 0.763123 | 0.763123 | 0.763123 | 0.401839 |
| 2.17778 | 0.211979 | 0.975102 | 0.975099 | 0.859983 |
| 12.1 | 0.0235532 | 0.998655 | 0.999989 | 0.999496 |
| 30.0444 | 0.00130851 | 0.999963 | 1. | 1. |
| 56.0111 | 0.0000363475 | 1. | 1. | 1. |
| 90 | $4.03861 \times 10^{-7}$ | 1. | 1. | 1. |

For $n = 5$ and $r = 2$, the compositions are $\{\{0, 5\}, \{1, 4\}, \{2, 3\}, \{3, 2\}, \{4, 1\}, \{5, 0\}\}$, and the moment-generating function is

$$\frac{1889568e^{5t/18}}{2476099} + \frac{524880e^{98t/45}}{2476099} + \frac{58320e^{121t/10}}{2476099} + \frac{3240e^{1352t/45}}{2476099} + \frac{90e^{5041t/90}}{2476099} + \frac{e^{90t}}{2476099}.$$

## 5 Concluding Remarks

As the number of replications becomes large, it was observed that, as expected, the parameters $\alpha$, $\theta$ and $\delta$ of the shifted gamma approximation respectively converge to $r/2$, 2 and 0. Moreover, as $n$ increases, overall, this approximation remains more accurate that the asymptotic chi-squared distribution. It should also be pointed out that the proposed methodology could readily be applied to other goodness-of-fit measures such as the Freeman-Tukey statistic or the sum of the squared deviations, which are not as sensitive as $\mathscr{P}$ to possible small values of $n\, p_i$, $i = 1, \ldots, r$.

The chi-squared approximation being inaccurate when the sample sizes are small or other conditions for its applicability are not satisfied, another continuous distribution, namely the shifted gamma distribution, is being proposed. Its parameters are determined from the exact distribution of $\mathscr{P}$, which is readily obtained by means symbolic computations. It is admittedly more convenient to resort to the chi-squared approximation; however, when it proves inadequate, the proposed methodology, which turns out to be easily implementable, provides a viable alternative for accurately determining specific critical values of Pearson's goodness-of-fit test statistic.

**Acknowledgments** The financial support of the Natural Sciences and Engineering Research Council of Canada is gratefully acknowledged. Thanks are also due to two referees for their valuable comments. This Festschrift, which was organized in recognition of Ian McLeod's significant contributions to Time Series as well as several other areas of Statistics, is indeed a fitting tribute to his scholarly accomplishments. Ian has truly been a valued colleague over the years.

## References

1. Andrés, A. M., & Tejedor, I. H. (2009). Comments on 'Tests for the homogeneity of two binomial proportions in extremely unbalanced 2×2 contingency tables'. *Statistics in Medicine, 28*, 528–531.
2. Cochran, W. G. (1952). The $\chi^2$ test of goodness of fit. *The Annals of Mathematical Statistics, 25*, 315–345.
3. Cox, M. K., & Key, C. H. (1993). Post hoc pair-wise comparisons for the chi-square test of homogeneity of proportions. *Key Educational and Psychological Measurement, 53*, 951–962.
4. DasGupta, A. (2008). *Asymptotic theory of statistics and probability, Springer Texts in Statistics*. New York: Springer.
5. Ferguson, T. S. (1996). *A course in large sample theory*. Boca Raton: Chapman & Hall.
6. Graybill, F. A. (1983). *Matrices with applications in statistics*. Belmont: Wadsworth.
7. Koehler, K. J., & Larntz, K. (1980). An empirical investigation of goodness-of-fit statistics for sparse multinomials. *Journal of the American Statistical Association, 75*, 336–344.
8. Mathai, A. M., & Pederzoli, G. (1996). A simple derivation of the chi-square approximation to Pearson's statistic. *Statistica, 56*, 407–413.
9. Moore, D. S., & Spruill, M. C. (1975). Unified large-sample theory of general chi-squared statistics for tests of fit. *Annals of Statistics, 3*, 599–616.
10. Nathan, G. (1972). On the asymptotic power of tests for independence in contingency tables from stratified samples. *Journal of the American Statistical Association, 67*, 917–920.
11. Pearson, K. (1900). On the criterion that a given system of deviations from the probable in the case of a correlated system of variables is such that it can be reasonably supposed to have arisen from random sampling. *Philosophical Magazine, 50*, 157–175.
12. Roscoe, J. T., & Byars, J. A. (1971). An investigation of the restraints with respect to sample size commonly imposed on the use of the chi-square statistic. *Journal of the American Statistical Association, 66*, 755–759.
13. Rosner, B. (2006). *Fundamentals of biostatistics. Chapter 10. Chi-square goodness-of-fit* (6th ed.). Duxbury, MA: Thomson Learning Academic Resource Center.
14. Tobin, J. (1958). Estimation of relationship for limited dependent variables. *Econometrica, 26*, 24–36.
15. Watson, G. S. (1959). Some recent results in chi-square goodness-of-fit tests. *Biometrics, 15*, 440–468.
16. Yates, F. (1934). Contingency table involving small numbers and the $\chi^2$ test. *Journal of the Royal Statistical Society, 1*, 217–235.

# Business Resilience During Power Shortages: A Power Saving Rate Measured by Power Consumption Time Series in Industrial Sector Before and After the Great East Japan Earthquake in 2011

**Yoshio Kajitani**

**Abstract** Many power crises have occurred in developing and developed countries such as through disruptions in transmission lines, excessive demand during heat waves, and regulatory failures. The 2011 Great Japan Earthquake caused one of most severe power crises ever recorded. This study measures the industry's ability to conserve power without critically reducing production ("power saving rate") as one of the indicator of resilience as a lesson of disaster. The quantification of the power saving rate leads to grasping the potential power reduction of industrial sector or production losses caused by the future incidents in many regions or countries. Using time series data sets of monthly industrial production and power consumption, this study investigates the power saving rate of Japanese industries during power shortages after the great earthquake. The results demonstrates the size of power saving rate right after the disaster, during the first severe peak demand season, as well as long-term continuous efforts of power saving in different business.

**Keywords** Power shortage · Resilience · Great East Japan Earthquake · Industrial sector

## 1 Introduction

Power shortages after the Great East Japan Earthquake on March 11, 2011 created prolonged impact on Japan's cities and businesses. Shortages during summer 2011 were serious in the Tohoku and Kanto regions, and power consumption by large businesses (with a maximum demand exceeding 500KW) was restricted by the Electricity Business Act (Article 27), which mandated each business to reduce peak

Y. Kajitani (✉)
Central Research Institute of Electric Power Industry,
1646 Abiko, Abiko City, Chiba 2701194, Japan
e-mail: y-kaji@criepi.denken.or.jp



W.K. Li et al. (eds.), *Advances in Time Series Methods and Applications*,
Fields Institute Communications 78, DOI 10.1007/978-1-4939-6568-7_12

power demand by 15 % from the previous year's peak. To achieve that goal, Japanese businesses undertook extensive efforts, such as shifting production before and after the summer season and reassigning weekday activities to weekends. Some installed power-saving machinery and equipment, such as LED (light-emitting diode). Their adaptations attained the 15 % targeted reduction, avoiding blackouts and major upsets in production.

The study defines their adaptive behavior as business resilience to power shortages. The key question is the degree to which business can reduce power consumption without reducing output. Research establishing the resilience of business to shortages is essential in preparing for future disasters. Previous research has provided extensive information regarding attempted adaptations and how much power is conserved (e.g., IEA [11]). This research extends that literature, focusing on the relationship between industrial production and power consumption. The quantification of resilience, in terms of power saving ability during the disaster with the consideration of production output, lead to grasping the potential power reduction of industrial sector or production losses caused by the future incidents in many regions or countries.

Time series analysis is appropriate for understanding the impact by removing the effects of seasonal trends and random errors observable even when disasters do not occur. This study adopts monthly power consumption for large business and an index of industrial production (IIP) as the most disaggregated datasets in Japan. The study detects changes in relationships between power consumption and IIP as adaptive behavior during power shortages. The study conducts a relatively disaggregated sector-by-sector analysis to identify the characteristics of each sector's resilience. Furthermore, it becomes milder, but the power shortage is an on-going issue in Japan (as of August 2015), and the how the Japanese industries adjust to this situation in a long-term basis is also an important issue for estimating not only short-term but also long-term energy saving potentials all over Japan and the world.

The study proceeds as follows Sect. 2 summarizes power demand forecasting models and conditions of power shortages after the Great East Japan Earthquake. Section 3 describes statistical models and the study's resilience index. Section 4 shows results from applying the model to Japan's Kanto (large power shortages) region. Section 5 summarizes.

## 2 Power Demand Forecasting and Industrial Adaptations to Shortages

### 2.1 *Power Demand Forecasting*

Models forecasting power demand are essential statistical tools, and many models have been developed in this area. In general, they estimate hypothetical demand if sufficient power is supplied and compare it with the actual demand. Forecasting approaches can be classified by their time scales and the statistical models adopted.

Autoregressive Integrated Moving Average (ARIMA), Artificial Neural Network (ANN), multiple regression, and combinations of all three are popular in statistical models.

In short-term forecasts, power demand per hour or less is modeled for real-time operations. Hilpert et al. [9] and An et al. [2] adopt ANN models; whereas, Tayler [20] employs ARIMA. Similar approaches are employed for mid-term daily or monthly forecasts, including Tayler and Buizza [21], Gonzalez-Romera [7], Pappas [19], Hyodo [10], and Vu et al. [24]. Weather-related parameters are central to the accuracy of estimates by these models and most differentiate on the basis of weekdays and weekends. Long-term forecasts, such as estimating annual power consumption, required more socio-economic variables (Fatai, [5]; Azadeh et al., [3]; Zahedi, [25]; Nawas, [17]; Kaytez et al., [16]). Population and GDP are typical exogenous variables because they correlate strongly with power consumption.

This study employs monthly power consumption time series data because a monthly production index is available in Japan. The model for forecasting power demand follows the general treatment of mid-term forecasts and partly that of long-term forecasts. The study selects the appropriate statistical models using the Akaike Information Criterion (AIC) and usual statistical tests on parameters.

## 2.2 *Power Shortages During Japan's 2011 Earthquake*

### 2.2.1 Damaged Power Plants and Recovery Situation

Figure 1 illustrates the damage and recovery status of supply capacity in regions supplied energy by Tokyo and Tohoku electric power companies. Operations of nuclear power plants were halted, and approximately 90 % of supply capacity depended on thermal power plants, which were massively damaged by earthquakes and tsunami. Many thermal power plants recovered by peak demand season in July and August, although risk of shortages remained high.

Figure 2 illustrates power supply and demand conditions on the peak demand day in both regions during summer 2011. Note that peak demand day differs in each region. During this period, Tokyo was expected to suffer the most severe supply and demand imbalances. Therefore, the central government asked most of industrial sectors to follow the 15 % mandatory reduction in power use. Considerable efforts by firms helped achieve the largest reduction of power consumption, and peak demand was covered by supply with a safe margin. As the figure shows, demand in Tohoku exceeded supply capacity; however, importing power from the regions outside Tohoku region avoided shortages. More severe conditions are evident in West Japan, where peak power consumption in most areas, excluding Chugoku, approached or exceeded capacity.

The remainder of this study primarily employs monthly demand (KWH) in its analysis because only monthly corresponding production index data are available. Considering that mandatory restriction of power usage is related to a peak demand

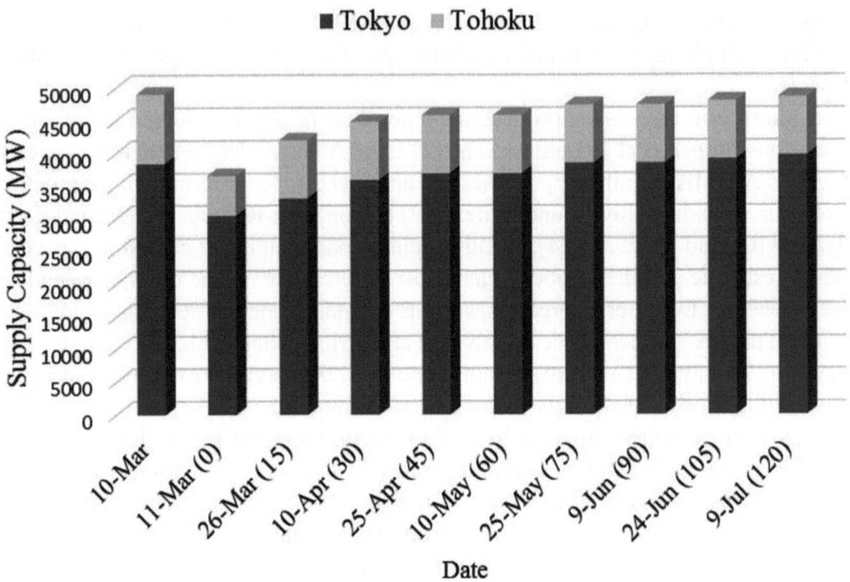


**Fig. 1** Power supply capacities after the 2011 earthquake and tsunami (*numbers in parenthesis* indicate the number of days after the previous event)

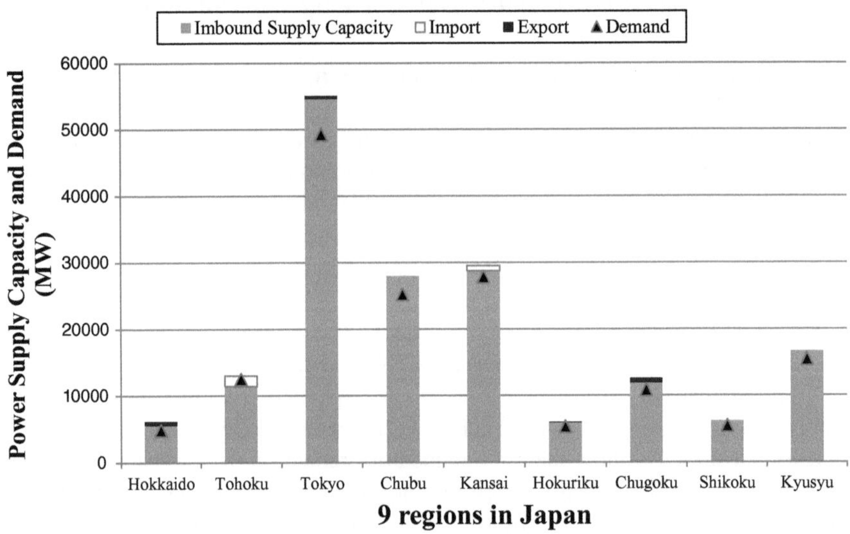


**Fig. 2** Power supply and demand after the 2011 earthquake and tsunami

(KW), it is necessary to clarify the relationship between peak demand (KW) and daily consumption (KWH) in advance. Figure 3 demonstrates the actual relationship between peak demand and daily consumption in the area served by Tohoku Electric

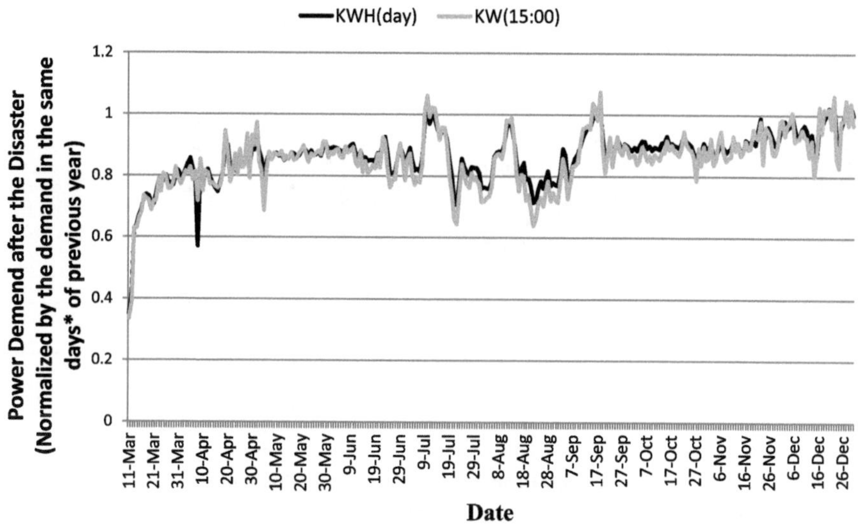


**Fig. 3** Time series plots of peak (KW) and daily (KWH) power demands after March 11

Power Co. [23]. This figure shows a linear relationship between these two measures. In summer 2011, many businesses reduced daily power consumption, which in turn reduced the peak power demand. For example, businesses took a day off or installed machinery that consumed less power.

### 2.2.2 Time Series Data of Power Consumption and IIP

Figure 4 plots power consumption among large businesses and the IIP in Kanto from March 2008 to December 2014 (METI-KANTO, [14]). Power consumption data (METI-KANTO, [15]) include only large customers; however, the share of total power supply is high. Consumption from generators owned by businesses is included. There are inconsistencies in regional definitions between datasets. The IIP covers 11 prefectures (Ibaraki, Tochigi, Gunma, Saitama, Chiba, Tokyo, Kanagawa, Niigata, Yamanashi, Nagano, Shizuoka); whereas, power consumption data target the area served by TEPCO. Its data exclude Nagano, Niigata, and part of Shizuoka, which are included in IIP.

Notwithstanding these slight regional inconsistencies, Fig. 4 shows that power consumption and IIP have an obvious linear relationship. This is because the regional overlap for these two indexes encompasses much of Japan's economy. Furthermore, it is likely that power consumption data capture a large share of total power consumption and display a synthetic trend with the remaining data because production trends are similar among small and large customers.

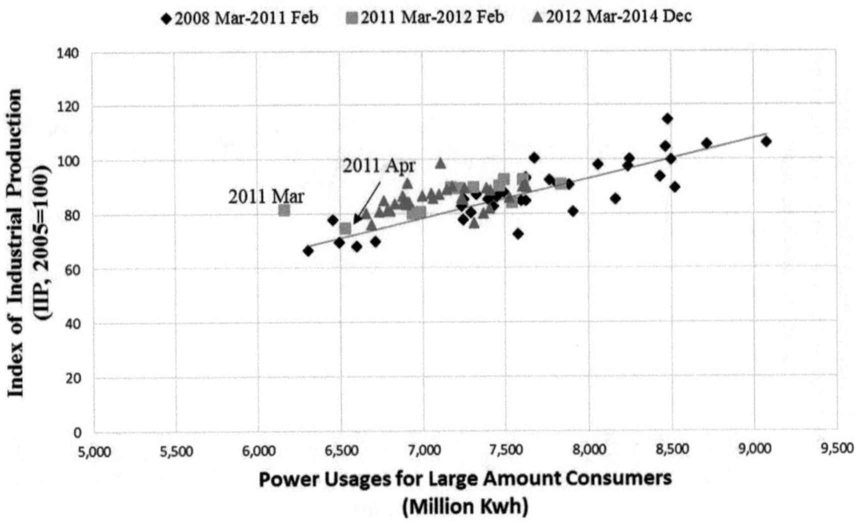


**Fig. 4** Relationships between power consumption by large businesses and index of industrial production in the Kanto Region

It is a slightly difficult to claim that a difference appears in earlier trends (March 2008 to February 2011) and trends during a year after the earthquake (March 2011 to February 2012). In other words, the earthquake produced no evident impact on the relationship between power consumption and production output. On the other hand, during the last period (March 2011 to February 2012), it is visible that the production is larger at the same level of electricity input, especially around 7100 million Kwh, compared with the production before the disaster. Again, it is also difficult to observe the difference between the two data sets at the larger input around 7500 million Kwh.

In fact, discussions of power consumption must consider temperature and seasons. In general, the cooling and heating demands are changed if the temperatures are changed. Whether the production is achieved with less electricity have to be inspected after the temperature effect is removed. Similarly, seasonal effects, such as changes in production items and systems, would affect the relationships between power consumption and productions. The modelling of power demand in industrial sector and deriving the indicator of quantifying the resilience is one of the main topics in this research and is described in Sect. 3.

### 2.2.3 Adaptation Behaviors in Business Sector and Related Research Questions

Nikkei [18] reports the elasticity of electricity input (production change under the change of electricity input) on March 24, 2011, estimated from the strong correlation between the IIP and power consumption for large businesses. Elasticity is 1.85 in

Kanto (TEPCO service region), highest among the nine regions in Fig. 2, followed by Chubu and Kansai, which are relatively industrialized. Average nationwide elasticity is 1, indicating that a 1 % electricity reduction (increase) yields a 1 % production increase (decrease). Based on elasticity in the Kanto region, the expected impact of the power shortage was extremely large.

A post-summer business survey by the Agency for Natural Resources and Energy [1] documents practical instances of countermeasures taken by businesses, their costs, and their benefits. For example, one large business in commodity resins and synthetic rubber shifted holidays to working days, day shifts to night shifts, and rescheduled production from mid-summer to early summer to reduce peak demand during the period of intensive power usage. As a result, 25 % of peak electricity demand was reduced from the previous year. The countermeasures that cost ¥180 million (¥80.75 = US$1 per Bank of Japan, 2011) help in avoiding losses worth ¥900 million, which were assumed to be created under the condition of 25 % power demand reduction without countermeasures. Other businesses adopted such countermeasures as moving periodic inspection days to peak demand season and shifting production from Kanto to other regions. Fujimi and Chang's [6] summary of 14 business surveys after the earthquake identify patterns of adaptation, revealing that manufacturers were more likely to implement schedule changes.

IEA [11] surveyed past major incidents of power shortages and summarizes adaptations.[1] The report notes three major strategies: raise electricity prices, encourage behavioral changes, and introduce energy-efficient technologies. Adaptations by Japanese business can be classified in greater detail, as follows.

- Restrictions (e.g., change air conditioner settings)
- Time Shifts (e.g., change production timing)
- Substitutions (e.g., change fuel types, imports from other regions)
- Relocations (e.g., change production locales, relocate data servers)
- Renewals (e.g., install of new machinery).

Restrictions and time shift can be instituted with little preparation, but they are grounded in human patience. Relocations and renewals require careful and lengthy preparation; however, their effects last over an extended period. The success of substitutions depends on costs and quality of substituted goods and services and can be temporary or semi-permanent. These countermeasures constitute the roots of resilience, and its characteristics can be exposed by investigating instances of power shortages that occasioned different durations of preparation. Detailed characteristics of resilience can be itemized by surveying individual business; however, overall resilience can be captured by monthly gross statistics in the following analysis. A research question regarding the types of adaptation is how long the power saving continues, which reflect the percentage of permanent adaptations undertaken by the business such as relocations and renewals. In contrast, the other adaptations would be revealed only temporarily during a severe condition. These question can be answered by analyzing relatively long time series data after the disaster.

[1]The report is updated in 2011.

# 3 Time Series Models and Index of Resilience

## 3.1 Time Series Model of Energy Demand

As noted, many statistical models forecast power demand. In these models, production outputs, temperatures, and seasonal factors are essential to estimate demand. Considering previous research, the following models are introduced as candidates for estimating power demand in this research. First, the basic structure of power demand function is represented as follows:

$$E(t) = a + bY(t) + \Gamma(t) + \Xi(t) + \varepsilon(t), \tag{1}$$

where $E(t)$ denotes power demand, $Y(t)$ production output, $\Gamma(t)$ temperature, $\Xi(t)$ remaining seasonal effects, and $\varepsilon(t)$ random errors at time $t$. For $\Gamma(t)$, either of two functional types are selected:

(Polygonal line type function)

$$\Gamma(t) = \begin{cases} \alpha - \beta_1(\gamma - T(t)) & (T(t) \leq \gamma) \\ \alpha + \beta_2(T(t) - \gamma) & (T(t) > \gamma) \end{cases} \\ s.t.\ \ \beta_1 \geq 0,\ \ \beta_2 \geq 0 \tag{2}$$

(Quadratic function)

$$\Gamma(t) = \alpha + \beta_1 T(t) + \beta_2 T(t)^2 \\ s.t.\ \ \beta_2 \geq 0,\ \ \beta_1 \leq 0 \tag{3}$$

$T(t)$ represents temperature at time $t$, and the other Greek letters indicate parameters. The restriction of parameters is determined so that the function satisfies downward convexity. This assumption comes from general observation of heating and cooling demand (i.e., demand increases as the temperature becomes apart from the most comfortable temperature to people). The model forms are selected to satisfy this condition with small parameters, but other appropriate model may exist.

For the seasonality term $\Xi(t)$, the dummy variables, Fourier series, or SARIMA models are candidates.

(Dummy variable)

$$\Xi(t) = \sum_{s=0}^{11} \eta_s D(t)\ ,\ where\ \ D(t) = \begin{cases} 1 & if\ t\,(\mathrm{mod}\ 12) = s \\ 0 & otherwise \end{cases}, \tag{4}$$

where $t(\mathrm{mod}\ 12)$ represents the remainder from dividing $t$ by 12. $s = 0$ indicates December, and other numbers (1 to 11) indicate corresponding months. $\eta_s$ is a parameter capturing dummy variables for month ($s$).

(Fourier series)

$$\Xi(t) = \sum_{i=1}^{k} a_i \sin\left(i\frac{2\pi}{K}t\right) + \sum_{i=1}^{k} b_i \sin\left(i\frac{2\pi}{K}t\right) \quad (K = 12) \tag{5}$$

(SARIMA)

$$\phi(B)\Phi(B^s)\nabla^d\nabla_s^D z_t = \theta(B)\Theta(B^s)a_t, \tag{6}$$

where $z_t$ is an original series, $\Phi(B^s)$ and $\theta(B)$ are seasonal autoregressive and moving average operators, respectively. $\nabla_s^D$ is seasonal differencing operator in $D$ times. $\varphi(B)$ and $\theta(B)$ are non-seasonal autoregressive and moving average operators, respectively. $\nabla^d$ is a non-seasonal differencing operator in $d$ times, and $a_t$ is white noise. More detailed explanations appear in Chap. 12 of Hipel and McLeod [8].

In total, six combinations (2 temperature effects × 3 remaining seasonal effects) are investigated. The best model is chosen based on goodness-of-fit defined by AIC and statistical tests on the significance of individual parameters.

## 3.2 *Approaches to Quantify the Resilience of Industrial Production During Power Shortages*

One of the simplest measures is how much business saved the power consumption which would have been used if the power crisis did not occur. In this study, the power consumption after the disaster in the pre-disaster condition is estimated by the power consumption time series model constructed by the pre-disaster data sets based on Eq. (1) as follow:

$$\widehat{E}_1(t_2) = \widehat{a}_1 + \widehat{b}_1\tilde{Y}(t_2) + \widehat{\Gamma}_1(t_2) + \widehat{\Xi}_1(t_2), \tag{7}$$

where the suffixes 1 and 2 indicate the parameters and variables before and after the disaster respectively. $\hat{E}_1(t_2)$ is a hypothetical power consumption estimated from the observed IIP and temperatures at time $t_2$ based on the time series model of pre-disaster structures. Furthermore, the power saving rate $ES(t_2)$ is defined by the hypothetical and the observed power consumptions $\widehat{E}_1(t_2)$ and $\widetilde{E}(t_2)$ as:

$$S(t_2) = 1 - \frac{\tilde{E}(t_2)}{\widehat{E}_1(t_2)}. \tag{8}$$

Note that Eq. (8) is the power saving rate of remained production because the cases of identical production outputs are compared before and after the disaster. In other words, more power reductions certainly entail the production decrease. The saving rate is also defined as the power reduction rate for producing a single unit of production.

Similar to the case of power saving rate, the increase in productivity during power shortages can be defined. In this case, production of time $t_2$ at the pre-disaster demand system is focused on and estimated as follows:

$$\hat{Y}_1(t_2) = \frac{1}{\widehat{b}_1}(\tilde{E}(t_2) - \widehat{a}_1 - \widehat{\Gamma}_1(t_2) - \widehat{\Xi}_1(t_2)). \tag{9}$$

Next, the productivity change rate $R$ is defined by

$$R(t_2) = \frac{\tilde{Y}(t_2)}{\widehat{Y}_1(t_2)} - 1. \tag{10}$$

Evidently, indices S and R have a clear relationship which is explained by a following formula based on the Eqs. (8)–(10).

$$\frac{S(t_2)}{R(t_2)} = \frac{\tilde{E}(t_2) - \widehat{a}_1 - \widehat{\Gamma}_1(t_2) - \widehat{\Xi}_1(t_2)}{\widehat{a}_1 + \widehat{b}_1\tilde{Y}(t_2) + \widehat{\Gamma}_1(t_2) + \widehat{\Xi}_1(t_2)}. \tag{11}$$

# 4 Case Study of the 2011 Great East Japan Earthquake

## *4.1 Analysis of Severely Affected Region (Kanto)*

First, total industrial sector in the Kanto area is analyzed. Each of industries were required to reduce peak power by 15 %. Time series data for three years before the earthquake (March 2008–February 2011) are used to select an appropriate statistical model and to estimate parameters. Temperature data are obtained from an observatory in center of Tokyo (JMA, [12]).[2] The monthly average daily maximum temperature is selected because the data fit better with the monthly average of daily average temperatures.

Table 1 shows the result of AIC comparisons among combinations of temperature and seasonality terms in Eqs. (1–6). The criterion supports a combination of Eq. (3), which employs a quadratic form for the temperature term, and Eq. (4) which employs a dummy variable. Parameter values appear in Table 2. This study imposes 5 % significance for selecting parameters. The dummy variables are significant in the parts of summer season (July, August), winter season (February and March) and autumn season (October). The signs of significant dummy variables in summer season are all positive. This is because of the remaining residuals of modeling nonlinear effects of temperatures on power demand by quadratic forms. The more cooling demand may be necessary in high temperature seasons, resulting in underestimations

[2]It is, of course, ideal to use detailed regional temperatures, especially when more disaggregated power consumption data are available.

**Table 1** AIC for different models

| Constant | Temperature | AIC |
|---|---|---|
| No | BiLinear | 507.3 |
| Yes | BiLinear | 492.9 |
| No | Polynomial | 511.0 |
| Yes | Polynomial | 487.5 |

| Seasonal Adj. | AIC |
|---|---|
| No | 487.5 |
| Fourier | 481.4 |
| Dummy (5) | 445.3 |
| SARIMA | 461.9 |

**Table 2** Estimated values of parameters (Kanto Area, March 2008–February 2011)

| Parameter | Estimates | SD | $t$ | $P$ |
|---|---|---|---|---|
| Constant | 3359.80 | 199.38 | 16.85 | 0 |
| $T$ | −47.03 | 19.85 | −2.37 | 0.03 |
| $T^2$ | 1.68 | 0.63 | 2.69 | 0.01 |
| IIP | 51.04 | 1.72 | 29.66 | 0 |
| Dummy (Feb) | −221.11 | 73.84 | −2.99 | 0 |
| Dummy (Mar) | −498.42 | 70.59 | −7.06 | 0 |
| Dummy (Jul) | 372.95 | 95.77 | 3.89 | 0 |
| Dummy (Aug) | 528.94 | 105.29 | 5.02 | 0 |
| Dummy (Oct) | 238.65 | 66.76 | 3.58 | 0 |
| Adjusted $R^2$ | 0.976 | | | |

of power demand. The power demand forecasts in the other dummy variables are also possibly affected by the residuals of modeling temperature effects. The other reason may be changes in monthly productivity. In general, scale effect should exist for production. That is, productivity increases if the amount of production increases. This is achieved by reducing idle time of production machinery and intensively inputting labor forces. This productivity change contributes to the power saving. In Japan, production increases in February and March mainly because these two months are the end of fiscal year and production orders are concentrated.

Table 3 indicates the result of applying the same procedure to the post-disaster data sets from March 2011 to December 2014. Basically, post-disaster time series is instable especially right after the event occurs. Therefore, the table is given mainly for reference purpose. Similar to Table 2, a quadratic form for the temperature term and a dummy variable for seasonal trend term are selected as a result of AIC comparisons and the criterial of 5 % significance level for choosing effective variables. Different from Table 3, only the dummy variable in March is available. This can be because of the effect of power saving efforts. The parameter value of IIP term is smaller than the value of pre-disaster case, which indicate that the effective production is conducted in average after the disaster. The parameters on quadratic term of temperature variable are similar in both cases; however, the parameter of the first-order term is different. There might be a possibility that patterns of cooling and heating demands changed

**Table 3** Estimated values of parameters (Kanto Area, March 2011–February 2014)

| Parameter | Estimates | SD | $t$ | $P$ |
|---|---|---|---|---|
| Constant | 3121.04 | 330.67 | 9.44 | 0 |
| $T$ | −34.22 | 12.62 | −2.71 | 0.01 |
| $T^2$ | 1.69 | 0.37 | 4.56 | 0 |
| IIP | 47.06 | 3.94 | 11.93 | 0 |
| Dummy (Mar) | −416.80 | 74.28 | −5.61 | 0 |
| Adjusted $R^2$ | 0.917 | | | |

after the disaster, or larger errors from regressed model may dominate the seasonality effects. A detailed analysis is required for understanding the change.

By eliminating temperature and seasonal effects from respective observed power consumptions before and after the disaster, Fig. 5 is illustrates the relationship between production and power consumption. Compared with Figs. 4 and 5 clearly reveals the effects of the earthquake.

Figure 6 illustrates a direct relationship between the rate of reduction in electricity input and production output. The baseline months are set from March 2010 to February 2011 and the reduction rates are calculated between the same months before and after the disaster (e.g., the reduction rate of productions outputs in March 2011 are estimated based on the change rate from the production output in March 2010.). The diagonal indicates the case in which the reduction rates of power and IIP are identical, and plots beneath it indicate that production is reduced at less reduc-

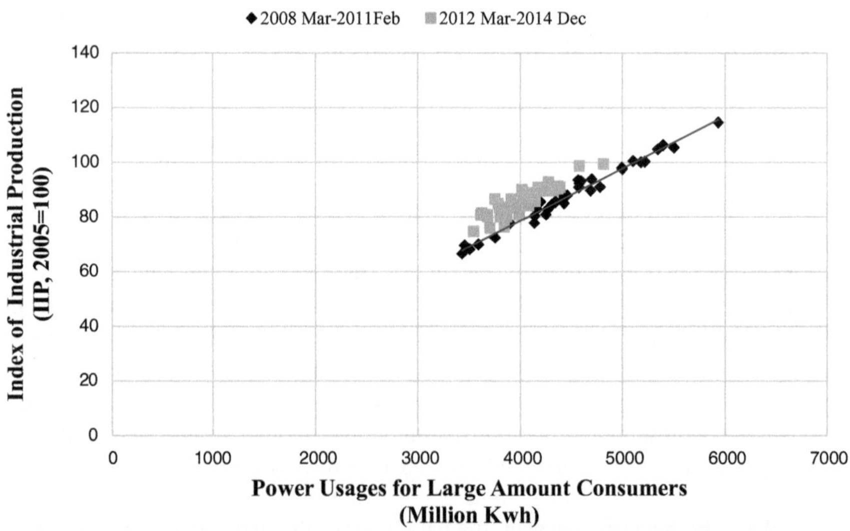


**Fig. 5** Relationships between output and power consumption in the Kanto region (temperature and remaining seasonality effects are adjusted.)

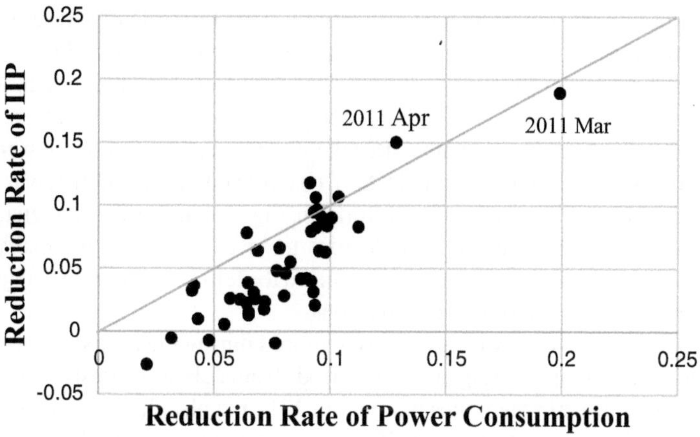


**Fig. 6** Rates of reduction in electricity input and production output for the Kanto region (from March 2011 to December 2014. Reduction rates are estimated by the comparison with the same month in the baseline year: March 2010–February 2011)

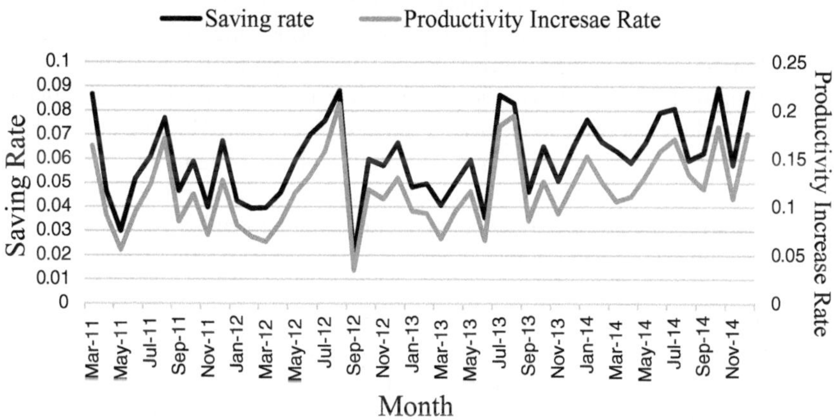


**Fig. 7** Estimated electricity saving rate and productivity increase rate from March 2011 to November 2014

tion rate of power. Reducing electricity consumption by reducing production starts at approximately 0.05. Incremental electricity reductions over 0.05 are achieved through reductions in production. The first and second largest curtails of electricity are seen in March and April 2011, where the production reduction rates are also large. It looks that the reduction of electricity requires the same or even larger rate of electricity input reduction right after the disaster.

Figure 7 plots the estimated electricity saving rates in Eq. (8) and productivity increase rates in Eq. (10). Both saving rates and productivity increase rates after the disaster are positive, which indicate that industries payed certain efforts to reduce energy without reducing outputs. Furthermore, both rates clearly have an almost identical moving pattern. This is apparent from the definition and Eq. (11), which determines the relationships between two rates. The following paragraph discusses the saving rates as same discussion holds for the production increase rates.

Observing the first three months after the disaster, saving rate is high in March (0.087); however, it becomes smaller in April (0.046) and May (0.030). It was difficult to know the types of power saving efforts in Fig. 7, but we can understand that the efforts were certainly taken by industries to save electricity, especially during March. In March, the industries may have utilized the leftover inventories for effective production with less electricity consumptions.

The largest efforts of power saving are required during the period of large power consumptions, normally during summer period (from July to August). These efforts are reflected in the estimated rates not only in 2011 but also during summer in other years. The saving rate in September 2012, 18 months after the earthquake, is small possibly because each business was released from the large restriction of power usages.

It should be also noted that the power saving rates look monotony increasing after the September 2012. This indicates that the electricity saving measures can be regularly practiced by employees and permanent countermeasures, such as renewals of old machineries, can be undertaken and continuously upgraded by the businesses.The other reason that may accelerate the power saving trend is increase in the price of electricity. In fact, the price of electricity increased in April 2012 by 2.58 yen/Kwh and 2.61 yen/Kwh for the customers with high and extra high voltage contracts, respectively (decreased by 0.25 yen in September 2012). These price increases may potentially affect the upward trend of saving rate after September 2012.

Overall, the saving rate without production decrease vary from 0.019 to 0.189. The average saving rate in the first six months after the earthquake is 0.059 and that in summer (July and August) increased to 0.068. The more reduction of electricity entailed the reduction of production. Finally, from the minimum saving rate during a first year after the disaster (appeared in May), at least 2–3 % of power saving in the first six months stems from the countermeasures of which effects last for a long period.

## *4.2 Analysis per Industrial Sector*

Resilience to power shortages likely differs among industrial sectors. For example, resilience could depend on the quantity of electricity consumed and the flexibility of production scheduling. The previous analysis is now applied to specific industrial sectors that exhibit clear relationships between production output and electricity consumption before the earthquake.

Figure 8 presents the electricity saving rates for Kanto's industrial sectors. The plot of productivity increase rates is omitted here because the rate has the similar movement with the electricity saving rate. Overall, the rate of each sector is similar

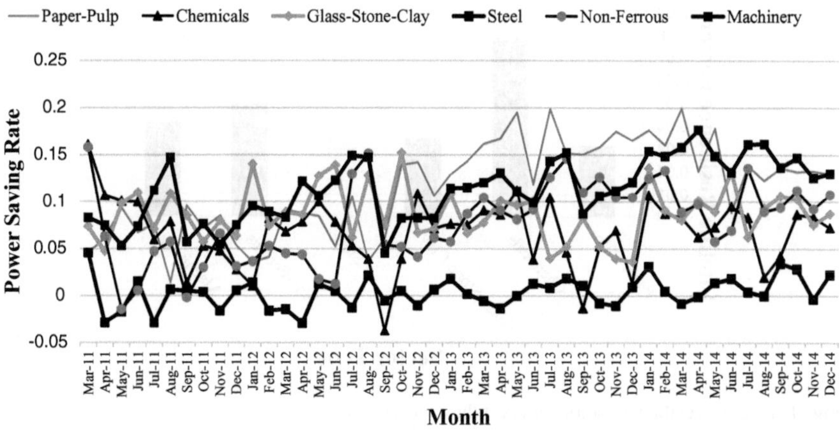


**Fig. 8** Monthly time series of estimated power saving rates for Kanto industrial sectors

to the case of aggregated industrial sector as shown in Fig. 7. That is, the saving rates during a few months after the earthquakes vary at the mean of approximately 0.05, and there is upward saving trend after the September 12. The machinery sector achieved the largest saving rate during the severest power shortage occurred in 2011 July and August.

In contrast, the resilience under power shortage and electricity saving capacity is inadequate during this period in the steel sector. The steel sector requires intensive electricity consumption and has difficulty in effectively adapting to shortages. In the application to future power shortages in many areas, the saving rates demonstrated in the sector can be adopted based on the structure of different economy. In each sector, sources of the adaptation for electricity saving come from temporal countermeasures similarly to the case of the overall sector.

Figure 9 illustrates the lifeline resilience factor, defined as production capacity remaining under power outages. Values are normalized between 0 and 1 and obtained from the past business survey in Aichi and Shizuoka prefectures, Japan (Kajitani and Tatano, [13]). 0 indicates that production capacity is lost and 1 indicates that production capacity remains at the same level when the outage does not occur. The steel industry has small resilience during complete and partial shortages. Chemicals display larger values, a finding consistent with instances of power shortages in 2011 March.

What would happen to total Japanese economy especially due to the small resilience in steel sector? Different from the automobile industries, which were required to recover for meeting the demands, the impacts of the steel sector on the Japanese economy were inferred as inadequate. In Fig. 10, which indicates the amount of steel imports from outside the country (The Japan Iron and Steel Federation, [22]), the imports increased in 2011 but decreased in 2012. The supply condition of steel may have been temporally adverse and the imports may have covered the

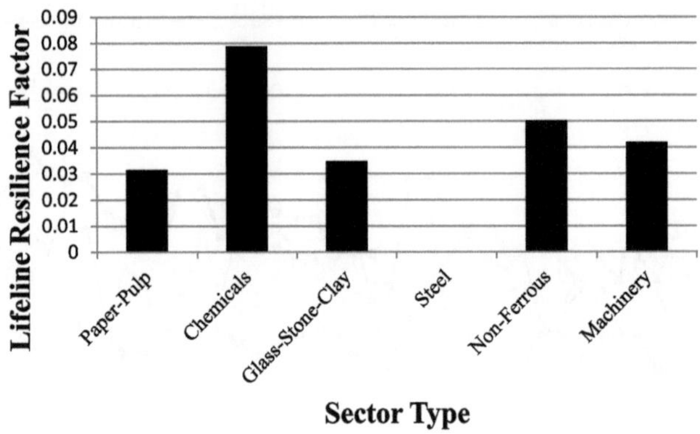


**Fig. 9** Lifeline resilience factor for six industrial sectors

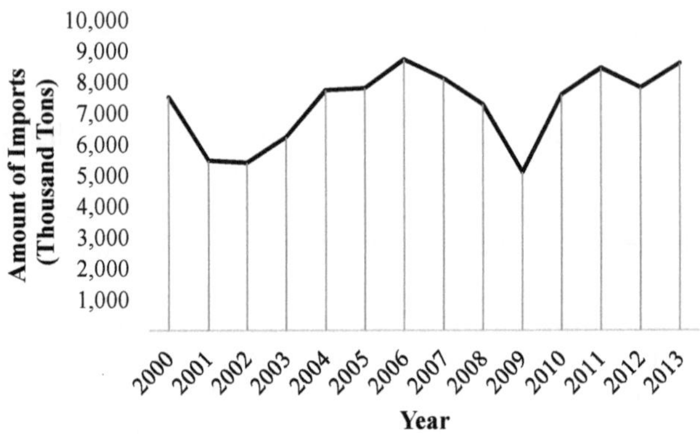


**Fig. 10** Amount of international imports of steel from 2000 to 2013

shortage; however, it is unlikely that the adverse conditions last for a long period considering that the imports decreased in 2012. However, what could be a bottleneck of the economy depends on the economic conditions. Significant impacts may have been created by the reduction of steel production if its demand is adverse worldwide, such as in the year of 2007. Furthermore, even if that demand is small, production reduction is a large risk for the sector in the competing market.

## 5 Conclusion

This study has investigated the resilience of Japanese businesses to power shortages by examining production output under differing degrees of shortages after the 2011 Great East Japan Earthquake. It has focused on the structural change in linear relationships between production output and electricity consumption using time series analysis to remove temperature and seasonal effects. Many studies quantitatively analyze the business impact of power blackouts; however, few establish metrics for resilience to shortages and power saving efforts following the shortages.

The following findings emerged from case study of Kanto region.

(a) Resilience characteristics during the first 6 months

The resilience of manufacturing in Kanto, which was represented by electricity saving rate and productivity increase rate, was high immediately after the earthquake (in March) even if a scheduled blackout was performed. The efforts such as utilizing inventories may have been effective. The average saving rate in first 6 months after the earthquake is 0.059 and that in summer (July and August) increased to 0.068. The more reduction of electricity entailed the reduction of production.

(b) Resilience characteristics after six months

From the minimum saving rate of time series during a first year after the disaster, at least 0.02–0.03 of power saving in the first six months stems from the countermeasures of which effects last for a long period. The power saving rates monotony increases after the September 2012. This indicates that the electricity saving measures can be regularly practiced by employees and permanent countermeasures, such as renewals of old machineries, can be undertaken and continuously upgraded by the businesses. Increases in electricity price in April 2012 may potentially accelerate this trend.

(c) Resilience characteristics of individual sector

Analysis of industrial sectors revealed differing degrees of resilience to power shortages. In the steel sector, production declined at rates greater than electricity conservation. These results reinforce previous survey results of business resilience during blackouts. Especially, it was shown that the resilience in steel sector is low both in the case of power shortage and blackout. These sectoral analyses provides the basic assumptions of production decrease caused by future power shortages occurred in the different business proportion environment.

In sum, this study determined the resilience of Japanese business to power shortages by documenting their responses to a real disaster. Resilience could differ whether the businesses have enough time to prepare for anticipated power shortages or other conditions such as accumulated inventories as well as the type of business. The long-term trend of power saving in industrial sector is a favorable point revealed in this study.

To apply our method of analysis to estimate risks of power shortages and enlarge it to examine other daily necessities, such as water and food, data need to be accumulated and the statistical model enhanced. Enhancements include comparing different techniques such as ARIMA and ANN for modeling time series, combining datasets with different spatial and temporal scales, and investigating other methods to detect structural changes before and after crises. Comparative analysis between different countries would help to extend the application to the shortages in the world.

## References


1. Agency for Natural Resources and Energy. (2011). *Follow-up of countermeasures to electricity power shortages in this summer (large business, small business, and household).* http://www.enecho.meti.go.jp/committee/council/basic_problem_committee/006/pdf/6-42.pdf. Accessed October 16 2015 (in Japanese).
2. An, N., Zhao, W., Wang, J., Shang, D., & Zhao, E. (2013). Using multi-output feedforward neural network with empirical mode decomposition based signal filtering for electricity demand forecasting. *Energy, 49*, 279–288. doi:10.1016/j.energy.2012.10.035.
3. Azadeh, A., Ghaderi, S. F., & Sohrabkhani, S. (2008). Annual electricity consumption forecasting by neural network in high energy consuming industrial sectors. *Energy Conversion and Management, 49*(8), 2272–2278. doi:10.1016/j.enconman.2008.01.035.
4. Bank of Japan. (2011). *Yen/dollar spot in the Tokyo market at 17:00 on July 11, 2011.* http://www.stat-search.boj.or.jp/ssi/mtshtml/d.html. Accessed October 4 2015.
5. Fatai, K., Oxley, L., & Scrimgeour, F. G. (2004). Modelling the causal relationship between energy consumption and GDP in New Zealand, Australia, India, Indonesia, The Philippines and Thailand. *Mathematics and Computers in Simulation, 64*(3–4), 431–445. doi:10.1016/s0378-4754(03)00109-5.
6. Fujimi, T., & Chang, S. E. (2014). Adaptation to electricity crisis: Businesses in the 2011 Great East Japan triple disaster. *Energy Policy, 68*, 447–457. doi:10.1016/j.enpol.2013.12.019.
7. González-Romera, E., Jaramillo-Morán, M. Á., & Carmona-Fernández, D. (2007). Forecasting of the electric energy demand trend and monthly fluctuation with neural networks. *Computers & Industrial Engineering, 52*(3), 336–343. doi:10.1016/j.cie.2006.12.010.
8. Hipel, K. W., & McLeod, A. I. (1994). *Time series modelling of water resources and environmental systems*. Amsterdam: Elsevier.
9. Hippert, H. S., Bunn, D. W., & Souza, R. C. (2005). Large neural networks for electricity load forecasting: Are they overfitted? *International Journal of Forecasting, 21*(3), 425–434. doi:10.1016/j.ijforecast.2004.12.004.
10. Hyodo, T. (2012). Demand analysis on electricity during energy crisis period after the earthquake 2011. *Transport Policy Studies' Review, 15*(1), 20–25 (in Japanese).
11. International Energy Agency (IEA). (2005). *Saving electricity in a hurry.* http://www.iea.org/publications/freepublications/publication/saving-electricity-in-a-hurry-2005.html. Accessed February 14 2015.
12. Japan Meteorological Agency (JMA). (2015). *Past climate information.* http://www.data.jma.go.jp/obd/stats/etrn/index.php. Accessed October 4 2015 (in Japanese).
13. Kajitani, Y., & Tatano, H. (2009). Estimation of resilience factors based on surveys of Japanese industries. *Earthquake Spectra, 25*(4), 755–776.
14. Kanto Bureau of Economy, Trade and Industry (METI-KANTO). (2015). *Result of electricity power demand (December).* http://www.kanto.meti.go.jp/tokei/denryoku/20130214index.html. Accessed February 14 2012 (in Japanese).
15. Kanto Bureau of Economy, Trade and Industry (METI-KANTO). (2015b). *Trend of industrial production.* http://www.kanto.meti.go.jp/tokei/kokogyo/kokogyo_index.html. Accessed February 14 2012 (in Japanese).
16. Kaytez, F., Taplamacioglu, M. C., Cam, E., & Hardalac, F. (2015). Forecasting electricity consumption: A comparison of regression analysis, neural networks and least squares support vector machines. *International Journal of Electrical Power & Energy Systems, 67*, 431–438. doi:10.1016/j.ijepes.2014.12.036.
17. Nawaz, S., Iqbal, N., & Anwar, S. (2014). Modelling electricity demand using the STAR (Smooth Transition Auto-Regressive) model in Pakistan. *Energy, 78*, 535–542. doi:10.1016/j.energy.2014.10.040.
18. Nihon Keizai Shinbun (Nikkei). (2011). Earthquake disaster and macro-economic. *Analysis, 24*, (in Japanese).

19. Pappas, S. S., Ekonomou, L., Karamousantas, D. C., Chatzarakis, G. E., Katsikas, S. K., & Liatsis, P. (2008). Electricity demand loads modeling using AutoRegressive Moving Average (ARMA) models. *Energy, 33*(9), 1353–1360. doi:10.1016/j.energy.2008.05.008.
20. Taylor, J. W. (2010). Triple seasonal methods for short-term electricity demand forecasting. *European Journal of Operational Research, 204*(1), 139–152. doi:10.1016/j.ejor.2009.10.003.
21. Taylor, J. W., & Buizzab, R. (2003). Using weather ensemble predictions in electricity demand forecasting. *International Journal of Forecasting*, *19*, 57–70.
22. The Japan Iron and Steel Federation. (2015). *Monthly steel supply and demand statistics*. http://www.jisf.or.jp/data/tokei/index.html. Accessed August 14 2015 (in Japanese).
23. Tohoku Electric Power Co. (2015). *Past power demand statistics*. http://setsuden.tohoku-epco.co.jp/download.html. Accessed August 14 2015 (in Japanese).
24. Vu, D. H., Muttaqi, K. M., & Agalgaonkar, A. P. (2015). A variance inflation factor and backward elimination based robust regression model for forecasting monthly electricity demand using climatic variables. *Applied Energy*, *140*, 385–394. doi:10.1016/j.apenergy.2014.12.011.
25. Zahedi, G., Azizi, S., Bahadori, A., Elkamel, A., & Wan Alwi, S. R. (2013). Electricity demand estimation using an adaptive neuro-fuzzy network: A case study from the Ontario province—Canada. *Energy*, *49*, 323–328. doi:10.1016/j.energy.2012.10.019.

# Atmospheric $CO_2$ and Global Temperatures: The Strength and Nature of Their Dependence

**Granville Tunnicliffe Wilson**

**Abstract** There is now considerable scientific consensus that the acknowledged increase in global temperatures is due to the increasing levels of atmospheric carbon dioxide arising from the burning of fossil fuels. Large scale global circulation models support this consensus and there have also been statistical studies which relate the trend in temperatures to the carbon dioxide increase. However, causal dependence of one trending series upon another cannot be readily proved using statistical means. In this paper we model the trend corrected series by times series methods which provide a plausible representation of their dependence. A consequence of trend correction and our use of relatively short series is that our model is unable to give precise long-term predictions, but it does illuminate the relationships and interaction between the series.

**Keywords** Time series prediction · Spectral coherency · Structural VAR · Graphical modeling

## 1 The Series

In a previous unpublished conference paper we modeled three series: (i) the atmospheric carbon dioxide ($CO_2$) concentration in parts per million (ppm) observed at Mauna Loa, (ii) the annual global mean temperature anomaly known as HadCRUT3 and (iii) the southern oscillation index (SOI). The HadCRUT3 series is a combination of the sea surface temperature (SST) anomaly series known as HadSST-gl and the land surface temperature (LST) anomaly series known as CRUTEM3. In the present paper we use these two separate series (with CRUTEM3 updated to CRUTEM4) in place of the combined series, which leads to simplification of the model.

G.Tunnicliffe Wilson (✉)
Department of Mathematics and Statistics, Lancaster University, Lancaster LA1 4YF, UK
e-mail: g.tunnicliffe-wilson@lancaster.ac.uk



W.K. Li et al. (eds.), *Advances in Time Series Methods and Applications*,
Fields Institute Communications 78, DOI 10.1007/978-1-4939-6568-7_13

The sources for the series are

(i) $CO_2$: ftp://aftp.cmdl.noaa.gov/products/trends/co2/
(ii) HadSST-gl and CRUTEM4: http://www.cru.uea.ac.uk/cru/data/temperature/
(iii) SOI: http://www.cpc.ncep.noaa.gov/data/indices/

The SOI is the observed sea level pressure difference between Tahiti and Darwin, Australia, and is strongly related to ocean temperatures across the Eastern Pacific Ocean. In our models we should consider the SOI series to be a proxy variable for these ocean temperatures, or possibly ocean temperature gradients across the region, because, unlike the temperature series, it has no visually evident trend. It is formed as the difference of the two standardized series, but to minimize preprocessing of the data we omit the standardization and we will also reverse the sign of the difference so our series is positively correlated with sea surface temperatures.

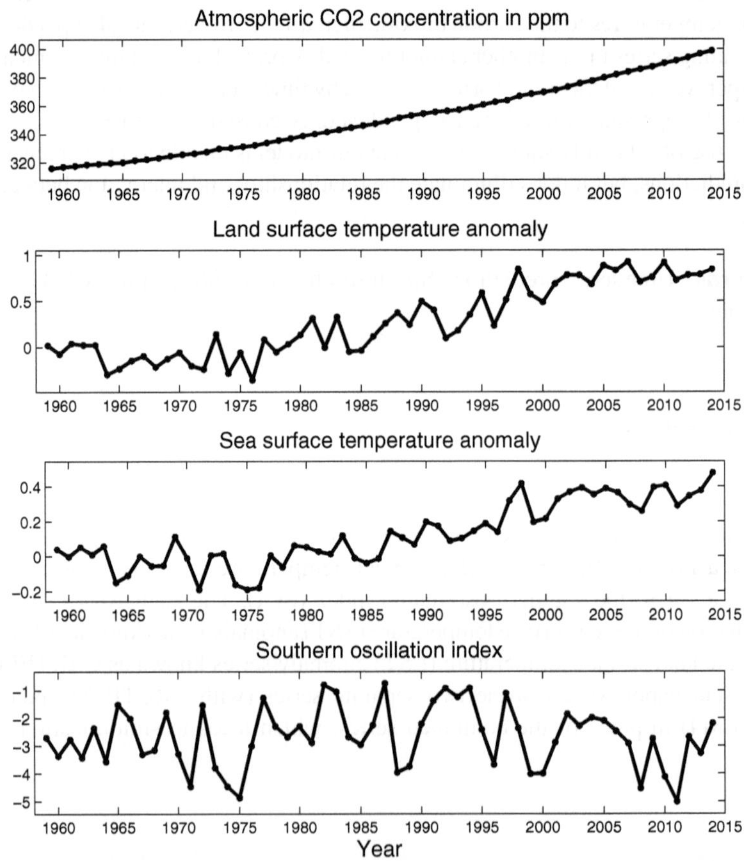


**Fig. 1** The four climate series analyzed and modeled in this paper

The $CO_2$ and SOI series are precisely defined and measured at respectively one and two stations. In contrast, the land and sea surface temperature anomalies are compiled from a large number of measurements around the globe. We use the global values but they are also available separately for the northern and southern hemispheres. However, we will be modeling annual mean values from 1959, the first full year of Mauna Loa records, to 2014, giving series of length 56. With 6 series, saturated forms of the vector autoregressive (VAR) models that we use would have an excessive number of parameters for models of more than very low order. We shall show in the next section that the spectral coherency between all the series is high, suggesting that there is little need for more information.

We model the annual series shown in Fig. 1 because of the substantial within-year variability of $CO_2$ and temperatures due to the natural seasonal influences.

Within 1 year we expect each variable to influence others, for example $CO_2$ levels influence air temperature and uptake of $CO_2$ in the oceans depends on sea surface temperatures. We will therefore develop structural forms of vector autoregressions with simultaneous equation relationships between the innovations in current annual values, to represent the net effect of these mutual influences.

## 2 Spectral Coherency Between the Series

Trends are evident in the raw series apart from the SOI. A time series model for the trending carbon dioxide and global temperature series over a longer time period is given by Young [18]. However, our intention as expressed in the abstract is to correct for these trends for which we use ordinary least squares (OLS) regression. The $CO_2$ series is corrected for a quadratic trend and the temperature series for linear trends which are visually evident. The SOI series is simply mean corrected; trend correction makes very little difference. Figure 2 shows these trend corrected series. The land and sea temperatures look very similar, but they have no obvious similarity with the other series.

However, the spectral coherency between all the series, which takes into account linear lagged dependence, is quite significant, as shown in the upper half of Fig. 3, in a form introduced by Dahlhaus [2, p. 167]. The strong coherency between the trend corrected monthly carbon dioxide and global temperature series was previously demonstrated by Kuo et al. [7], for records over the shorter period from 1958 to 1988.

The lower half of Fig. 3 shows the partial coherencies which measure, for a pair of series, the further dependence of the one upon the second, given its dependence upon the other two. These are seen to be much weaker than the pairwise coherencies. In all these plots the significance limits are those which are exceeded with 5 % probability when there is no coherency at a particular frequency. The coherencies may appear significant at no, some or all frequencies. The limits are higher at the end points of the frequency range because the band of frequencies over which smoothing is applied then includes sample spectral values outside the frequency range from 0 to 0.5 which are correlated with those within this range. The shape of the smoothing window is

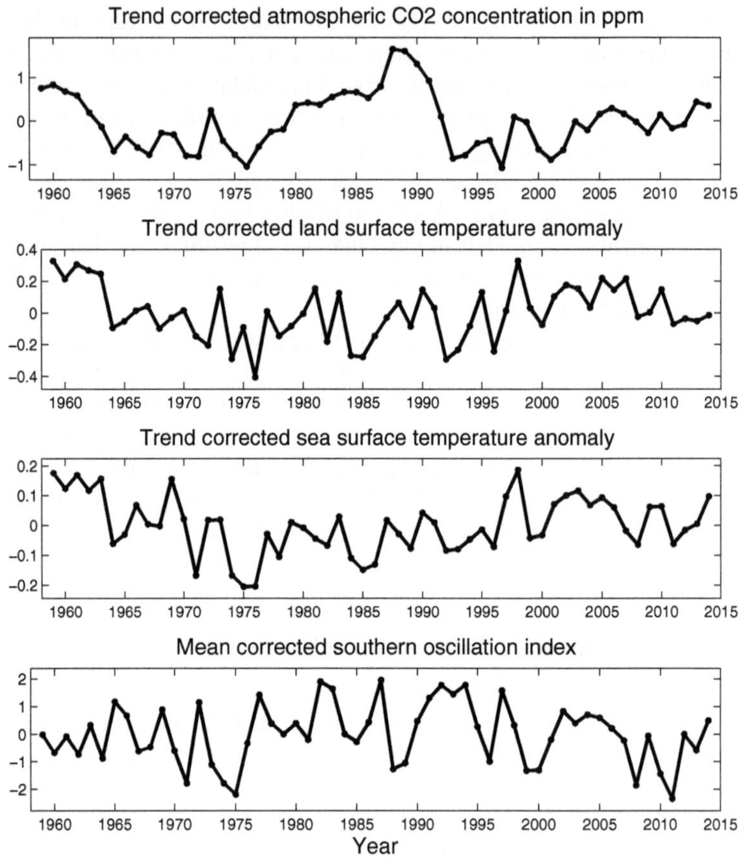


**Fig. 2** The four climate series after correction for mean and trend components

that of the sum of 4 uniform random variables. We summarize the partial coherencies in Fig. 4 in which the links between the series reflect the significance observed in the lower half of Fig. 3. A solid line corresponds to significant coherency over some part of the frequency range and a broken line corresponds to marginal significance, otherwise no link is shown.

This graph provides only a limited description of the dependence between the series but it can be related to their causal dependence as described by Dahlhaus and Eichler [3]. Subject, of course, to statistical uncertainty, the graph implies that in a structural VAR model for the series there should be no explicit dependence of either $CO_2$ or LST on present or past values of the other. We will comment further on this after we have built such a model.

Spectral analysis can also estimate the lagged response of one series to another, and we show these for selected pairs of the climate series in Fig. 5. Thus the uppermost plot of this figure shows the regression coefficients of SOI on SST at positive and negative

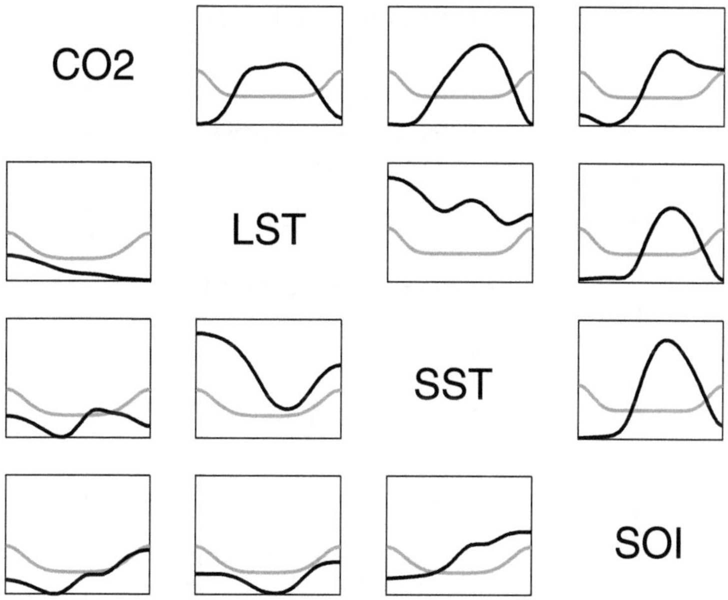


**Fig. 3** An array of plots showing the pairwise squared coherencies above the diagonal and partial coherencies below the diagonal, between the four climate series indicated on the diagonal. In each plot the *horizontal axis* is the frequency range from 0 to 0.5 cycles per annum and the *vertical axis* is the squared correlation range from 0 to 1. The proportion of tapering applied to the series is 0.1, the squared coherencies are smoothed using a bandwidth of 0.2 and the partial coherencies using a bandwidth of 0.3. The coherencies are shown in the *solid black line* and their 5 % significance limits by the *solid gray line*

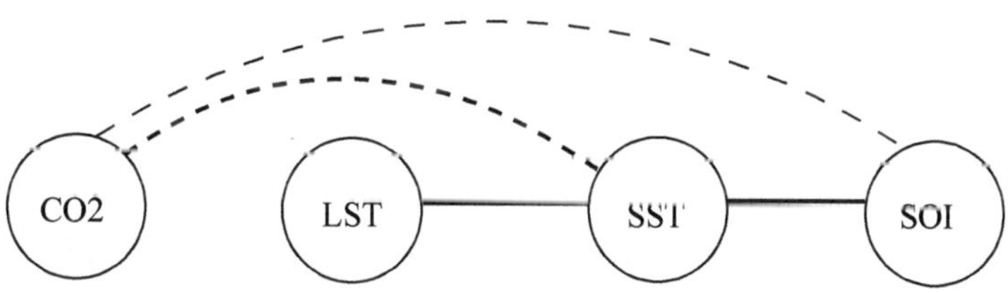


**Fig. 4** The partial coherency graph between the four climate series

lags. At positive lags the coefficients describe the effect of the current value of SST on future values of SOI—the response of SOI to SST. The regressions are estimated, however, in the presence of feedback between the series, so the coefficients do not correctly estimate the causal response, and significant coefficients may be observed at negative lags. For the correct causal response we will use a VAR model.

The response of SOI to SST shows the acknowledged strong positive relationship (with our sign convention) in the same year, but this is followed by a strong negative relationship in the subsequent year. This reflects the eponymous oscillatory nature of the SOI, with a period of between 2 and 3 years. The response of $CO_2$ to SST is

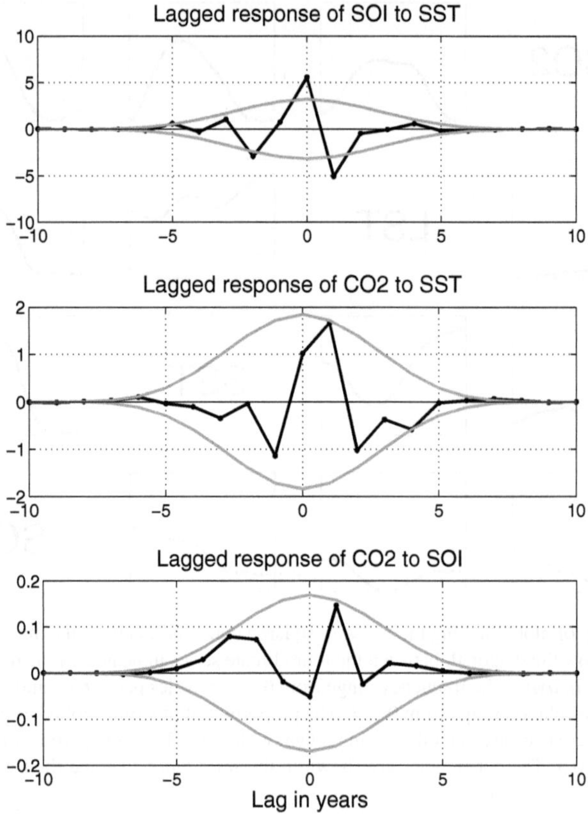


**Fig. 5** The lagged responses of selected pairs of climate series, with the estimated coefficients shown by the *points* connected by *solid black lines* and the significance limits shown by *gray lines*. The estimates lie outside these limits with 5 % probability if there is no lagged partial regression relationship between the series

positive in the same and subsequent year, though only marginally significant when estimated by spectral means. This reflects the fact that at higher temperatures the sea cannot absorb so much $CO_2$, so there is an apparent consequent net increase. Of particular interest is the response of $CO_2$ to SOI at a lag of 1 year. A low SOI is associated with cooler sea temperatures in the Eastern Pacific with upwelling of cold seawater close to the South American west coast. A high SOI (with our sign convention) is associated with the El Niño effect in which this upwelling fails and warmer water floods into that area. This warmer water can absorb much less $CO_2$ than the cold upwelling. The response shown in Fig. 5 suggests that actually our annual SOI series is predictive of this effect by 1 year.

Spectral estimation of these responses are of course limited in their causal interpretation and as semi-parametric estimators they are not as statistically efficient as the parametric models we consider next. They do, however, give useful predictive information and insight into the relationships between the series.

## 3 A Standard VAR Model for the Series

As a preliminary step in developing an empirically identified structural VAR model we first identify a saturated VAR model of order $p$, in the standard form which may be found, for example in Lütkepohl [11] or Reinsel [13]:

$$x_t = \Phi_1 x_{t-1} + \Phi_2 x_{t-2} + \cdots + \Phi_p x_{t-p} + e_t, \tag{1}$$

where $x_t$ is the vector of series values at time $t$ and $\Phi_1, \ldots, \Phi_p$ are the matrix coefficients of dependence upon lagged series values. The error or innovation vector $e_t$ has, in general, correlated elements with covariance matrix $V_e$. It is also multivariate white noise and uncorrelated with all past observations $x_{t-k}$ for $k > 0$.

The series are all zero mean as a consequence of mean and trend correction and we assume they are second order stationary. The VAR model is therefore fitted for increasing orders of $p$ and the AIC, Akaike [1], plotted as in the left hand plot of Fig. 6. The model is fitted by exact maximum likelihood under the normal assumption. This gives very similar estimates to the use of lagged OLS regression but uses information

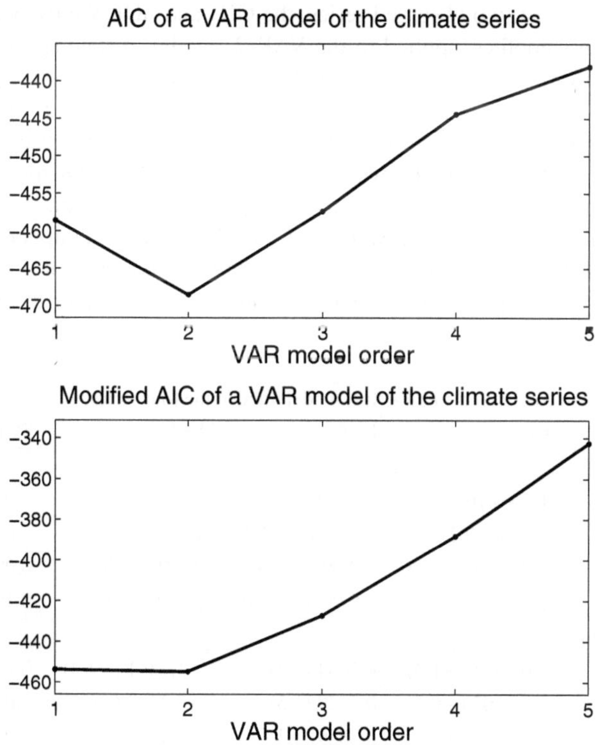


**Fig. 6** Plots of the AIC and modified AIC of a VAR model for the climate series

in the full length of series, which is preferable for short series. Lagged regression of order $k$ necessarily uses the reduced series length of $n - k$. The plot on the right of Fig. 6 shows the AIC as modified by Hurvich and Tsai [6] to improve the order selection by this criterion in small samples. In fact both have their minimum for the model of order 2, though the modified AIC is only marginally less than its value at order 1. For univariate series the AIC is known not to be consistent for order selection, with a small probability of selection of an order somewhat higher than the true one. However, the situation is not so simple for our multivariate series because 16 new coefficients are introduced when the order is increased by 1. The probability of overestimating the true order, $p$, is then quite low, and on the contrary, in small samples the true order may be underestimated if only a small number of the 16 coefficients are non-zero at lag $p$. Other criteria such as that due to Schwarz [14] may be asymptotically consistent in selecting the true order, but are even more prone to underestimation of the order in this circumstance. In fact, for the model of order 2, as shown in Table 1, the dependence of LST on SOI at lag 2 has $t$-value 2.0 and the dependence of SST on LST at lag 2 has $t$-value 2.3. This supports the selection of the order $p = 2$ indicated by the modified AIC.

We now demonstrate the extent to which this VAR(2) model for the four climate series captures their dependence by using it to estimate the whole record of the SST series from just two observed series, the $CO_2$ and SOI. We use what is known as the Kalman smoother, applied to the VAR(2) model in state space form. To be precise, the information used to estimate the SST series and the error limits of this estimate, is just the pair of full $CO_2$ and SOI series and the VAR(2) model (fitted to the four full series). The only information otherwise used from observations of the SST series is the linear trend by which it was corrected for VAR model estimation and which was used to restore that linear component after estimation. Neither was any information used from observations of the LST series. The upper plot in Fig. 7 shows the observed SST series, its estimate and two standard error limits. Note that the estimates are not predictions from past values. Each value of the SST series is

**Table 1** Estimated coefficients of a VAR(2) model fitted to the four series, with $t$-values in brackets

| | $CO_2$ | LST | SST | SOI |
|---|---|---|---|---|
| Lag 1 coefficients | | | | |
| $CO_2$ | 1.0061 (6.8) | 0.0402 (0.1) | 0.9272 (0.8) | −0.1249 (−1.8) |
| LST | 0.0106 (0.2) | −0.2262 (−1.1) | 1.2352 (2.7) | −0.0211 (−0.8) |
| SST | −0.0377 (−1.1) | −0.0937 (−0.8) | 0.7718 (2.9) | 0.0032 (0.2) |
| SOI | 0.4336 (1.0) | −0.0010 (−0.0) | 3.6441 (1.1) | 0.3945 (1.9) |
| Lag 2 coefficients | | | | |
| $CO_2$ | −0.1931 (−1.3) | 0.1634 (0.3) | −1.7341 (−1.4) | 0.1265 (1.9) |
| LST | −0.0105 (−0.2) | 0.3229 (1.5) | −0.2382 (−0.5) | 0.0519 (2.0) |
| SST | 0.0214 (0.6) | 0.2859 (2.3) | −0.4309 (−1.5) | 0.0119 (0.8) |
| SOI | −0.5969 (−1.4) | −1.1557 (−0.7) | 1.3989 (0.4) | −0.0025 (−0.0) |

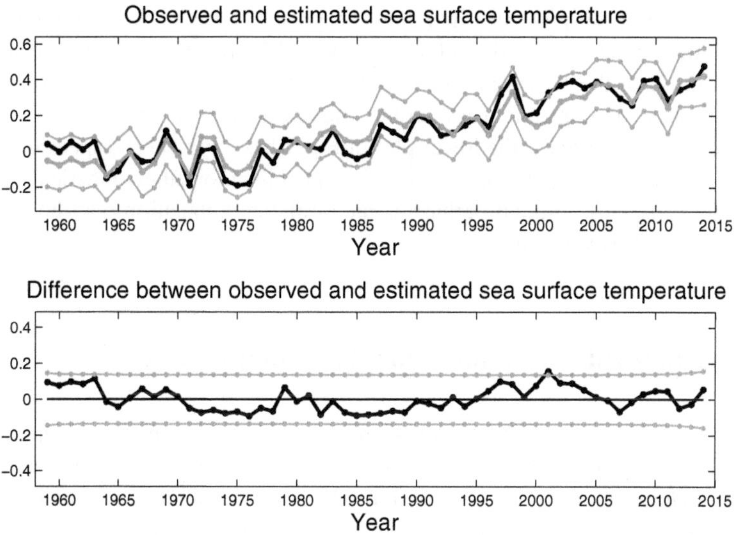


**Fig. 7** The *upper plot* shows the observed SST series (*black line*) and its estimate and two standard error limits (*gray lines*), derived from observations only of the $CO_2$ and SOI series, using a VAR(2) model fitted to the four climate series. The *lower plot* shows the difference between the observed and estimated series with the error limits

estimated from the whole sequence of the $CO_2$ and SOI series. The lower plot in Fig. 7 shows the difference between the observed and estimated series with the error limits.

We make the following remarks on these plots.

1. The estimated SST follows the pattern of the observed SST remarkably well, and even over periods where it is generally lower or higher, it follows the year to year movements well.
2. This similarity does not imply causality in either direction between the predicting series of $CO_2$ and SOI and the predicted series of SST, because they may be related by mutual dependencies which we aim to model in later sections.
3. Though the observed SST varies by only a few tenths of a degree over the whole record, it is compiled from a large number of temperature measurements which are subject to much greater diurnal and annual variation. The precisely and objectively defined nature of the predicting series gives strong support to the claim that the well predicted observed temperature series is similarly well defined.
4. When the VAR model is fitted only to the observations of the four climate series before the year 2000, there is no visual difference in the plots of Fig. 7 when they are constructed using this restricted model. In particular there is no visual difference in the SST estimated from the $CO_2$ and SOI series over the years 2000–2014. This suggests that there is no essential change in the times series nature of the four climate series over this period.

5. We note from the plots in Fig. 7 that the general level of the observed SST is above that of the estimated level over the period from 1995 to 2005. The apparent leveling off of the observed SST over the first decade of the new century is due to this surge, and its conclusion around the middle of that decade.
6. The difference series in the lower plot of Fig. 7 has a typical red spectrum, rising at lower frequencies, and has no significant cyclical, or other statistical, features worthy of remark, though perhaps some climatic explanation may be found by inspection of its variation.

# 4 A Structural VAR Model for the Climate Series

By a structural VAR (SVAR) model we mean one in which each current value of the series may depend upon other current values besides a set of lagged values. This dependence explains the correlation which would otherwise remain between the innovation series if dependence was allowed only upon past values, as in the standard VAR. The structural VAR model therefore results in structural innovations which are uncorrelated. The model equation is now

$$\Phi_0 x_t = \Phi_1 x_{t-1} + \Phi_2 x_{t-2} + \cdots + \Phi_p x_{t-p} + a_t, \tag{2}$$

where $\Phi_0$ represents dependence between current values and the structural innovations $a_t$ have diagonal variance matrix. Our aim is to identify and estimate this model with sparse forms of the coefficient matrices $\Phi_k$ which represent the structure of the dependence within and between the series.

We have previously propounded methods of identifying SVAR models, for example in Reale and Tunnicliffe Wilson [12] and Tunnicliffe Wilson et al. [16]. This methodology is very much motivated by and based upon the graphical modeling procedures used to identify relationships between a general set of variables represented by directed acyclic graphs (DAGs). This is clearly set out in the books Whittaker [17], Lauritzen [8] and Edwards [4]. One of the main statistical tools used in this identification is the conditional independence graph (CIG) between the variables. A simple rule determines how the CIG for a set of variables may be derived from the DAG representing their dependence. A given CIG does not, however, necessarily determine uniquely the structure of this DAG. In many cases, however, a small number of possible DAG representations can be determined, one of which may be selected as the best, following their estimation by maximum likelihood.

The idea of acyclic dependence in a DAG, is that the variables may be ordered so that each is dependent only on a subset of those which are previous in the ordering. More traditionally, in econometrics, see Zellner and Theil [19], a set of simultaneous equations, suggested by economic theory, may be used to represent the relationships between variables. Each equation may only involve a small subset of all the variables, upon which restrictions are imposed to ensure that the relationships may be uniquely identified and estimated. However, there is no requirement that the dependence be acyclic. These methods have been extended to represent simultaneous equation relationships between current values of a structural VAR model, see for example the much cited paper of Sims [15].

In our earlier exposition, Reale and Tunnicliffe Wilson [12], of graphical modeling as applied to identifying SVAR models, we restricted ourselves to acyclic relationships between current variables. However, in Tunnicliffe Wilson et al. [16] we explored the possibility of using cyclic simultaneous equation representations, which we believe may be appropriate for the climate series which are the subject of this paper. Fortunately, the CIG between the variables, determined empirically by statistical analysis, may still be used in the identification of their dependence, whether or not this is acyclical.

Because our interest is restricted to linear relationships between variables described by their second order statistical moments, we can construct their CIG from their sample partial correlation graph (PCG). As described in Tunnicliffe Wilson et al. [16], the PCG is formed from the data matrix **X** whose columns are the four mean and trend corrected climate series and their values to lag 2—the order of their standard VAR representation. Their sample covariance matrix is then $\widehat{V} = \frac{1}{n}\mathbf{X}'\mathbf{X}$ and their sample inverse covariance matrix is computed as $\widehat{W} = \widehat{V}^{-1}$. From its entries, $\widehat{W}_{i,j}$, the sample partial correlations can be calculated as

$$\widehat{\pi}(X_i, X_j) = \frac{-\widehat{W}_{i,j}}{\sqrt{\widehat{W}_{i,i}\widehat{W}_{j,j}}}. \tag{3}$$

A sample partial correlation between two variables is closely related to the $t$-value of the coefficient of one of them, in the regression of the other upon the whole set of variables. Under the hypothesis that $\pi(X_i, X_j) = 0$, the ratio

$$\frac{\widehat{\pi}(X_i, X_j)\sqrt{n - m + 1}}{\sqrt{1 - \widehat{\pi}(X_i, X_j)^2}} \tag{4}$$

is distributed as a $t_{n-m+1}$ variable where $n - m + 1$ is the number of degrees of freedom. We therefore reject the null hypothesis that $\pi(X_i, X_j) = 0$ at level $\alpha$ if

$$\left|\widehat{\pi}(X_i, X_j)\right| > \frac{t_{\alpha/2,n-m+1}}{\sqrt{t^2_{\alpha/2,n-m+1} + (n - m + 1)}}. \tag{5}$$

where $t_{\alpha/2,n-m+1}$ is the corresponding critical value of the $t_{n-m+1}$ distribution.

Our interest is purely in the relationships between current variables and between current and past variables, not between past variables. On applying this procedure to our climate variables we show in Fig. 8 the CIG of their dependence. No link is shown between two variables for which the associated $t$-value defined in (4) is less than 1.645 in absolute value. The strength of a link is shown also by the style of line:

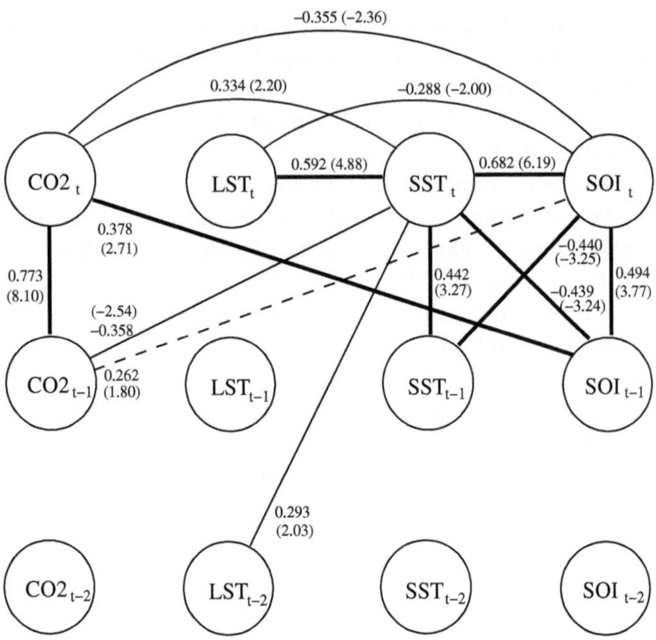


**Fig. 8** The sample CIG between the four climate series and their values to lag 2. Significant partial correlation between two variables is shown by a link, against which is shown the value of the partial correlation with its associated $t$-value in *brackets*

either broken, thin solid or thick solid, if the $t$-value is greater than 1.645, 1.96 and 2.575, corresponding to test levels of 10, 5 and 1 %.

Our next step is to postulate an SVAR for which the corresponding CIG (or PCG) is given by that in Fig. 8. We will also present this SVAR graphically using a similar diagram to that shown in the figure, but with directions attach to the links, indicated by arrow heads, showing which variables are used to explain the current variables at time $t$. All links from the past are naturally directed towards current variables. A link from one current variable to another specifies them as respectively explanatory and dependent variables. If the relationship so specified between the current variables is acyclic, then the SVAR may be estimated by OLS linear regression of each current variables on its specific set of explanatory variables. Otherwise, the Gaussian likelihood of the model is evaluated, by transforming it to the standard VAR form, and maximized numerically.

To illustrate the rule by which the CIG of an SVAR may be derived, we use that represented by the diagram in Fig. 9.

The relationship between current variables in this figure is *not* acyclic. If it were we would call it a directed acyclic graph (DAG). As it is we note the cycle of links $CO_{2_t} \rightarrow SST_t \rightarrow SOI_t \rightarrow CO_{2_t}$, and we will call it a directed structural graph (DSG). However, for both types of graph the implied CIG (or PCG) between the variables can be derived using the moralization rules of Lauritzen and Spiegelhal-

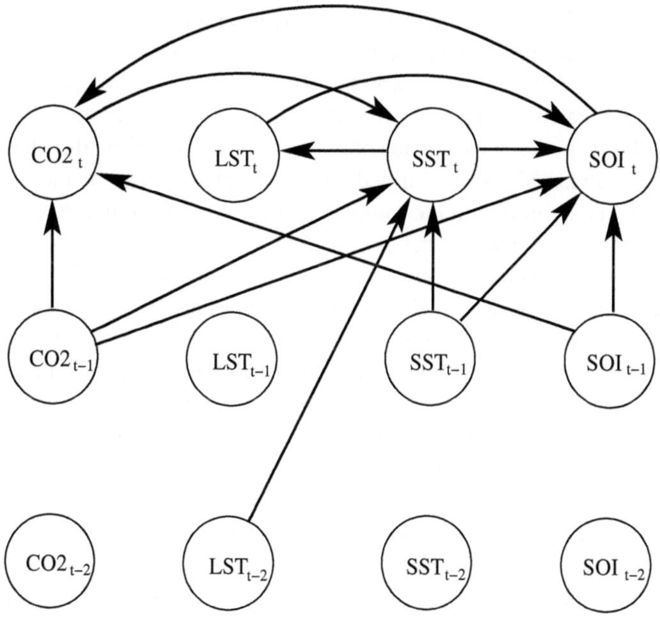


**Fig. 9** A diagram representing an SVAR which might explain the CIG shown in Fig. 8

ter [9]. These are expressed in the language of graphical modeling in which each of the 12 variables shown in Fig. 9 are referred to as nodes and for a given node its parent nodes are those from which it receives a directed link—i.e. its explanatory variables. The rule is then:

1. For each node of the DAG or DSG insert an undirected edge between all pairs of its parent nodes, unless they are already linked by an edge. This is called marrying the parents (to make them moral).
2. Replace each directed edge in the DAG or DSG by an undirected edge.

Doing this for the graph in Fig. 9 gives the graph in Fig. 8, but only on omitting a few extra links for reasons on which we comment shortly. Note that moralization in this case introduces the link between $\mathrm{SOI}_{t-1}$ and $\mathrm{SST}_t$ in Fig. 8 because in Fig. 9 these are both parents of $\mathrm{SOI}_t$. Also the choice of direction for the link $\mathrm{SST}_t \rightarrow \mathrm{LST}_t$ avoids the introduction of a moralization link between $\mathrm{SST}_{t-1}$ and $\mathrm{LST}_t$. Reversing the direction of the postulated link $\mathrm{SST}_t \rightarrow \mathrm{LST}_t$ would lead to the introduction of this moralization link, in conflict with the graph in Fig. 8. Such considerations help to specify the postulated model. There are some other moralization links such as between $\mathrm{SST}_{t-1}$ and $\mathrm{LST}_t$ that should be added because these are also both parents of $\mathrm{SOI}_t$. However, moralization links generally have a lower associated partial correlation and may not appear as significant in the PCG. In Tunnicliffe Wilson et al. [16] we present some quantitative rules which can be applied in restricted contexts. For example, if we were given just the three nodes $\mathrm{LST}_t$, $\mathrm{SOI}_t$ and $\mathrm{SST}_{t-1}$ related as in Fig. 9, the moralization partial correlation between the parents $LST_t$ and $SST_{t-1}$ would be minus the product of their partial correlations with their children, i.e. $-(-0.288 \times -0.440) = -0.126$, which is well below the threshold of significance. Such simplified calculations are of initial value in assessing whether moralization links might be seen as a consequence of a particular choice of the directed links. On fitting the postulated SVAR its implied partial correlations can be more accurately calculated and compared with the sample values used to form the PCG. We do this as a form of model checking in the next section.

Of course, just because a given DAG or DSG (from here on we will just write DSG for this pair) is consistent with a given CIG does not mean that it is the correct model. For example a link in the CIG that might possibly be ascribed to moralization is not necessarily so explained. However, the true DSG will not contain links that are not present in the CIG, except by possible numerical coincidence, for example when a true link and a moralization link contribute canceling effects to give a zero partial correlation which would otherwise appear as a link in the CIG.

Identifying which of the links between current variables are not due to moralization, and the directions of the remaining links, is the key to identifying the DSG. All the links from past values that appear in the CIG can then be included in the DSG. Any due solely to moralization should be found, on fitting the model, to have coefficients that are not significant.

We have, in fact, for this example used a strategy which may be viewed as exploiting this last point. It is certainly not universally applicable, because it relies on the CIG having a high level of sparsity in its links. Even then it may not be successful because it relies on the fitting of what may be an over-specified model, for which there may be a range of likelihood equivalent parameter values, i.e. no unique set of estimates. We have, however, used it before with success in modeling term rates series in Tunnicliffe Wilson et al. [16].

The strategy is to fit a DSG which includes every link in the CIG, with the directions from past to current naturally given by the arrow of time, but with every link between current variables being made bidirectional. On fitting this model by Gaussian estimation the coefficients of low significance can be removed successively, with the level of significance being confirmed by differences in the Gaussian likelihood. This procedure was followed without difficulty, leading to the DSG identical to that shown in Fig. 9 except for the bidirectional links $LST_t \leftrightarrow SST_t$. The link to the right, $LST_t \rightarrow SST_t$, has a low $t$-value of $-1.43$. However, this is based on a local quadratic approximation of the likelihood and on removing this term the deviance (minus twice the log likelihood) increases by 9.02. This suggests that both this term is significant and that the $t$-value approximation is poor in this case. We have however, on removing other terms in the model, found that their $t$-values were consistent with differences in deviance, and all the remaining terms are significant.

## 5 The Final Model, Its Interpretation and Properties

The model so far identified is now subject to diagnostic checking of the residual series which in this case are the estimated structural model innovations. The largest sample cross-correlation between these at lag zero is 0.104, much less than the nominal two standard error limit of 0.267. However, there is a lagged cross-correlation with the value of −0.370 between the residuals of $CO_{2_t}$ and $SST_{t-2}$. We therefore introduced a corresponding further term into the model to explain this dependence and on estimation this term had a $t$ value of −2.93. Although there remained several lagged cross-correlations on or just below their two standard error limits, this is to be expected among the 250 cross-correlations plotted up to lag 10 in Fig. 11. The model with this extra term has a deviance in excess of that of the saturated standard VAR model by only 26.20, with 22 fewer coefficients, suggesting that we have not sacrificed any significant goodness of fit by using the sparsely parameterized structural model. The final model is displayed in Fig. 10 with the coefficients displayed adjacent to the links and their $t$-values in brackets. The $t$-values are derived from local approximations of the deviance, except for the link $LST_t \rightarrow SST_t$ for which it is derived from the deviance difference.

Because of the cyclical nature of the relationship between current variables, we have to be careful how we interpret this diagram. It appears to present, for each current series value, its prediction given all the other current series values and the

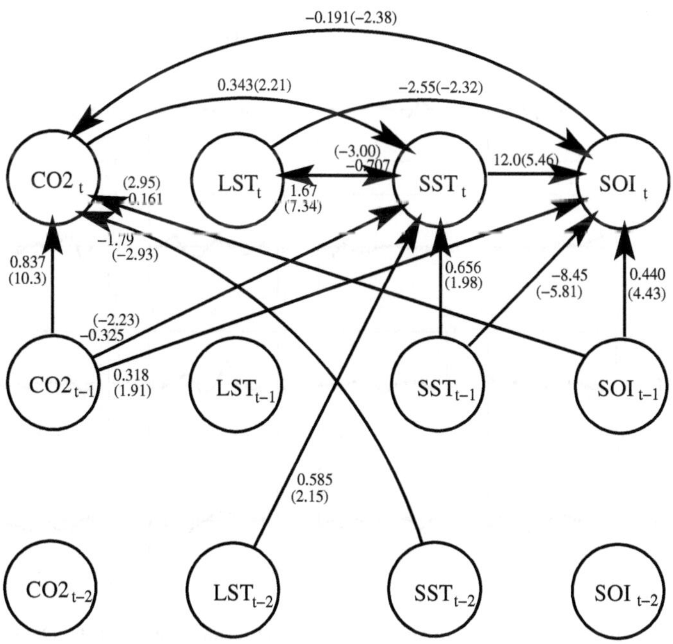


**Fig. 10** The final DSG fitted to the four climate series

past values to lag 2. This is not true. If it were, this prediction would be the same as that of standard VAR model given the same series values. We will give further consideration of the properties of this model, but first present evidence that the model adequately represents the series.

The sample correlation matrix of the model residuals, or structural innovations, is:

$$\begin{pmatrix} 1.000 & & & \\ -0.089 & 1.000 & & \\ 0.044 & 0.009 & 1.000 & \\ -0.045 & -0.005 & 0.081 & 1.000 \end{pmatrix}. \quad (6)$$

All these entries are small and give no reason to doubt the assumption that the residuals are uncorrelated. The lagged cross-correlations of the residuals are shown in Fig. 11.

Again, these generally lie within their bounds except that there appears to be some significant negative autocorrelation in SST at lag 4, which might be removed by a further term. The overall measure of the magnitude of the lagged cross-correlations, the sum of squares of all 160 values scaled by the series length 56, is 130.52. This is a form of the multivariate portmanteau statistic, Hosking [5], Li and McLeod [10]. It does not even exceed the expected value of $145 = 160 - 15$ (the number of estimated coefficients) and again gives no evidence to doubt the model. The squared residuals

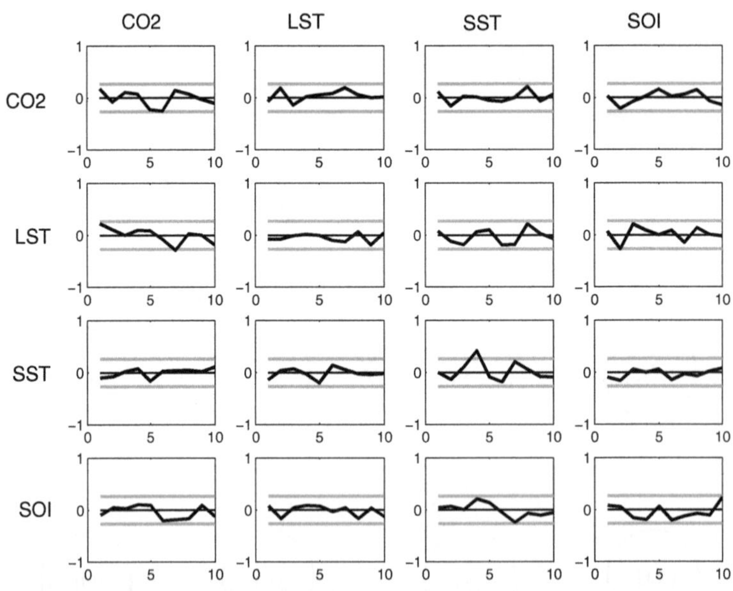


**Fig. 11** The cross-correlations up to lag 10 between the residual series of the DSG model. The *gray bands* show their approximate 2 standard error limits

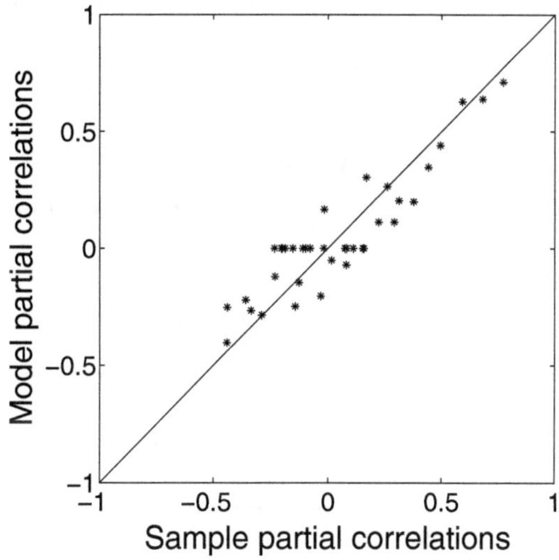


**Fig. 12** The model partial correlations up to lag 2 plotted against the corresponding sample partial correlations

were also checked with no evidence of significant autocorrelation that might have indicated heteroscedasticity.

Finally, we compare the partial correlation properties of the fitted model with the sample values used to identify the model. These are compared in the plot of Fig. 12.

There are 14 zero values implied by the model, as seen in the horizontal line of points in the center. The lowest of these has a corresponding sample value of −0.23 and the largest a sample value of 0.15, so lie within their two standard error limits of ±0.27. There are 23 non-zero model values arising from the 15 links in Fig. 10 and their moralization links. There is a good correspondence between the model and sample values of these, subject to sampling fluctuation. Most of the sample partial correlations corresponding to moralization links are small and do not appear in Fig. 8. Again, this plot generally supports the model.

It is appropriate here to refer back to the partial coherency graph of Fig. 4. The model in Fig. 10 implies that the partial coherency graph should be of the form shown in Fig. 4, except for the addition of a link between LST and SOI. The main point is the separation between $CO_2$ and LST which have no connecting links in Fig. 10, or consequent moralization links. And there is no link between these series in Fig. 4. The comparison of these two graphs could highlight deficiencies in the SVAR, though not in this case. The partial coherency graph also requires little effort to compute and present.

We return to interpretation of the model. We have said that we cannot, for example, take the predictor of $SST_t$ in terms of all the other variables to be the linear combination of $CO_{2_t}$, $LST_t$, $CO_{2_{t-1}}$, $SST_{t-1}$ and $LST_{t-2}$ as indicated by its parents in Fig. 10. However, we *could* do so if at any time the cyclical feedback links from $SST_t$ to $LST_t$ and $SOI_t$ to $CO_{2_t}$ were broken. As it is we have to solve the simultaneous equations relating the contemporary variables to determine this predictor.

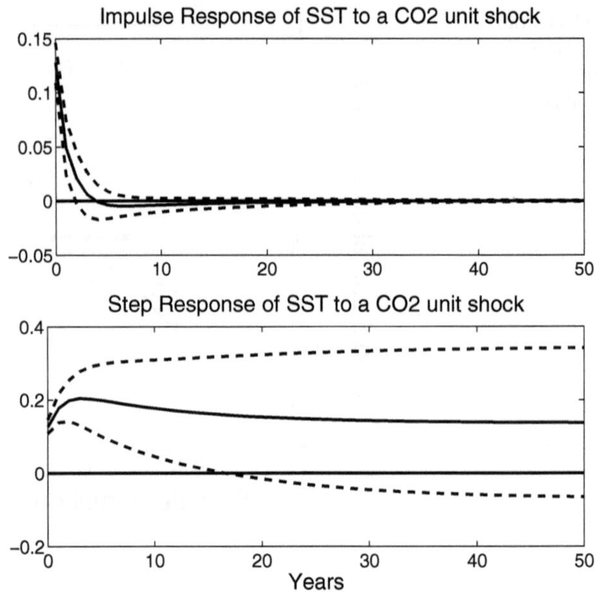


**Fig. 13** The impulse and step responses of SST to a unit shock in $CO_2$, with one standard error limits

A widely used property of an econometric time series model is its impulse response function. This plots the future effect on the different series of a unit shock to the innovation in one of the series—in our case a structural innovation. Figure 13 shows the impulse response, and its cumulative value the step response, of the SST series, given a shock in $CO_2$.

One standard deviation error limits are shown, and even these are very wide for high lead times, in consequence of the fact that long-term information about the model has been removed by trend correction. Before commenting on these we show also, in Fig. 14, the impulse and step responses of $CO_2$ to a unit shock in $CO_2$.

A unit innovation shock to $CO_2$, even in the year zero at which it enters the system, leads to a less than unit increase in $CO_2$ in that year. This is because of the negative feedback within that year from SOI to $CO_2$, an effect which was previewed in the lowest plot of Fig. 5. That plot shows the much larger positive response of $CO_2$ occurring in the following year, which is supported in the final model of Fig. 10 by the link $SOI_{t-1} \rightarrow CO_{2_t}$. This positive feedback results in the step response to a series of unit shocks in $CO_2$ leading to a continuing build up of $CO_2$, and at the same time the greenhouse effect of the $CO_2$ leads to a sea temperature increase. The model represents the effect that warm seas can absorb less $CO_2$, leading to further net increase in the level of $CO_2$.

Although the error limits on the responses are wide, the ratio of eventual level of the step response in SST to that of $CO_2$ is approximately 0.033. The ratio in the trend slope of SST to that of $CO_2$ seen over the period 1975–2005 where the slopes

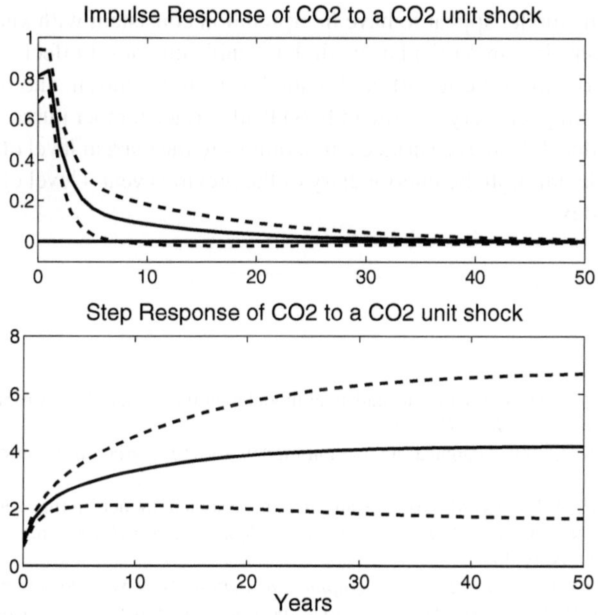


**Fig. 14** The impulse and step responses of $CO_2$ to a unit shock in $CO_2$, with one standard error limits

are steepest, in Fig. 1, is somewhat less at 0.012. However, given the uncertainty in the model properties at higher lead times, these are not wildly inconsistent.

We also remark on the signs of some of the links between current variables. With our definition of the sign of SOI the current value is strongly positively dependent on the current value of SST, though the immediate past value of SST has a largely compensating negative effect—again as previewed in the upper plot of Fig. 5. A substantial decrease in SST would have a negative effect on SOI and after a delay of 1 year this will led to an increase in $CO_2$ associated with an El-Niño event.

Finally, the reciprocal roots of the SVAR model operator have maximum modulus 0.9133. This corresponds to a real root, with the next largest in absolute value being a pair of complex conjugate roots with modulus less than 0.5. This suggests that there may be a unit root process underlying the series, most likely deriving from the trend like behavior of $CO_2$, with LST and SST being co-integrated with this.

## 6 Conclusion

Our analysis of these four climate series lead directly to a structural SVAR representation of the trend corrected series. The model is consistent with the trending appearance of the original carbon dioxide and sea surface temperature series and the terms in the model appear to have interpretation consistent with known climatological relationships. In particular we find a significant causal effect on sea surface temperatures of atmospheric carbon dioxide levels in the current and previous year, and this effect impacts very significantly on land surface temperatures. There is also a positive feedback from sea surface temperatures to the current level of atmospheric carbon dioxide, through the intermediary of the previous year's level of the southern oscillation index.

## References

1. Akaike, H. (1973). A new look at statistical model identification. *IEEE Transactions on Automatic Control, AC–19*(2), 716–723.
2. Dahlhaus, R. (2000). Graphical interaction models for multivariate time series. *Metrika, 51*, 157–172.
3. Dahlhaus, R., & Eichler, M. (2003). Causality and graphical models in time series analysis. In P. J. Green, N. L. Hjort, & S. Richardson (Eds.), *Highly structured stochastic systems*. Oxford: Oxford Universiity Press.
4. Edwards, D. (2000). *Introduction to graphical modelling*. New York: Springer.
5. Hosking, J. R. M. (1980). The multivariate portmanteau statistic. *Journal of the American Statistical Association, 75*, 602–608.
6. Hurvich, C. M., & Tsai, C.-L. (1989). Regression and time series model selection in small samples. *Biometrika, 76*(2), 292–307.
7. Kuo, C., Lindberg, C., & Thomson, D. J. (1990). Coherence established between atmospheric carbon dioxide and global temperature. *Nature, 343*, 709–714.
8. Lauritzen, S. L. (1996). *Graphical models*. Oxford: Oxford University Press.
9. Lauritzen, S. L., & Spiegelhalter, D. J. (1988). Local computations with probabilities on graphical structures and their application to expert systems. *Journal of the Royal Statistical Society: Series B, 50*, 157–224.
10. Li, W. K., & McLeod, A. I. (1981). Distribution of the residual autocorrelations in multivariate arms time series models. *Journal of the Royal Statistical Society: Series B, 43*, 231–239.
11. Lütkepohl, H. (1993). *Introduction to multiple time series analysis*. New York: Springer.
12. Reale, M., & Wilson, G. T. (2001). Identification of vector AR models with recursive structural errors using conditional independence graphs. *Statistical methods and applications, 10*, 49–65.
13. Reinsel, G. C. (1993). *Elements of multivariate time series analysis*. New York: Springer.
14. Schwarz, G. (1978). Estimating the dimension of a model. *The Annals of Statistics, 6*(2), 461–464.
15. Sims, C. A. (1996). Are forecasting models usable for policy analysis. *Federal Reserve Bank of Minneapolis Quarterly Review, 10*, 2–16.
16. Tunnicliffe Wilson, G., Reale, M., & Haywood, J. (2015). *Models for dependent time series*. New York: CRC Press.
17. Whittaker, J. C. (1990). *Graphical models in applied multivariate statistics*. Chichester: Wiley.
18. Young, P. C. (2014). Hypothetico-inductive data-based mechanistic modelling, forecasting and control of global temperature. Technical report, Lancaster Environment Center, Lancaster University. http://captaintoolbox.co.uk/Captain_Toolbox.html/Captain_Toolbox.html.
19. Zellner, A., & Theil, H. (1962). Three-stage least squares: Simultaneous estimation of simultaneous equations. *Econometrica, 30*, 54–78.

# Catching Uncertainty of Wind: A Blend of Sieve Bootstrap and Regime Switching Models for Probabilistic Short-Term Forecasting of Wind Speed

**Yulia R. Gel, Vyacheslav Lyubchich and S. Ejaz Ahmed**

**Abstract** Although clean and sustainable wind energy has long been recognized as one of the most attractive electric power sources, generation of wind power is still much easier than its integration into liberalized electricity markets. One of the key obstacles on the way of wider implementation of wind energy is its highly volatile and intermittent nature. This has boosted an interest in developing a fully probabilistic forecast of wind speed, aiming to assess a variety of related uncertainties. Nonetheless, most of the available methodology for constructing a future predictive density for wind speed are based on parametric distributional assumptions on the observed wind data, and such conditions are often too restrictive and infeasible in practice. In this paper we propose a new nonparametric data-driven approach to probabilistic wind speed forecasting, adaptively combining sieve bootstrap and regime switching models. Our new bootstrapped regime switching (BRS) model delivers highly competitive, sharp and calibrated ensembles of wind speed forecasts, governed by various states of wind direction, and imposes minimal requirements on the observed wind data. The proposed methodology is illustrated by developing probabilistic wind speed forecasts for a site in the Washington State, USA.

**Keywords** Renewable energy · Resampling · Power grid · Predictive density · Sustainability · Nonparametrics

Y.R. Gel (✉)
University of Texas at Dallas, Richardson, TX, USA
e-mail: ygl@utdallas.edu

V. Lyubchich
University of Maryland Center for Environmental Science, Cambridge, MD, USA
e-mail: lyubchic@umces.edu

S.E. Ahmed
Brock University, Saint Catharines, ON, Canada
e-mail: sahmed5@brocku.ca

 
W.K. Li et al. (eds.), *Advances in Time Series Methods and Applications*,
Fields Institute Communications 78, DOI 10.1007/978-1-4939-6568-7_14

## 1 Introduction

As impacts of global warming get increasingly alarming, renewable energy and energy efficiency technologies are now fully recognized as the key components for limiting environmental footprint and making our communities sustainable for future generations. Many countries and regions are now accelerating their policies through comprehensive climate and renewable energy packages, aiming to reduce greenhouse gas emissions and pollution. For example, recently the European Union endorsed its ambitious 20-20-20 strategic plan, with a target to unilaterally cut at least 20 % in greenhouse gas emissions by 2020 (relative to 1990 levels), to increase a proportion of renewable energy to 20 % of the overall energy supply, and to reduce primary energy use by 20 % by improving energy efficiency. The United States and Canada are also in the process of developing a number of landmark policies that enhance and stimulate renewable energy systems at both national and local levels, e.g., the US Western Renewable Energy Zones Initiative and the Standard Offer Contract in Canada.

In the midst of recent proliferation of alternative power sources, wind power is one of the most attractive options. Indeed, in the past decade, wind power has been the fastest growing renewable power source around the globe, with an average annual growth rate more than 30 % and an exponential increase in some countries, e.g., Germany, Spain, Portugal, UK, and Italy [22, 28]. Nevertheless, as the New York Times mentioned, "the dirty secret of clean energy is that while generating it is getting easier, moving it to market is not" [37]. Efficient and reliable integration of wind farms into the electric power system is considered to be one of the main obstacles on the way of renewable wind power to an end user. In particular, power markets need to deal with variability and uncertainty of wind energy resources that are highly intermittent and volatile in nature. Hence, in the advent of renewable energy sources, accurate, reliable and timely forecasting of wind conditions is viewed as a key tool for profitable and safe management of wind power and improving its position in a liberalized electricity market. This in turn has boosted a major interest in development of new methodology for wind speed forecasting. While a substantial part of studies still focus on various point forecasting procedures (see detailed overviews by [10, 28, 30–32, 36, 38, 39, and references therein], there is an increasing demand in producing a probabilistic forecast of wind power, especially providing a full predictive probability density for each horizon rather than only predefined quantiles or intervals, which enables to enhance risk management and facilitate decision-making. Among such statistical developments on probabilistic wind power prediction are, for example, the regime-switching space-time diurnal (RSTD) approach based on an autoregressive conditionally heteroscedastic (ARCH) model with Gaussian innovations [17]; a nonparametric approach based on kernel density estimation [21, 22]; a fuzzy inference model with adapted resampling [33]; a family of autoregressive methods [31, 34]; quantile regression [5] and high-frequency methodology [2].

In this paper we propose a new nonparametric data-driven approach to probabilistic wind speed forecasting, adaptively combining sieve bootstrap and regime switching models. Our new bootstrapped regime switching (BRS) model releases a number of restrictive assumptions imposed by previous studies and, hence, generalizes the methodology proposed by Gneiting et al. [17], Pinson and Madsen [34], and Jeon and Taylor [21]. In particular, most of the earlier approaches to modeling predictive density of wind power are developed under the restrictive parametric conditions on wind speed (usually, the normality assumption) that are typically unjustifiable and unrealistic [21, 25, 31]. Hence, a nonparametric resampling can be viewed as a preferred method for constructing an ensemble of future wind speed scenarios. We propose to employ a flexible and parsimonious technique of sieve bootstrap [4, 7, 9] in an adaptive autoregressive setting with regime switches, which enables to accurately and robustly assess a full probabilistic structure of wind speed, governed by wind direction. In contrast to a regime-switching model proposed by Gneiting et al. [17], the new BRS model constructs future states of wind directions using solely on-site historical observations, rather than requiring recent off-site information from other neighboring wind farms. This is especially important as the lack of expertise and associated costs in wind farm site selection still remain among the key barriers to untapping considerable wind resources in many regions. Particularly, in the developing countries, wind farms, if any, are very sparse [1], therefore there exists no off-site information from adjacent wind farms. Thus, our new nonparametric and robust approach, with minimal model and data assumptions, provides an added degree of flexibility and generality and is of a particular interest to various regions, especially developing countries, where a new era of sustainable wind energy is yet to see its dawn.

The paper is organized as follows. In Sect. 2, we provide a brief overview of wind data. In Sect. 3, we present the new bootstrapped regime switching model. Section 4 describes performance measures that are used for assessment of probabilistic and point forecasts of wind speed. Predictive performance of the new BRS model vs. the benchmark model with normal innovations is reported in Sect. 5. The paper concludes by discussion in Sect. 6.

## 2 Data Description

In this paper we consider wind speed and wind direction data from the Kennewick wind tower in the Pacific Northwest of the United States, collected by the Energy Resources Research Laboratory (ERRL) at the Oregon State University. The ERRL wind data archive represents the largest wind data base in the Pacific Northwest and one of the oldest and most comprehensive wind data archives in the United States. The ERRL wind data are now widely used as benchmark for a variety of environmental and statistical studies. The data are recorded from August 2002 till December 2008. Wind speed is measured in meters per second ($ms^{-1}$) and wind direction is recorded in degrees. The data are available in two forms: observations taken every 10 min,

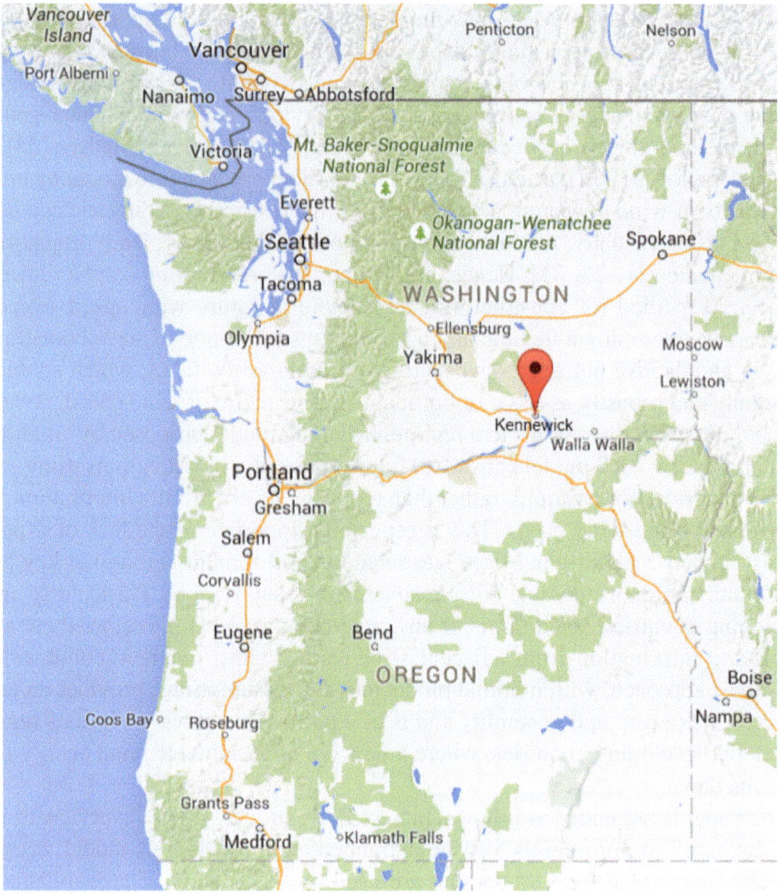


**Fig. 1** Location of Kennewick in the US Pacific Northwest

and the 10-min data averaged to hourly data series. Since our purpose is to produce 1- and 2-h ahead forecasts, we employ only the aggregated version.

The map in Fig. 1 shows geographical location of Kennewick. A detailed discussion of the wind data archive and properties of wind measurements at this location can be found on http://mime.oregonstate.edu/ERRL/WRC/.

## 3 Bootstrapped Regime Switching Model

Let $x_t$ denote a wind speed measurement at a time point $t$. We assess dynamics of $x_t$ using a two-state regime switching model

$$x_t = \begin{cases} c_1 + \sum_{i=1}^{p_1} \phi_{1i} x_{t-i}^{(1)} + \varepsilon_{1t} & \text{if } S_t = 1, \\ c_2 + \sum_{i=1}^{p_2} \phi_{2i} x_{t-i}^{(2)} + \varepsilon_{2t} & \text{if } S_t = 2, \end{cases} \tag{1}$$

where $S_t$ represents the underlying state at a time point $t$ ($t = 1, \ldots, T$) and is defined by the corresponding wind direction of $x_t$, i.e., $S_t = 1$ and $S_t = 2$ are easterly and westerly regimes, respectively; $c_1$ and $c_2$ are constants corresponding to mean levels for each regime; $\{\varepsilon_{1t}\}$ and $\{\varepsilon_{2t}\}$ are independent and identically distributed (i.i.d.) random variables with $E\varepsilon_{jt} = 0$ and $E\varepsilon_{jt}^2 = \sigma_j^2$ for $j = 1, 2$ and are also independent of each other. Here $x_t^{(1)}, \ldots, x_{t-n_1}^{(1)}$ and $x_t^{(2)}, \ldots, x_{t-n_2}^{(2)}$ are historical wind speed observations for easterly and westerly regimes, respectively, that are available at a time point $t$, and $n_1$ and $n_2$ are respective sample sizes. Although $x_t^{(1)}, \ldots, x_{t-n_1}^{(1)}$ and $x_t^{(2)}, \ldots, x_{t-n_2}^{(2)}$ are generally irregularly sampled observations, we treat them, for simplicity, as evenly observed data. One can overcome this limitation by employing, for example, an irregularly-spaced autoregressive (IS-AR) model of Erdogan et al. [13] or spectral estimation of autoregressive (AR) model for non-equidistant time series, suggested by Bos et al. [6]. The orders $p_1$ and $p_2$ of AR models in (1) can be estimated with one of the information criteria. We also assume that the AR parameters $\phi_{1,i}$ and $\phi_{2,i}$ satisfy the weak stationarity condition, i.e., roots of the corresponding AR characteristic equations for $\phi_{1,i}$ and $\phi_{2,i}$ lie strictly outside of the unit circle. Note that the model (1) is not identifiable for data modeling purposes since there is no unique way to identify the states and the two sub-equations are interchangeable. To avoid this problem, following McCulloch and Tsay [27], we assume without loss of generality that $c_2 > c_1$; indeed, as shown in our case study, the average wind speed in the westerly regime is higher than in the easterly regime. We assume that no additional off-site information is available, and define the state at $t + h$ as the most recently observed state, i.e., $S_{t+h} = S_t$. Since we are focusing on short-term forecasts ($h = 1$ or 2h), such naive selection of $S_{t+h}$ is appropriate.

Now, we proceed to the sieve bootstrap procedure [4, 7, 8, 23] to develop a full predictive distribution of wind speed. This resampling approach is robust, because it allows us to generate a probabilistic forecast of wind speed without imposing any parametric distributional assumptions on observed data.

The employed sieve bootstrap procedure is given by Algorithm 1. Given the likeliest state of a future wind direction, we first decide which AR equation in (1) to use (step 1). Then, we estimate the AR model parameters and obtain the residuals (steps 2 and 3). At each bootstrap replication $b$, we sample with replacement $T + h$ values of the residuals and plug them back into the AR model to obtain bootstrap values $x_T^*(h)$ (steps 4–8). Note that if $h \leq 0$, $x_T^*(h) = x_{t+h}$, which corresponds to the actual wind speed observations. Finally, the unknown distribution of future wind speed is approximated by $F^*(x)$ (step 9), which can be used to construct prediction intervals and carry out further inference on the future wind speed dynamics.

**Algorithm 1**: Sieve bootstrap for a regime switching model

**Input** : Observed time series $x_t$, $t = 1, \dots, T$; the most recent state $S_T = j$, $j = 1, 2$.
**Output**: Bootstrap distribution of the future wind speed $F^*(x)$.

$S_{T+h} = S_T$;
estimate parameters $\hat{c}_j$ and $\hat{\phi}_{j1}, \dots, \hat{\phi}_{jp_j}$ in (1);
$r_t^{(j)} = x_t^{(j)} - \hat{c}_j - \sum_{i=1}^{p_j} \hat{\phi}_{ji} x_{t-i}^{(j)}$;
**for** $b = 1, \dots, B$ **do**
  sample with replacement $\left\{r_1^{(j)*}, \dots, r_{T+h}^{(j)*}\right\}$ from $\left\{r_t^{(j)}\right\}_{p+1}^{T}$;
  $x_T^*(h)[b] = \hat{c}_j + \hat{\phi}_{j1} x_T^*(h-1) + \dots + \hat{\phi}_{jp} x_T^*(h-p) + r_{T+h}^{(j)*}$,
  where $x_T^*(h) = x_{t+h}$ if $h \leq 0$;
**end**
$F^*(x) = \sum_{b=1}^{B} \mathbb{1}\{x_T^*(h)[b] < x\}$.

## 4 Performance Measures for Probabilistic Forecasts

Let us briefly outline measures for assessment of forecast ensembles. To evaluate a probabilistic forecast, we employ standard criteria and diagnostic tools, such as coverage probability, length of developed prediction intervals (PI), verification rank histograms, and continuous rank probability score (CRPS) [12, 16, 29]. Coverage probability for a $100(1-\alpha)\%$-prediction interval is defined by a relative proportion of observations, falling within $Q_{\alpha/2}$ and $Q_{1-\alpha/2}$-quantiles, and measures *calibration* of a probabilistic forecast. In turn, length of the developed prediction intervals provides assessment of ensemble *sharpness*. A forecast ensemble is called *calibrated* if estimated coverage is close (by absolute value) to a declared $100(1-\alpha)\%$-confidence level, and a calibrated ensemble of forecasts with shorter lengths of prediction intervals is preferred.

To assess how well ensemble spread represents true variability of wind speeds, we utilize a Talagrand rank histogram [19]. A probabilistic forecast with appropriate spread is characterized by a flat rank histogram; while $\bigcup$- or $\bigcap$-shapes indicate underdispersed or overdispersed ensembles, respectively; a skewed ranked histogram implies that ensemble contains bias.

The CRPS is defined as

$$CRPS(F, x) = \int_{-\infty}^{\infty} \left(F(y) - \mathbb{1}\{y \geq x\}\right)^2 dy, \tag{2}$$

where $\mathbb{1}\{y \geq x\}$ is the indicator function that attains 1 if $y \geq x$ and 0 otherwise, and $F$ is the forecast distribution. The CRPS evaluates the predictive skill of a probabilistic forecast in terms of the entire predictive distribution and simultaneous assesses sharpness and calibration [12, 16, 29]. Let $F_{ens}$ be a discrete predictive distribution from a forecast ensemble of size $B$. The predictive cumulative distribution function $F_{ens}$ has $B$ jumps of size $1/B$ at the respective $B$ ensemble member values $y_1, \dots, y_B$. Then, the empirical CRPS score can be calculated as follows (see [15] for more details):

$$CRPS(F_{ens}, x) = \frac{1}{B}\sum_{i=1}^{B}|y_i - x| - \frac{1}{2B^2}\sum_{i=1}^{B}\sum_{j=1}^{B}|y_i - y_j|, \quad (3)$$

where $y_1, \ldots, y_B$ are the ensemble members, i.e., independent random samples from the predictive distribution $F_{ens}$. Note that in the case of a bootstrap-based predictive distribution, $y_1, \ldots, y_B$ are only conditionally independent.

## 5 Case Study

The main goal of our study is to produce 1- and 2-h ahead probabilistic forecasts of wind speed and to evaluate performance of the new bootstrapped regime switching (BRS) model that allows to utilize only on-site wind data. We use the 2008 data set for Kennewick that consists of 8784 data points. Similarly to the approach by Gneiting et al. [17], we adopt the sliding window method for parameter estimation. The method by Gneiting et al. [17], however, requires the last 45 days of observations due to the complexity of the employed autoregressive conditionally heteroscedastic (ARCH) model. In contrast, our sliding window is noticeably shorter and consists of only 25 days (600 h). Also, in view of a relatively short window, we find that seasonal adjustments yield no gain in predictive performance.

Before actual implementation of BRS, we perform an exploratory analysis of wind speed and wind direction. We define wind measurements corresponding to wind direction falling within the range of 90°–270° as the westerly regime and the rest of the wind data as corresponding to the easterly regime. Wind speeds in the westerly regime are higher on average and more dispersed than their eastern counterpart (see summary statistics in Table 1, box plots in Fig. 2, and time series plot in Fig. 3). Thus, we can conclude that the underlying dynamics in these two regimes is different, and different models are to be employed to assess westerly and easterly wind speeds.

*Remark* We select states based on easterly and westerly regimes to obtain a homogeneous variance within each state. More generally, Pinson and Madsen [34] suggest to select states based on homoscedasticity of data within each state.

**Table 1** Summary statistics for the hourly wind speed (m $s^{-1}$) at Kennewick, observed in January 1–December 31, 2008

| Wind direction | Number of observations | Min. | 1st quartile | Median | Mean | 3rd quartile | Max. |
|---|---|---|---|---|---|---|---|
| Westerly | 5808 | 0.0 | 5.8 | 9.7 | 10.4 | 14.2 | 35.6 |
| Easterly | 2976 | 0.0 | 3.2 | 4.5 | 4.9 | 6.1 | 18.6 |
| All | 8784 | 0.0 | 4.2 | 7.0 | 8.5 | 11.8 | 35.6 |

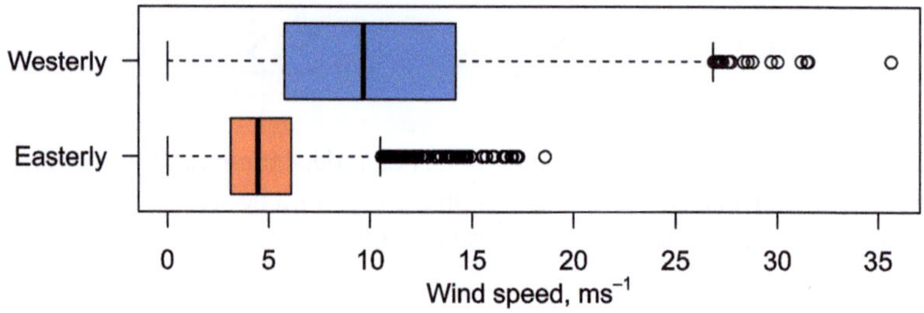


**Fig. 2** Box plot for westerly and easterly wind speed observations at Kennewick, January 1–December 31, 2008

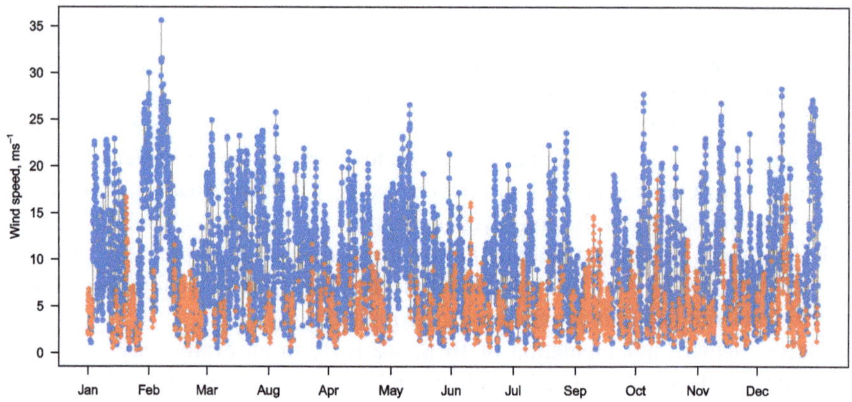


**Fig. 3** Dynamics of hourly wind speed at Kennewick, January 1–December 31, 2008. *Blue circles* denote westerly wind, *red diamonds* stand for easterly wind

The initial model training set starts from the 24th hour of December 6, 2007 up to the 23rd hour of December 31, 2007. Hence, the first 25 days of data are used to produce probabilistic BRS forecasts, which are then compared with observations in the evaluation set, in terms of various performance measures. Then, data points in the evaluation set are moved into the estimation set with the first value in the estimation set discarded, keeping the size of the window constant at 600 data points. This process is repeated until all data points in the evaluation set are exhausted. To illustrate difference of wind dynamics among months in the evaluation set of 2008, all performance measures are calculated separately for each month. Since we advocate a nonparametric and data-driven approach to developing probabilistic wind forecasts, in contrast to Gneiting et al. [17], we employ only raw wind data and do not pre-process wind measurements by discarding or trimming any stretches of the data.

*Remark* Selection of a number of regimes and driving forces for regimes is an open question which is specific for a particular set of observations. For example, Gneiting et al. [17] select regimes based on wind direction, while Pinson and Madsen [34] suggest to select regimes based on volatility of wind speed. In our study we investigated both approaches and found that both selection methods yield a similar predictive performance, slightly better for regimes driven by the wind direction.

We now apply the new BRS model driven by the easterly and westerly regimes to forecast wind speed at Kennewick. We choose the same benchmark model as suggested by Gneiting et al. [17]. Namely, we consider an autoregressive model with innovations following a truncated normal distribution (AR-TN) and select its optimal order using Akaike information criterion (AIC), with a maximum order 4.

For both models, we construct 1000-member forecast ensembles, which are bootstrap-generated for the BRS model and simulated from a truncated normal distribution for the AR-TN model. Tables 2 and 3 show monthly performance measures, evaluated for Kennewick in the verification period January–December, 2008. The model with the lowest value for each measure for every month is shown in bold. For the coverage probability measure, the bold values are those values that are the closest by absolute value to the target coverage of 90 %.

Tables 2 and 3 indicate that the new BRS model provides equally calibrated but sharper probabilistic forecasts than the benchmark AR-TN model, with the improvement up to 10 % in the length of 90 % prediction intervals (CL for January in Table 2). The BRS ensembles of forecasts also yield consistently lower CRPS than the AR-TN forecasts (except of January and March where both methods provide equal CRPS results). In general, the AR-TN model tends to generate overdispersed ensembles with a higher coverage than expected and, hence, wider prediction intervals. The $\bigcap$-shape of verification rank histogram confirms these findings (see the right panel of Fig. 4). This is due to the fact that the AR-TN model based on a parametric truncated normal distribution is more sensitive and less robust to winds that are higher than usual. Remarkably, the new BRS model is capable to capture well a wide range of possible wind events and its ensemble spread is close to the true variability of wind speeds, as also suggested by a flat rank histogram (see the left panel of Fig. 4).

Overall, the new data-driven nonparametric bootstrap approach of the BMRS model provides more robust and sharp probabilistic short-term forecasts of future wind speeds, while requires only on-site wind observations. The proposed methodology can be readily extended to incorporate off-site observations (if available) in a form of exogenous regressors. In addition, the proposed new full density prediction based on a data-driven nonparametric sieve bootstrap can be extended to more complicated ARCH structures using sieve linearization procedures [9].

## 6 Discussion

While demand for renewable and clean wind power is gaining an impressive momentum across the world, still only a small fraction of available natural wind resources is utilized. One of the key challenges on the way of wind power to end-users is its thorny integration into electricity markets, due to a highly volatile and intermittent nature of winds. In addition, most of wind power production is limited only to developed parts of the world (mainly Europe and North America), while being a completely new

**Table 2** Monthly coverage probability (CP) and coverage length (CL) for 90 %-prediction interval and continuous ranked probability score (CRPS) evaluated at Kennewick in 2008 for 1-h forecasts and based on the proposed bootstrapped regime switching (BRS) model and the benchmark autoregressive model with truncated normal innovations (AR-TN)

| Measure | Model | Jan | Feb | Mar | Apr | May | Jun | Jul | Aug | Sep | Oct | Nov | Dec |
|---|---|---|---|---|---|---|---|---|---|---|---|---|---|
| CP (%) | BRS | 89.48 | 89.28 | 89.04 | 89.69 | 90.21 | 90.21 | 90.13 | 90.28 | 90.39 | 90.43 | 90.35 | 90.24 |
| | AR-TN | 91.50 | 90.71 | 90.57 | 90.39 | 91.09 | 91.12 | 91.34 | 91.44 | 91.61 | 91.44 | 91.45 | 91.34 |
| CL | BRS | **5.17** | **5.41** | 5.26 | **5.27** | **4.76** | **5.07** | **5.24** | **4.66** | **4.44** | **4.59** | **4.57** | **4.93** |
| | AR-TN | 5.74 | 5.79 | **5.03** | 5.69 | 4.91 | 5.19 | 5.58 | 4.91 | 4.49 | 4.60 | 4.89 | 5.20 |
| CRPS | BRS | 0.93 | **1.32** | 1.45 | **1.40** | **1.11** | **1.34** | **1.28** | **1.23** | **0.99** | **1.28** | **1.19** | **1.22** |
| | AR-TN | 0.93 | 1.36 | 1.45 | 1.41 | 1.14 | 1.37 | 1.30 | 1.27 | 1.02 | 1.31 | 1.20 | 1.23 |

**Table 3** Monthly coverage probability (CP) and coverage length (CL) for 90 %-prediction interval and continuous ranked probability score (CRPS) evaluated at Kennewick in 2008 for 2-h forecasts and based on the proposed bootstrapped regime switching (BRS) model and the benchmark autoregressive model with truncated normal innovations (AR-TN)

| Measure | Model | Jan | Feb | Mar | Apr | May | Jun | Jul | Aug | Sep | Oct | Nov | Dec |
|---|---|---|---|---|---|---|---|---|---|---|---|---|---|
| CP (%) | BRS | 88.02 | 88.23 | 89.04 | 89.42 | 89.01 | 89.39 | 90.06 | 89.85 | 89.12 | 89.16 | 89.23 | 89.26 |
| | AR-TN | 90.83 | 90.24 | 90.24 | 90.01 | 90.56 | 90.42 | 90.53 | 90.42 | 90.37 | 90.27 | 90.33 | 90.32 |
| CL | BRS | **7.57** | **7.67** | 7.69 | **8.10** | **6.91** | **7.51** | **7.43** | **6.63** | **6.42** | 6.81 | **6.99** | **6.91** |
| | AR-TN | 8.39 | 8.26 | **7.51** | 8.57 | 7.15 | 7.68 | 7.68 | 6.94 | 6.56 | 6.81 | 7.19 | 7.40 |
| CRPS | BRS | 1.41 | **1.57** | 1.81 | **1.64** | **1.35** | **1.63** | **1.51** | **1.43** | **1.23** | **1.57** | **1.47** | **1.46** |
| | AR-TN | 1.41 | 1.61 | 1.81 | 1.66 | 1.41 | 1.65 | 1.54 | 1.47 | 1.27 | 1.60 | 1.48 | 1.47 |

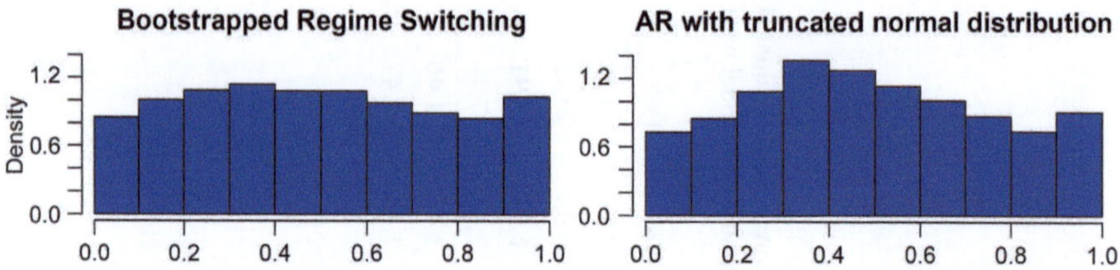


**Fig. 4** Verification rank histograms for the BRS and AR-TN benchmark models, evaluated at Kennewick, over 2008 for 2-h ahead forecast

and unexplored energy source in developing countries, where wind data archives are very sparse or do not exist at all. This ignites an interest in developing new reliable and robust statistical models for wind speed prediction, with minimal assumptions on available wind data.

In this paper we propose a new bootstrapped regime switching (BRS) model, aiming to produce fully probabilistic forecasts of wind speed. In particular, the dynamics of wind speed is modeled by two autoregressive structures, switched in accordance to a state of wind direction. Future states are determined by the most recently observed wind direction at the site of interest. Ensembles of future wind speeds are then developed with a data-driven sieve bootstrap [4, 7, 8, 23] without imposing restrictive and often infeasible parametric assumptions on wind speed data. The new BRS model further extends some of the previously suggested approaches for wind speed prediction in a number of ways. First, since the BRS ensembles of wind speed forecasts are generated using a nonparametric bootstrap techniques and do no impose any distributional assumptions on wind measurements, BRS is more robust in modeling a wider range of winds, including extremes, which enables us to accurately capture the true wind uncertainty and produce substantially sharper and more calibrated probabilistic wind forecasts. Note that in contrast, the RSTD approach of Gneiting et al. [17] and the geostatistical output perturbation (GOP) method for probabilistic surface temperature forecasts of [14] may be viewed as parametric bootstrap-based counterpart procedures since they generate prediction ensemble from a hypothesized normal distribution [11]. Another advantage of the BRS model is that it has minimal demands on data availability. It allows developing reliable wind forecasts with only on-site wind measurements, which is especially useful for regions with limited wind data archives and no history of wind power generation. If off-site data become available, the BRS methodology can be easily extended to a spatio-temporal setting.

The new model can be advanced in a number of ways. First, currently we disregard for simplicity that observations in easterly and westerly wind regimes are naturally unevenly spaced time series. Clearly, such a limitation leads to biases and some informational losses. Instead, to release this assumption and generalize our approach, we can employ, for example, an irregularly-spaced autoregressive (IS-AR) model of Erdogan et al. [13] or spectral estimation of autoregressive model for non-equidistant time series, suggested by Bos et al. [6]. Second, following the ideas of Hering and Genton [20], in modeling wind direction we can recognize more regimes as well as consider Markov processes and related Markovian local bootstrap of higher orders. In addition, as a possible refinement, instead of the sieve bootstrap for generating wind speed scenarios, we can utilize a more general technique of overlapping block bootstrap of a moving length [35]. Another alternative to account for local variations at a particular station is to employ a random effect model with bootstrap of homoscedastic blocks followed by bootstrap within blocks [18, 26]. All the new information, e.g., off-site wind data or other weather variables, if available, can be readily incorporated into the model as exogenous variables, which will likely further enhance accuracy of the resulting wind speed prediction, as mentioned by Alexiadis et al. [3] and Larson and Westrick [24].

Given the competitive performance, flexibility and minimal data requirements of the new BRS model, we anticipate it will find its place in a variety of wind farms worldwide and will be widely used to deliver reliable and robust operational wind power forecasts, especially in regions with limited wind data, which yet to see a dawn of clean and sustainable wind energy.

**Acknowledgments** The authors would like to thank the Bonneville Power Administration and the Oregon State University Energy Resources Research Laboratory, particularly, Stel Walker, for the generosity in providing the wind speed and direction data. We would like to thank William Weimin Yoo and Kimihiro Noguchi for the help at the initial stage of this project. Research of Ejaz Ahmed is supported in part by the Grant from the Natural Sciences and Engineering Research Council of Canada, and Yulia R. Gel is supported in part by the National Science Foundation Grant DMS1514808.

## References

1. Abramowski, J., & Posorski, R. (2000). Wind energy in developing countries. *DEWI Magazine*, *16*, 46–53.
2. Agrawal, M. R., Boland, J., & Ridley, B. (2013). Analysis of wind farm output: Estimation of volatility using high-frequency data. *Environmental Modeling & Assessment*, *18*(4), 481–492.
3. Alexiadis, M. C., Dokopoulos, P. S., & Sahsamanoglou, H. S. (1999). Wind speed and power forecasting based on spatial correlation models. *IEEE Transactions on Energy Conversion*, *14*(3), 836–842.
4. Alonso, A. M., Peña, D., & Romo, J. (2002). Forecasting time series with sieve bootstrap. *Journal of Statistical Planning and Inference*, *100*, 1–11.
5. Anastasiades, G., & McSharry, P. (2013). Quantile forecasting of wind power using variability indices. *Energies*, *6*(2), 662–695.
6. Bos, R., De Waele, S., & Broersen, P. M. T. (2002). Autoregressive spectral estimation by application of the Burg algorithm to irregularly sampled data. *IEEE Transactions on Instrumentation and Measurement*, *51*(6), 1289–1294.
7. Bühlmann, P. (1997). Sieve bootstrap for time series. *Bernoulli*, *3*(2), 123–148.
8. Bühlmann, P. (2002). Bootstraps for time series. *Statistical Science*, *17*(1), 52–72.
9. Chen, B., Gel, Y. R., Balakrishna, N., & Abraham, B. (2011). Computationally efficient bootstrap prediction intervals for returns and volatilities in ARCH and GARCH processes. *Journal of Forecasting*, *30*(1), 51–71.
10. Costa, A., Crespo, A., Navarro, J., Lizcano, G., Madsen, H., & Feitosa, E. (2008). A review on the young history of the wind power short-term prediction. *Renewable and Sustainable Energy Reviews*, *12*(6), 1725–1744.

11. Efron, B. (1982). *The jackknife, the bootstrap and other resampling plans* (Vol. 38). Society of Industrial and Applied Mathematics CBMS-NSF Monographs.
12. Epstein, E. S. (1969). A scoring system for probability forecasts of ranked categories. *Journal of Applied Meteorology*, *8*(6), 985–987.
13. Erdogan, E., Ma, S., Beygelzimer, A., & Rish, I. (2004). Statistical models for unequally spaced time series. In *Proceedings of the Fifth SIAM International Conference on Data Mining, SIAM*, pp. 626–630.
14. Gel, Y., Raftery, A. E., & Gneiting, T. (2004). Calibrated probabilistic mesoscale weather field forecasting: The geostatistical output perturbation method. *Journal of the American Statistical Association*, *99*(467), 575–583.
15. Gneiting, T., & Raftery, A. E. (2007). Strictly proper scoring rules, prediction, and estimation. *Journal of the American Statistical Association*, *102*(477), 359–378.
16. Gneiting, T., Raftery, A. E., Westveld, A. H, I. I. I., & Goldman, T. (2005). Calibrated probabilistic forecasting using ensemble model output statistics and minimum CRPS estimation. *Monthly Weather Review*, *133*(5), 1098–1118.
17. Gneiting, T., Larson, K., Westrick, K., Genton, M. G., & Aldrich, E. (2006). Calibrated probabilistic forecasting at the Stateline wind energy center: The regime-switching space-time method. *Journal of the American Statistical Association*, *101*(475), 968–979.
18. Gray, B. R., Lyubchich, V., Gel, Y. R., Rogala, J. T., Robertson, D. M., & Wei, X. (2016). Estimation of river and stream temperature trends under haphazard sampling. *Statistical Methods & Applications*, *25*(1), 89–105.
19. Hamill, T. M. (2001). Interpretation of rank histograms for verifying ensemble forecasts. *Monthly Weather Review*, *129*(3), 550–560.
20. Hering, A. S., & Genton, M. G. (2010). Powering up with space-time wind forecasting. *Journal of the American Statistical Association*, *105*(489), 92–104.
21. Jeon, J., & Taylor, J. W. (2012). Using conditional kernel density estimation for wind power density forecasting. *Journal of the American Statistical Association*, *107*(497), 66–79.
22. Juban, J., Siebert, N., & Kariniotakis, G. N. (2007). Probabilistic short-term wind power forecasting for the optimal management of wind generation. In *Proceedings of 2007 IEEE Lausanne Power Tech*, pp. 683–688.
23. Kreiss, J. P., & Franke, J. (1992). Bootstrapping stationary autoregressive moving-average models. *Journal of Time Series Analysis*, *13*(4), 297–317.
24. Larson, K. A., & Westrick, K. (2006). Short-term wind forecasting using off-site observations. *Wind Energy*, *9*(1–2), 55–62.
25. Lau, A., & McSharry, P. (2010). Approaches for multi-step density forecasts with application to aggregated wind power. *The Annals of Applied Statistics*, *4*(3), 1311–1341.
26. Lyubchich, V., Gray, B. R., & Gel, Y. R. (2015). Multilevel random slope approach and nonparametric inference for river temperature, under haphazard sampling. In *Machine learning and data mining approaches to climate science* (pp. 137–145). Cham: Springer.
27. McCulloch, R. E., & Tsay, R. S. (1994). Statistical analysis of economic time series via Markov switching models. *Journal of Time Series Analysis*, *15*(5), 523–539.
28. Monteiro, C., Bessa, R., Miranda, V., Botterud, A., Wang, J., & Conzelmann, G. (2009). *Wind power forecasting: State-of-the-art 2009*. Tech. Rep. ANL/DIS-10-1, Decision and Information Sciences Division, Argonne National Laboratory (ANL).
29. Murphy, A. H. (1969). On the "ranked probability score". *Journal of Applied Meteorology*, *8*(6), 988–989.
30. Palomares-Salas, J. C., de la Rosa, J. J. G., Ramiro, J. G., Melgar, J., Agüera, A., & Moreno, A. (2009). Comparison of models for wind speed forecasting. In *Proceedings of The International Conference on Computational Science (ICCS)*.
31. Pinson, P. (2012). Very-short-term probabilistic forecasting of wind power with generalized logit-normal distributions. *Journal of the Royal Statistical Society: Series C (Applied Statistics)*, *61*(4), 555–576.
32. Pinson, P. (2013). Wind energy: Forecasting challenges for its operational management. *Statistical Science*, *28*(4), 564–585.

33. Pinson, P., & Kariniotakis, G. (2010). Conditional prediction intervals of wind power generation. *IEEE Transactions on Power Systems*, *25*(4), 1845–1856.
34. Pinson, P., & Madsen, H. (2012). Adaptive modelling and forecasting of offshore wind power fluctuations with Markov-switching autoregressive models. *Journal of Forecasting*, *31*(4), 281–313.
35. Politis, D. N. (2003). The impact of bootstrap methods on time series analysis. *Statistical Science*, *18*(2), 219–230.
36. Tascikaraoglu, A., & Uzunoglu, M. (2014). A review of combined approaches for prediction of short-term wind speed and power. *Renewable and Sustainable Energy Reviews*, *34*, 243–254.
37. Wald, M. L. (2008). *The energy challenge: Wind energy bumps into power grid's limits*. New York: New York Times.
38. Wan, C., Xu, Z., Pinson, P., Dong, Z. Y., & Wong, K. P. (2014). Optimal prediction intervals of wind power generation. *IEEE Transactions on Power Systems*, *29*(3), 1166–1174.
39. Wang, X., Guo, P., & Huang, X. (2011). A review of wind power forecasting models. *Energy Procedia*, *12*, 770–778.

MIX
Papier aus verantwortungsvollen Quellen
Paper from responsible sources
FSC® C105338
FSC
www.fsc.org

If you have any concerns about our products,
you can contact us on
**ProductSafety@springernature.com**

In case Publisher is established outside the EU,
the EU authorized representative is:
**Springer Nature Customer Service Center GmbH**
**Europaplatz 3, 69115 Heidelberg, Germany**

Printed by Libri Plureos GmbH
in Hamburg, Germany